Fortschritte der Chemie organischer Naturstoffe

Progress in the Chemistry of Organic Natural Products

45

Founded by L. Zechmeister
Edited by W. Herz, H. Grisebach, G. W. Kirby

Authors:
J. A. Elix, M. V. Sargent, Y. Shimizu,
D. A. H. Taylor, A. A. Whitton

Springer-Verlag
Wien New York 1984

Dr. W. Herz, Professor of Chemistry, Department of Chemistry,
The Florida State University, Tallahassee, Florida, U.S.A.

Prof. Dr. H. Grisebach, Biologisches Institut II, Lehrstuhl für Biochemie der Pflanzen,
Albert-Ludwigs-Universität, Freiburg i. Br., Federal Republic of Germany

G. W. Kirby, Sc. D., Regius Professor of Chemistry, Chemistry Department,
The University, Glasgow, Scotland

With 2 Figures

ISSN 0071-7886

ISBN-13:978-3-7091-8719-7   e-ISBN-13:978-3-7091-8717-3
DOI: 10.1007/978-3-7091-8717-3

Contents

List of Contributors

ELIX, Dr. J. A., Faculty of Science, Department of Chemistry, The Australian National University, GPO Box 4, Canberra, ACT 2601, Australia.

SARGENT, Professor M. V., Department of Organic Chemistry, The University of Western Australia, Nedlands, WA 6009, Australia.

SHIMIZU, Professor Dr. Y., Department of Pharmacognosy and Environmental Health Sciences, College of Pharmacy, University of Rhode Island, Kingston, RI 02881, U.S.A.

TAYLOR, Professor D. A. H., Department of Pure and Applied Chemistry, University of Natal, King George V Avenue, Durban, Natal, South Africa.

WHITTON, Dr. A. A., Faculty of Science, Department of Chemistry, The Australian National University, GPO Box 4, Canberra, ACT 2601, Australia.

The Chemistry of the Limonoids
from Meliaceae

By D. A. H. TAYLOR, Department of Pure and Applied Chemistry,
University of Natal, Durban, South Africa

Contents

I. Introduction

The purpose of this article is to review the chemistry of the limonoids isolated from plants of the natural order Meliaceae; limonoids from the Rutaceae and Cneoraceae are not included. The chemistry of the limonoids was last reviewed in this series by DREYER (84); a more recent review is by CONNOLLY, OVERTON and POLONSKY (73). Taxonomic aspects have been reviewed recently by the present author (199), and chemistry and phytochemistry have been discussed at a recent meeting (81). The present article sets out to review advances in limonoid chemistry since the last review in this series (84); the tables however are intended to be comprehensive.

The limonoid chemistry of the Meliaceae began in 1960 with the isolation of gedunin (1) from the West African timber tree *Entandrophragma angolense* (19); the proof of structure (20) was dependent on that of limonin (2) (23, 24) which had recently been published. Subsequently many limonoids have been isolated; the tables in this review list 280 compounds several of which occur as different esters in different plants.

(1)   (2)

The separation of these compounds which are often very similar, frequently unstable and may interconvert on thin layer chromatography, is sometimes very difficult. High pressure liquid chromatography has been much used, but is not free from problems of isomerisation. Separation is further complicated by the occurrence of mixtures of esters; mixed esters of isobutyric and 2-methylbutyric acids are particularly common and difficult to separate.

Certain compounds of the prieurianin group were reported to be separable by t.l.c. in Glasgow, but to isomerise on attempted separation in Durban (*35, 136*). This was eventually traced as an effect caused by the thickness of the absorbent layer. Thin plates, which run rapidly, separate the mixtures while thicker plates, which run more slowly, isomerise them. Not all commercially-available plates are suitable for these separations. Particular difficulty was experienced with the limonoids of *Trichilia emetica*. These proved inseparable by conventional methods (*200*) but were separated by a special technique of high pressure liquid chromatography (*138*). Because of this, some compounds which have been isolated may be artefacts; a case in point is the limonoids of *Aphanamixis polystacha* (*35*) which were isolated only after repeated column chromatography.

When isolated, the limonoids may be of considerable molecular complexity, and the structure determination of compounds such as azadirachtin (**3**) (*202*) has presented formidable problems. In most cases however, the structure elucidation has been fairly straightforward using physical methods, and as a result of this the chemistry of many limonoids has been little investigated and is poorly known.

(**3**)

Even in simple cases, the structure determination may present unexpected pitfalls for the unwary, and several methods of determining stereochemistry have failed when applied to limonoids. The molecular conformation is not always readily predictable, and in n.m.r. spectroscopy of limonoids use of the Karplus equation and of solvent shift methods has led to mistaken conclusions (*98, 152*). More reliable has been the effect of substituents on the chemical shift of nearby methyl groups (*105, 151*). It has been shown that the presence of a hydroxy substituent causes a considerable downfield shift of methyl substituents in a 1,3 diaxial relationship, while an acetoxy or other carbonyl containing substituent may or may not produce a similar shift depending on orientation effects. This can produce good evidence of the stereochemistry, and in suitable cases, of the position of a substituent (*98, 152*).

These considerations show that the structure assigned to dysobinin (*181*) is incorrect. Dysobinin was considered to be 6β-acetoxyazadirone (**4**). However, the methyl group farthest downfield resonates at δ 1.3, while use of model compounds (*152*) shows that an axial acetoxy group at C-6 should produce a downfield shift of the C-10 methyl group to approximately δ 1.5. The 6α-configuration is predicted to lead to the observed shift (*64*). The structure of 6β-acetoxyobacunol (**5**) (*201*), although unusual, appears to be correct as the methyl shifts recorded agree with those calculated.

(**4**)   (**5**)

Many limonoids are potentially available in very large quantities. The timber of some species may yield 1% of an isolated crystalline limonoid; while a single tree of *E. angolense* may contain more than 100 kg. of gedunin, much of it easily recoverable from timber mill offcuts. The biological advantage for the plants of the production of such large amounts of material has always been a problem. Some time ago it was shown that certain limonoids were active as insect anti-feedants (*123*). Recently, it has been shown that many, perhaps even most, limonoids are active as insect antifeedants, although most are not directly insecticidal (*119, 120*). The effect may be so powerful that insects will starve rather than eat leaves treated with limonoids, and it is probable that this is the biological advantage of the limonoids to the Meliaceae. At present commercial applications are being investigated; unfortunately indications are that insects may adapt to limonoids rather quickly. Thus African insects eat African Meliaceae in preference to similar South American species, while the reverse is true for similar South American insects (*95*).

Some limonoids have also been found to be active against some types of cancer (*108*); to date activity is confined to certain compounds of the havanensin and prieurianin groups containing 14,15 epoxide rings. It is as yet too soon to say if the effect will be significant; if it is, obtaining a supply of the rather rare and complex compounds involved will not be easy.

II. Classes of Limonoids

Structurally, the limonoids are derived from tetracyclic triterpenes similar to euphol or tirucallol by a series of oxidative changes interspersed with molecular rearrangements. The oxidations are either epoxidations of double bonds or Baeyer-Villiger attacks on ketones and are all of the type to be expected from a biological per-acid equivalent, presumably a peroxidase. Some of the oxidations are subject to extreme steric hindrance *in vitro* and laboratory duplication is sometimes difficult to bring about. In contrast, the rearrangements are very ready and often spontaneous. These transformations have been studied by examination of the minor naturally-occurring constituents, and many have now been repeated in the laboratory. There is as yet no direct proof of the hypothetical biosynthesis by tracer incorporation, but the schemes are sufficiently well supported by circumstantial evidence to make it unlikely they are greatly in error. The details will be discussed later.

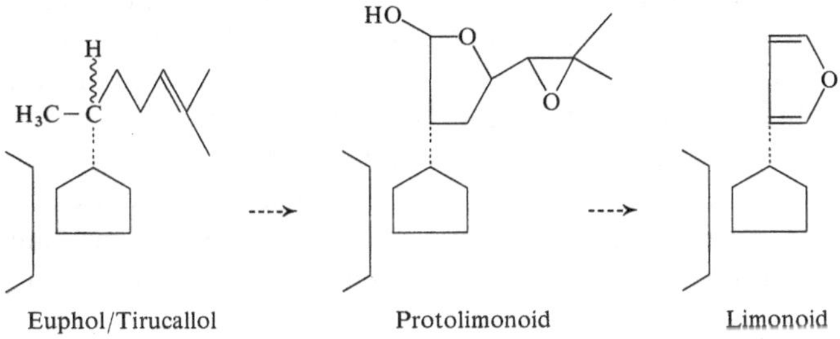

Scheme 1. Side chain oxidation of protolimonoids

In the course of these changes (Scheme 1), the triterpene side chain is first oxidised, eventually to a β-substituted furan ring. The usual definition of a limonoid is that it is a triterpene derivative in which the side chain has become a furan ring by the loss of four carbon atoms, hence an alternative name, tetranortriterpenoids, for the limonoids. In a few compounds which are obviously related a fifth carbon atom has been lost by decarboxylation at C-4, and in a few the furan ring is replaced by a different five membered heterocyclic ring. Some of these last derivatives appear to be oxidation products of the limonoids (*40*); others may be intermediates in the production of the furan ring (*64*).

There are also a number of oxidised intact triterpenes known which by their biological occurrence and oxidation pattern appear to be biochemical precursors of the limonoids. Such compounds are therefore known as protolimonoids. All these groups are included in this review. When the C-20 stereochemistry of triterpenes is preserved in protolimonoids, it has usually been found to be the 20α-H configuration characteristic of tirucallol, but there are also several examples of the 20β-H configuration found in euphol, and it is not known whether the biosynthesis uses only one of these isomers or can start from either. Feeding evidence suggests that in *Melia* both tirucallol and euphol can be incorporated, but that euphol is more readily used (*87*); it may be significant that the known protolimonoids with the euphol configuration all occur in *Melia,* while the tirucallol derivatives occur in other genera.

Recent research has found limonoids co-occurring with triterpenes of the dammarane class which show no obvious structural relationship. The significance of this is not known; these triterpenes which frequently also occur in other plants are not included in this review.

The limonoids have been classified on the basis of which of the four rings in the triterpene nucleus have been oxidised. If the protolimonoids are included sixteen groups can be recognised in this way of which ten are now known. It has been suggested that 12-ketones in which ring D is a lactone and which would be the precursors of C, D-*bis-seco*limonoids are unstable and give rise to the quassinoids (*81*) (see Scheme 2). If this is correct, this would reduce the number of possible groups of limonoids to thirteen, which suggests that nearly all may now have been found. Groups which are missing are those in which rings A and C, B and C, and A, B, and C are oxidised. Compounds in which rings B and D are oxidised are common in the tribe Swietenioideae and it is convenient to divide them into sub-groups which depend on whether further transformations have occurred. In subgroup (a) rings B and D are opened, in subgroup (b) a new ring has been formed between C-2 and C-30, while in subgroup (c) compounds of subgroup (b) are further modified by bridging of ring A. These groups are used to classify the limonoids in the present review; their possible taxonomic significance has been considered (*81*).

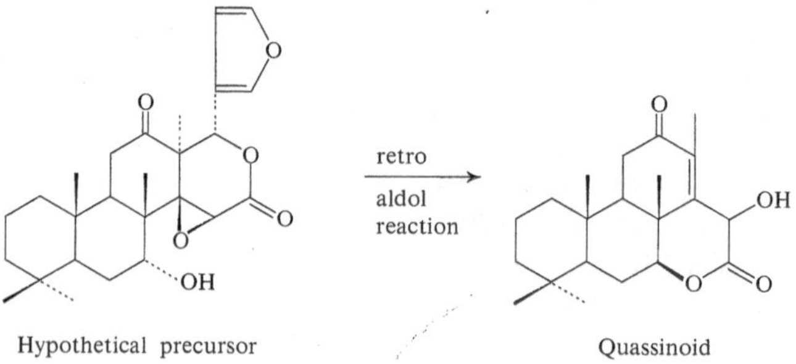

Hypothetical precursor   Quassinoid

retro
aldol
reaction

Scheme 2. Hypothetical biogenesis of quassinoids

1. Group I. Protolimonoids and Related Triterpenes

In these compounds the triterpene side chain is intact, but usually highly oxidised and often cyclised to form an ether ring.

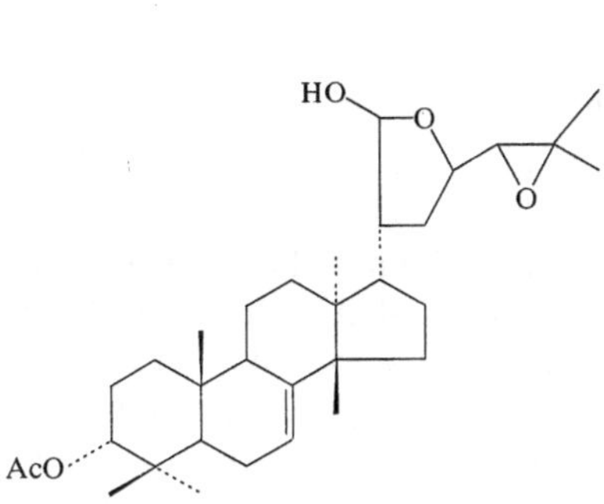

Turreanthin

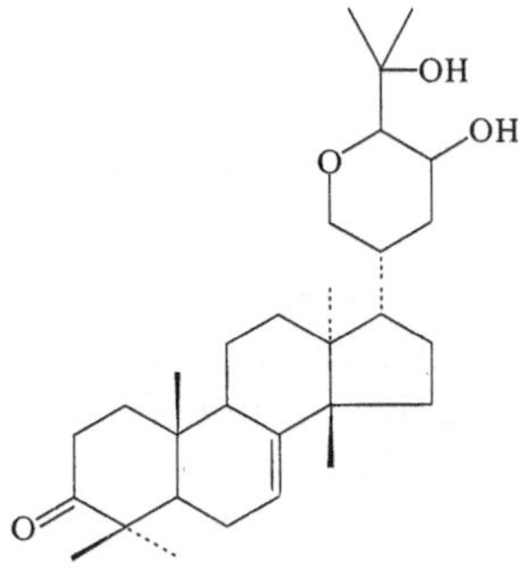

Bourjotinolone A

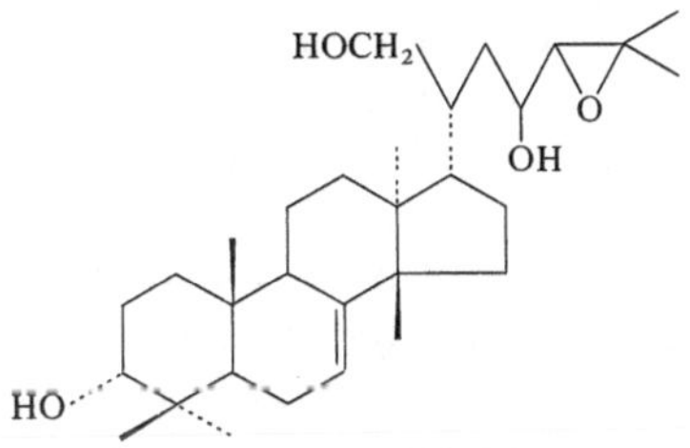

Entandrophragma triol

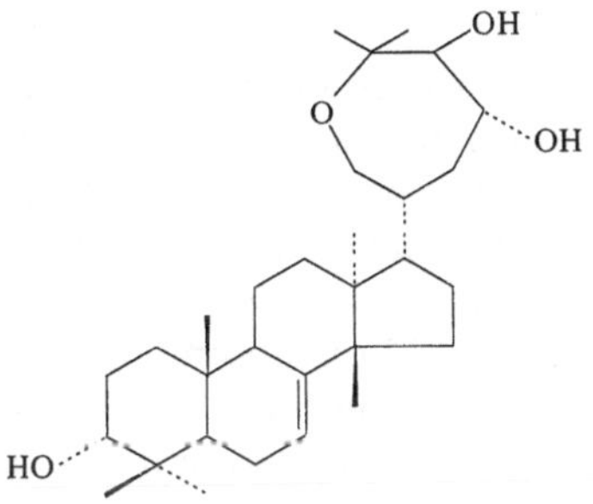

Sapelin B

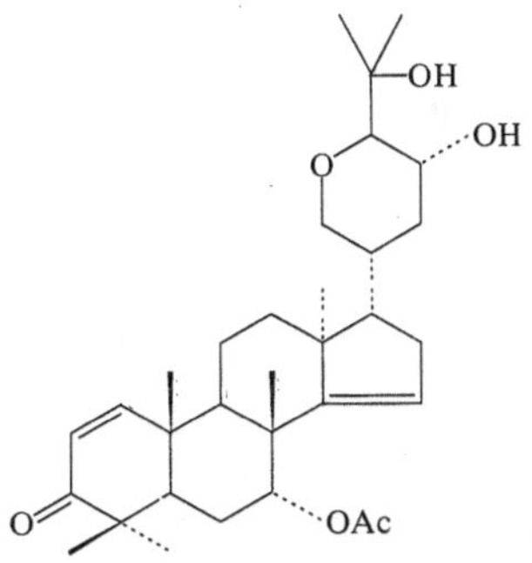

Grandifoliolenone

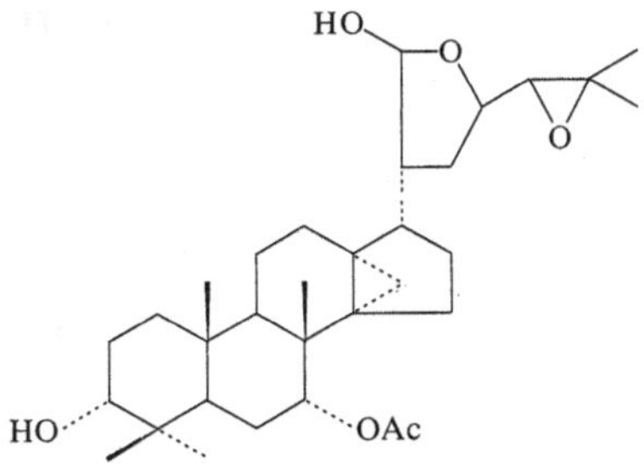

Glabretal

Types of Protolimonoids

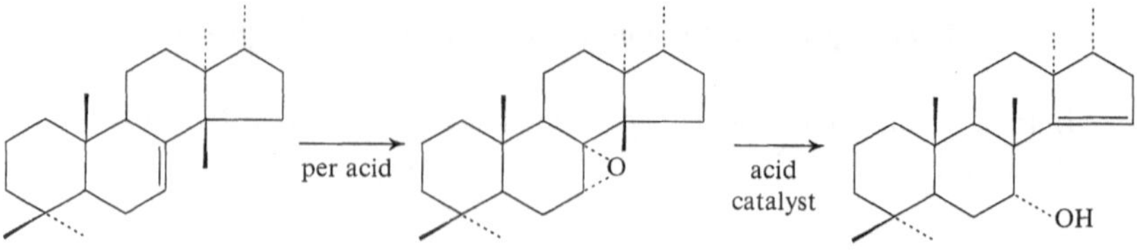

Scheme 3. Apo-change

All the main types of protolimonoids, apart from glabretal, had been discovered at the time of the last review, and although new ones have been added they are not of special interest. The protolimonoids fall in one of two classes. The first, like euphol, have a β-methyl group at C-14 and a double bond at $\Delta^{7,8}$; while in the second, the so-called *apo* group, there is an α-hydroxyl group at C-7, the double bond has moved to $\Delta^{14,15}$ and the β-methyl group to C-8. It is believed that this change is dependent on the opening of a 7,8α-oxide (Scheme 3), and this has been demonstrated in the laboratory (*37, 39, 82, 83, 124, 127*). The side chains appear to be the same in both groups of protolimonoids so there is apparently no specific stage at which the *apo* change occurs. Glabretal (*92*) occupies an intermediate position between these two groups, as it has the 7α-hydroxyl and the 8β-methyl, but the 14,15-double bond is replaced by a 13,14-cyclopropane. It is possible to imagine a common carbocation which can either lose a proton from C-15 or capture the C-13 angular methyl. It is also possible that the cyclopropane may be a normal intermediate in the *apo*-change, under laboratory conditions it would not be likely to survive the conditions of acid catalysis used to open the 7,8-oxide ring. The structure of glabretal was elucidated by x-ray analysis; no other compounds of this type are yet known from the Meliaceae, but several have been found in the Cneoraceae (*185*).

2. Group II. Havanensin Group

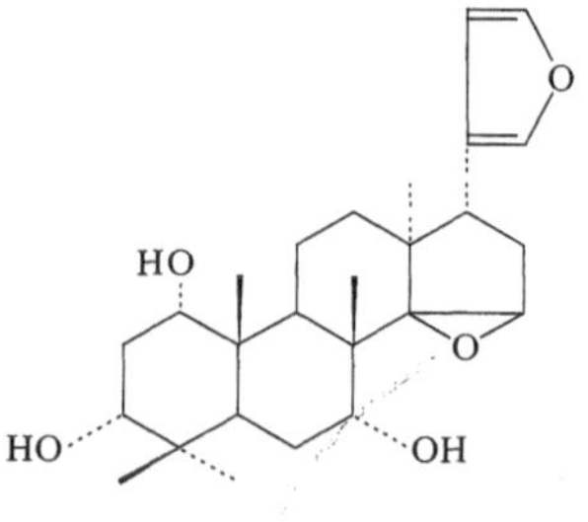

Havanensin

This group consists of compounds with a furan side chain in which all rings of the nucleus remain intact. Many new compounds of this group have recently been discovered. They consist of two kinds, the first being minor constituents of comparatively simple structure found during extensive re-working of extracts from *Melia* and *Azadirachta* in the search for compounds with anti-insect activity, the second consisting of highly oxidised compounds, mainly from *Trichilia, Azadirachta* and *Melia*.

Simple ones which have been recently discovered include 7-benzoates of known limonoids (*114*), 17-*iso*-compounds (*111*), and 17-hydroxy compounds. 17-*Epi*azadiradione (**6**), obtained from *Azadirachta indica* (*111, 114*) had spectral properties similar to those of azadiradione (**7**), the main difference being shifts of some methyl groups, particularly that at C-13 which is shifted downfield in the epimer from δ 1.03 to δ 1.44. The structure was proved by the use of the n.o.e., which showed that H-17 was *cis* to the 13-methyl group.

(**6**) 17β
(**7**) 17α
Azadiradione

(**8**)

17β-Hydroxyazadiradione (**8**), obtained in the same work, was found to have an extra tertiary hydroxyl group. The location of this was shown by the [13]C n.m.r. spectrum; in comparison with that of azadiradione, the resonance assigned to C-17 became a singlet and shifted sharply downfield (δ 60.7 d to δ 80.7 s). The stereochemistry was again shown by the n.o.e., by which the furan ring was found to be *cis* to the 13-methyl group.

Azadiradione has a carbonyl group at C-16 and hence can isomerise at C-17. The configuration of azadiradione itself was therefore uncertain until the present work. The configuration of azadiradione derivatives cannot be regarded as certain unless confirmed, preferably by the n.o.e.

A compound which was said to be isomeric with 17-hydroxyaza-diradione has been isolated from *Azadirachta indica* (*180*). The recorded [1]H n.m.r. spectra of the two 17-hydroxy compounds are closely similar while the [13]C n.m.r. spectra differ considerably and unpredictably. It seems probable to the present reviewer that the two compounds are the same, but that the ^{13}C n.m.r. spectrum of the second compound (*180*) has been recorded incorrectly.

Vilasinin (**9**), which is interestingly related to the more complex compounds of group VII such as salannin (**10**) by the presence in both of a 28,6α-oxide, has been isolated from *Azadirachta indica* (*112, 164*) and from *Chisocheton paniculatus* (*64*). The stereochemistry of this compound has been fully confirmed by KRAUS (*112*) by use of the n.o.e., which has shown that the three methyl groups at C_4, C_{10} and C_8 are all *cis*. So far, all known natural limonoids substituted at 6 or 7 have the 6α-configuration, except 6β-acetoxyobacunol (*201*).

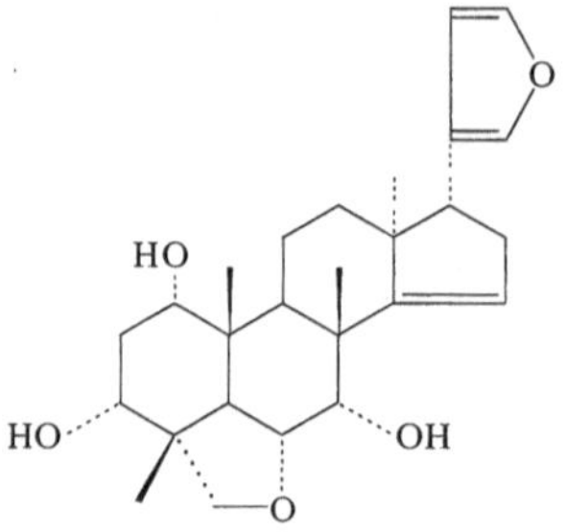

Vilasinin
(**9**)

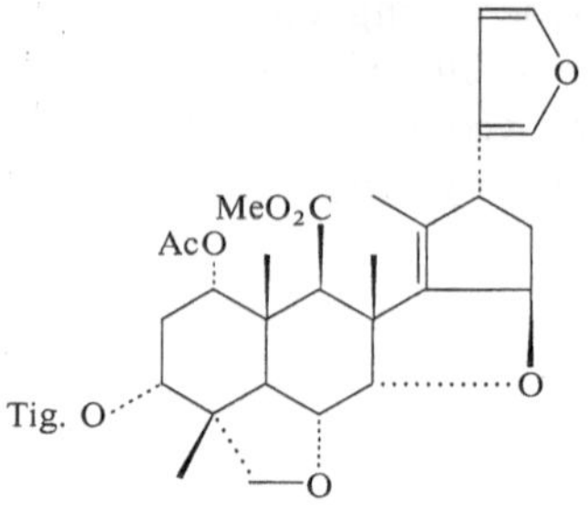

Salannin
(**10**)

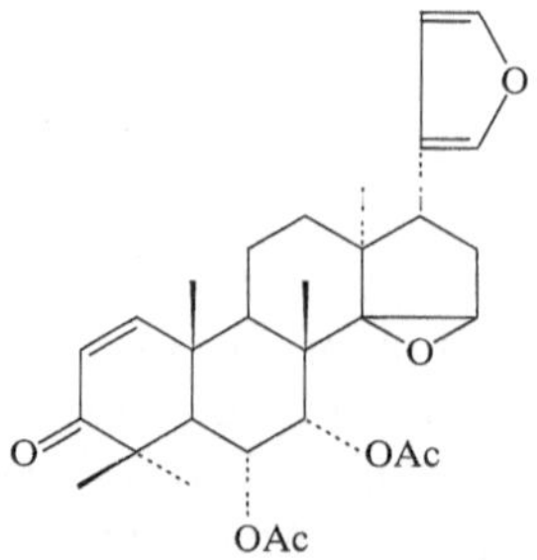

(**11**)

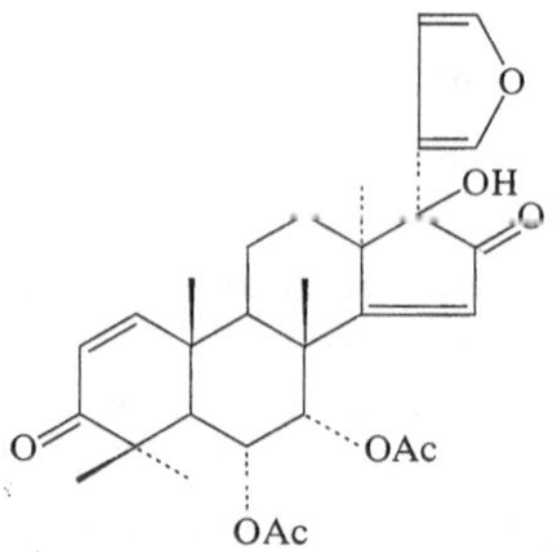

(**12**)

Chisocheton paniculatus (*64*) yielded 6α-acetoxyepoxyazadiradione (**11**), previously isolated from *Carapa guyanensis* (*129*), and four new azadirone derivatives, only one of which had a furan ring. This was identified as 6α-acetoxy-17β-hydroxyazadiradione (**12**), from the similarity of the spectrum to that of 17β-hydroxyazadiradione described by KRAUS (*111*).

The other three have the 6α-hydroxyazadirone nucleus but lack the 16-oxo group and have modified side chains. One is the saturated γ-lactone

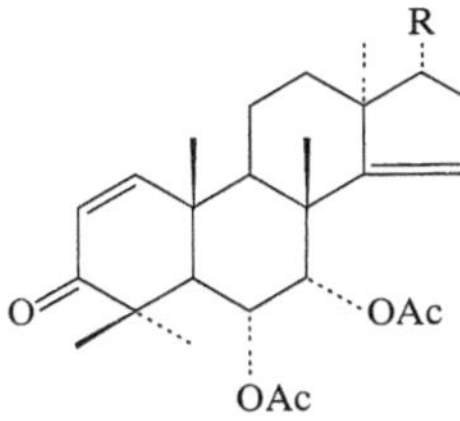

(13)　R =

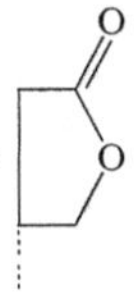

(14)　R =

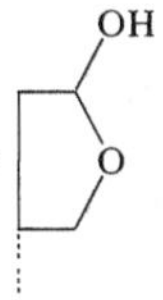

(15)　R =

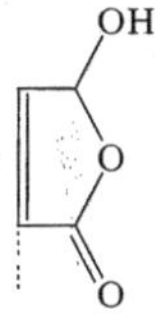

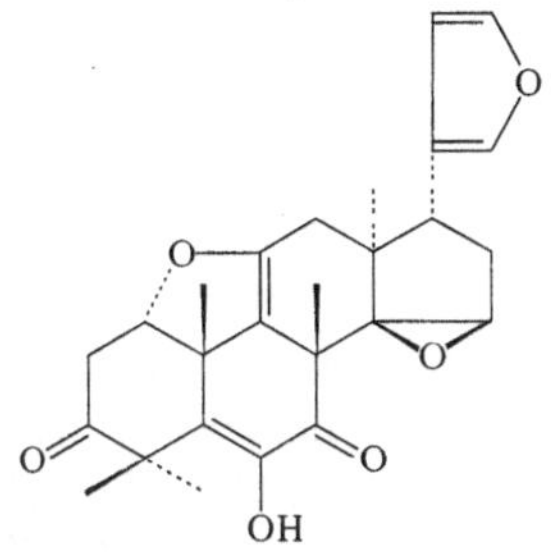

(16)

(13), identified spectroscopically, the second the related lactol (14), identified by oxidation to the lactone. The third contained a γ-hydroxy-α,β-unsaturated lactone ring (15), a well known system which will be discussed later in reference to photogedunin. An unusual compound obtained by HALSALL from *Khaya anthotheca* (16) (*98*) has a 1,11-ether and a 9,11-double bond. The structure was determined by spectroscopy and comparison with anthothecol. The enol ether grouping is stable to both acid and alkali, as is the isomeric ketone 11-oxocedrelone. This unreactivity is probably due to steric hindrance toward attack on the enol system. The biosynthesis of this unusual compound is quite unknown.

At the time of the last report the limonoid nucleus of the heudelottins (17) isolated from *Trichilia heudelotti* had been identified, but the nature of the esterifying acids was unknown. This has now been determined (*160*).

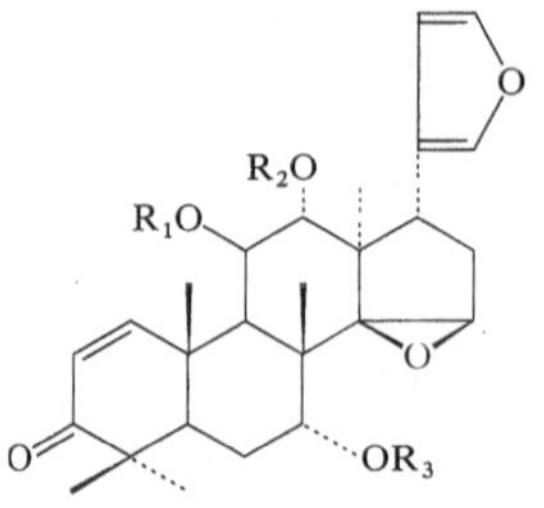

(17) R_1, R_2, $R_3 = H$

(18) Heudelottin E $R_1 = $

$R_2 = CH_3$

$R_3 = $

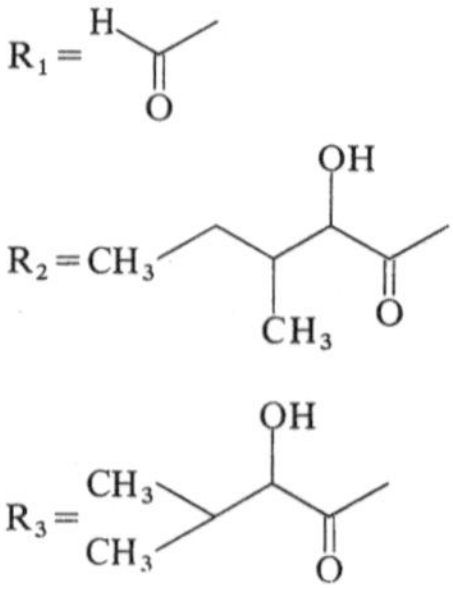

(19) Heudelottin F $R_1 = H-C-$

$R_2 = CH_3$

$R_3 = $

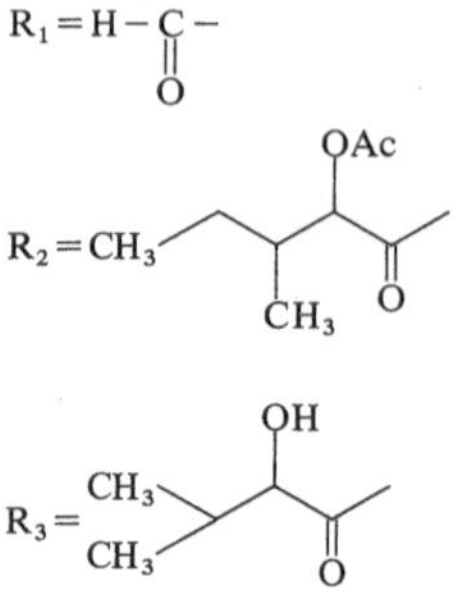

(20) Heudelottin C $R_1 = H$

$R_2 = CH_3$

$R_3 = $

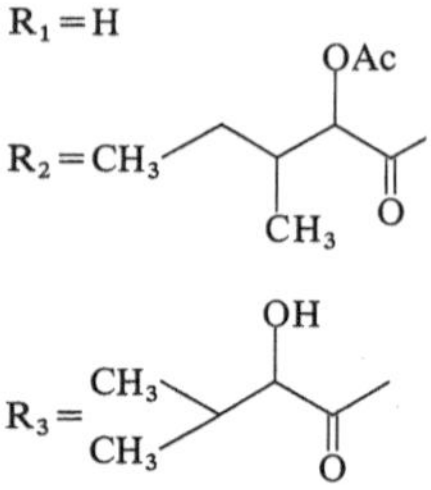

(21) R_1, R_2, $R_3 = Ac$

Bicarbonate hydrolysis removed first the 11- and then the 12-substituent; it was shown by mass spectroscopy that heudelottin E (18) had a formate at C-11, a 2-hydroxy-3-methylvalerate at C-12 and a 2-hydroxy-3-methylbutyrate at C-7. Heudelottin F (19) was acetylated in the side chain at C-12, while heudelottin C (20) was the 11-deformyl derivative of F. These compounds are of interest as they are the simplest ones containing the 11β-formyloxy-12α-(2-hydroxy-3-methylvaleryloxy) system which has been found very commonly in the complex compounds of the prieurianin group. Acid treatment of heudelottin, like that of the simpler havanensin, very readily isomerises the oxide to a 15-ketone (22), but by alkaline hydrolysis of heudelottin it is possible to obtain the epoxy alcohol (17)

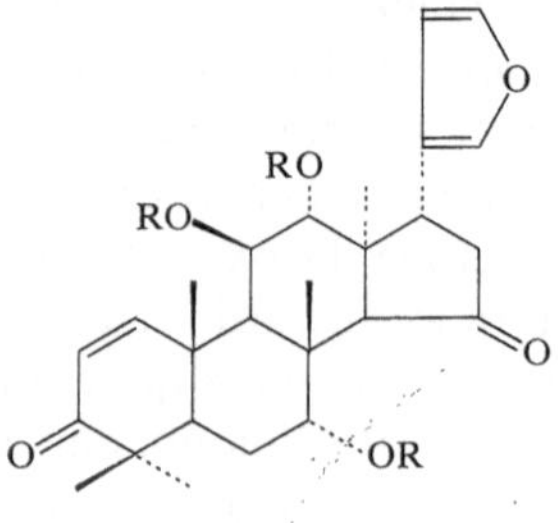

(22)

References, pp. 93—102

$(R_1, R_2, R_3 = H)$, which on acetylation gives the corresponding triacetate (21). The 12α-acetate in this compound shows a methyl resonance at δ 2.0, unlike the corresponding ring D lactones, where a similar 12α-acetate experiences a powerful shielding influence from the furan ring, and resonates at about δ 1.6 (77) (c.f. p. 20). Closely related to the heudelottins are some compounds obtained from *Turrea floribunda* (18) (23, 24), which have a carbomethoxy group at C-4. The C-8, C-10 and C-13 methyl group shifts of these compounds agree with those calculated from heudelottin, while the remaining methyl group at C-4 is shifted downfield by δ 0.4. This is consistent with the presence of a C-28 carbomethoxy group, but not with the presence of a C-29 ($= 4\beta$) carbomethoxy, which would be expected to produce a big shift of the C-10 methyl resonance.

(23)   (24)   (25)
                                                Hirtin

Hirtin (25), from *Trichilia hirta* (47), has been known since 1966, but the stereochemistry at C-4 is unknown. The methyl shifts agree with those calculated for a C-28 carbomethoxy structure using the *Turrea* compounds and cedrelone as references, so it is likely this is the configuration.

More complex representatives of this group are a number of closely similar compounds obtained from *Trichilia emetica* (138), *Aphanamixis grandiflora* (166) and *Melia azedarach* (145). The basic nucleus of these is the hemiacetal of 19-hydroxyhavanensin-29-aldehyde (26), although this substance itself is unknown. Amoorastatin (27) (167), the 11-oxo derivative; the 12α-hydroxy derivative of this (168), and the 2α,12α-dihydroxy derivative (28) (aphanastatin, 166) have all been isolated by POLONSKY from *Aphanamixis grandiflora*, while sendanin (29) (145), the 12,29 diacetate of 12α-hydroxyamoorastatin has been isolated by OCHI from *Melia azedarach*. The structures of all these compounds have been determined by x-ray crystallography. The related trichilins (138) were isolated by NAKANISHI from *Trichilia emetica*.

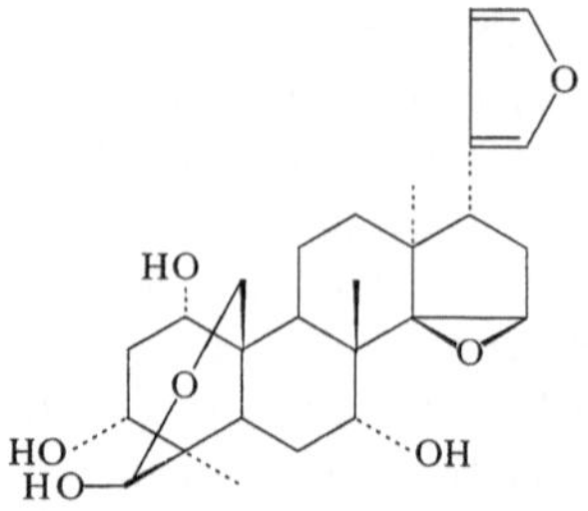

(26)

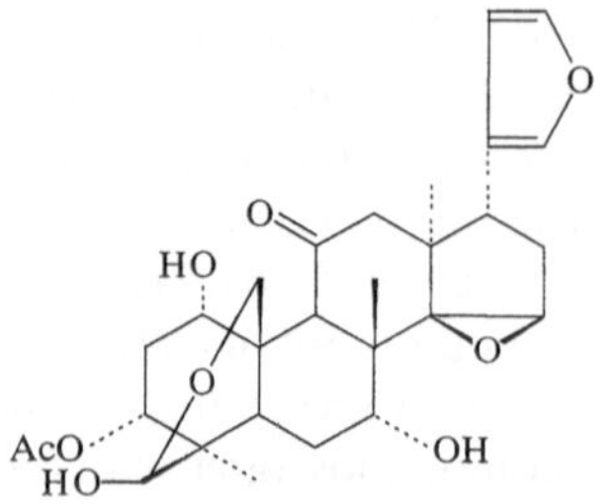

(27)
Amoorastatin

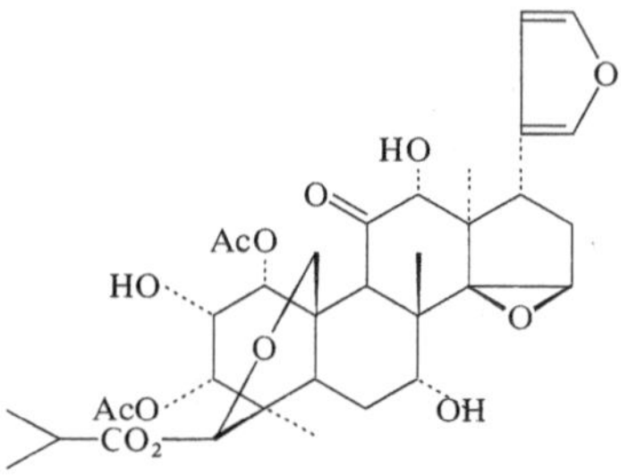

(28)
Aphanastatin

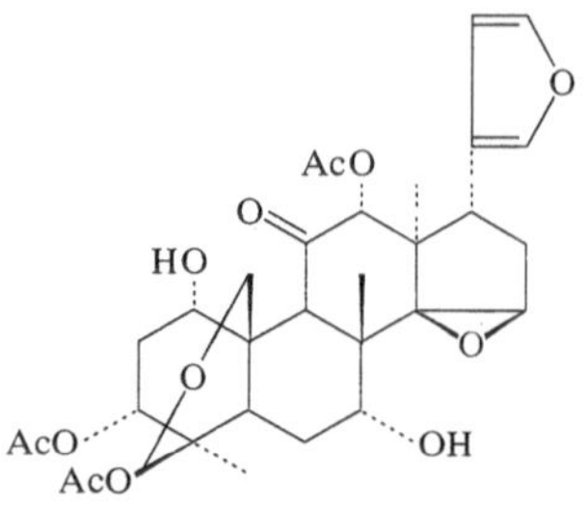

(29)
Sendanin

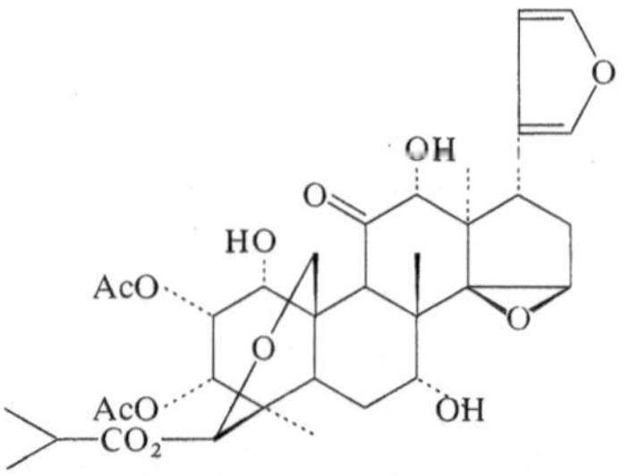

(30)
Trichilin B

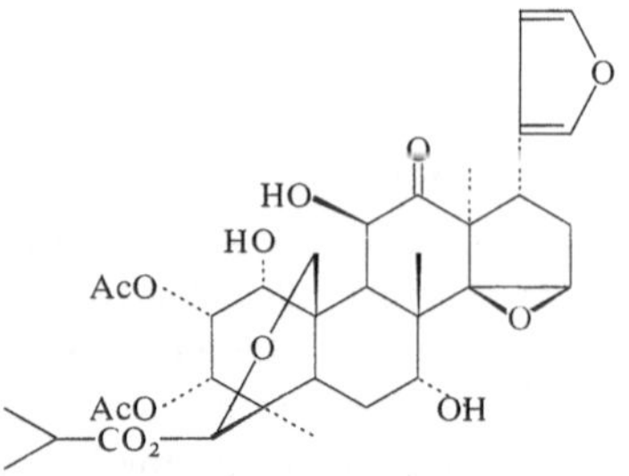

(31)
Trichilin C

Of these, trichilin B (30) differs from aphanastatin by having an acetate at C-2 instead of C-1, trichilin A is the 12β-isomer of this, trichilin D is the 12-deoxy derivative, trichilin C the 11β-hydroxy-12-oxo-isomer (31), and trichilin E the 12β-isomer of aphanastatin. The structures of these were proved by n.m.r. methods, and by correlation with aphanastatin which is produced by a unique isomerisation of trichilin B on treatment with zinc borohydride. Dissolved in chloroform, A slowly isomerised to C in which the H-9 resonance is shifted upfield (δ 4.72 to δ 3.25) but remains a singlet. The location and configuration of the ring C hydroxyl in compound B is shown by the upfield shift in the furan and benzene protons of the 12-p-bromobenzoate, in which the two aromatic rings lie side by side.

It was found by NAKANISHI that the trichilins isomerised in chloroform and on contact with silica gel. Using these reagents the present author failed to isolate any pure compounds from *Trichilia emetica*.

Sendanal (**32**), also obtained from *Melia azedarach* (*149*) belongs to a stereoisomeric series, being a C-28 aldehyde.

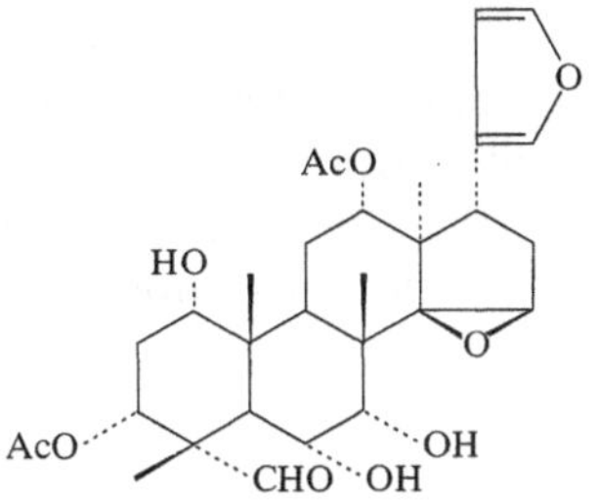

(**32**)
Sendanal

The partial synthesis of the furan ring in havanensin-type compounds has been accomplished by sodium metaperiodate oxidation of turreanthin in the presence of perchloric acid (Scheme 4). Execution of the *apo* rearrangement as previously described then led to the partial synthesis of a simple havanensin derivative (*36, 39*). It is not known if the biosynthesis goes by this route or by an alternative following the more usual limonoid pattern of an oxide rearrangement (*81*).

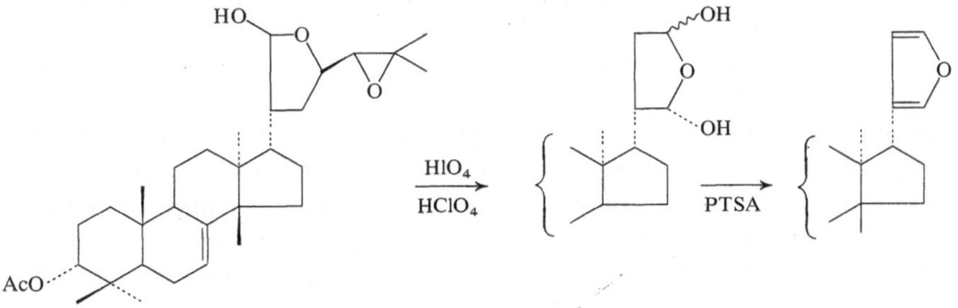

Turreanthin

Scheme 4. Side chain oxidation of turreanthin

3. Group III. Gedunin Group (Ring D Opened)

Gedunin

This group contains the first limonoids to be discovered in the Meliaceae and little has been added to it in recent years. 11β-Acetoxykhivorin (**33**) has been isolated from *Khaya* species (*188, 190, 191*) and it has been shown that nyasin (*187*) is 11β-hydroxykhivorin (**34**) (*80*). Mild hydrolysis of 11β-acetoxykhivorin gives either the 7-deacetyl or the 11-deacetyl derivative, apparently depending on the conditions employed.

(33) R=Ac   **(35)**
(34) R=H   Photogedunin

Recently KRAUS (*114*) has isolated the benzoate corresponding to gedunin; this is of interest since the 7α-hydroxyl shows considerable steric hindrance to acylation. It has been shown that gedunin can be photochemically oxidised in the furan ring to give a 4-hydroxy-2,3 unsaturated γ-lactone named photogedunin (**35**) which has been isolated from *Cedrela odorata* (*40*). A number of other compounds with the same ring system have been isolated. It is not known whether these are natural products or artefacts, and it has been suggested that the presence of these compounds in *Chisocheton* extract may be due to the influence of Glasgow sunshine (*64*). However LAVIE has drawn attention to the fact that photooxidation produces only those isomers in which the terpenoid residue is attached to C-2 of the lactone ring, while compounds which have been isolated from plants include both these and the C-3 isomers (*175, 203*). It seems certain therefore that at least some of these compounds are genuine natural products (*62, 117, 170*).

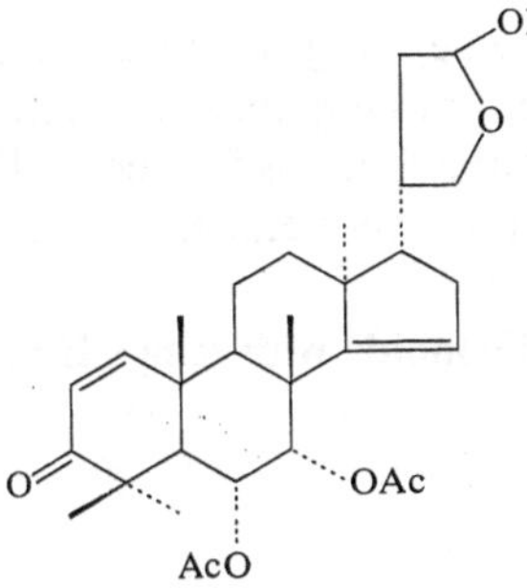

(36)

It is not known what relevance this has to the biosynthesis of the furan ring. Certain compounds such as **(36)** from *Chisocheton* (*64*) are at a lower oxidation level than furan and therefore could be biological intermediates. In this case, the biosynthesis must be quite different from either of the possibilities discussed above, since both of these produce the five membered ring at the oxidation level of furan. On the whole it seems probable that these compounds belong in a different biosynthetic chain; one possibility would be that the furan ring results from biological degradation of a compound similar to turreanthin, with a C-21 aldehyde already present, while the *Chisocheton* compounds result from a similar degradation applied to a 21-hydroxy derivative.

The biosynthesis of the lactone ring D in gedunin seems to be well understood, a series of potential intermediates between the cyclopentene ring of deoxyhavanensin and the epoxy lactone having been isolated in both the gedunin and the khivorin series, and the hypothetical linking stages have been carried out in the laboratory (*38, 121, 126, 128*) (Scheme 5).

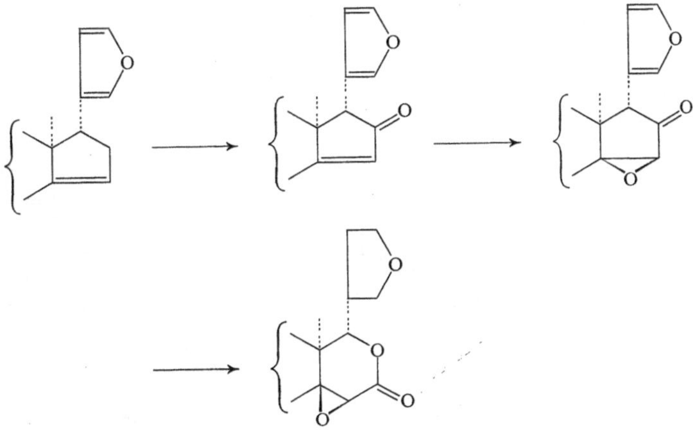

Scheme 5. Partial synthesis of gedunin

It seems possible that the key stage is the allylic oxidation of the double bond to a 16-ketone. Once this happens, formation of the lactone ring follows; if, however, the double bond is first epoxidised, then formation of the ketone is blocked and the five-membered ring remains intact.

4. Group IVa. Limonoids with Rings B and D Opened

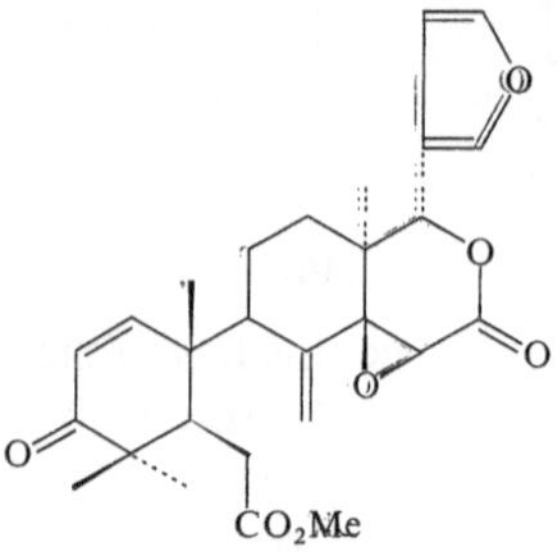

(37)
Andirobin

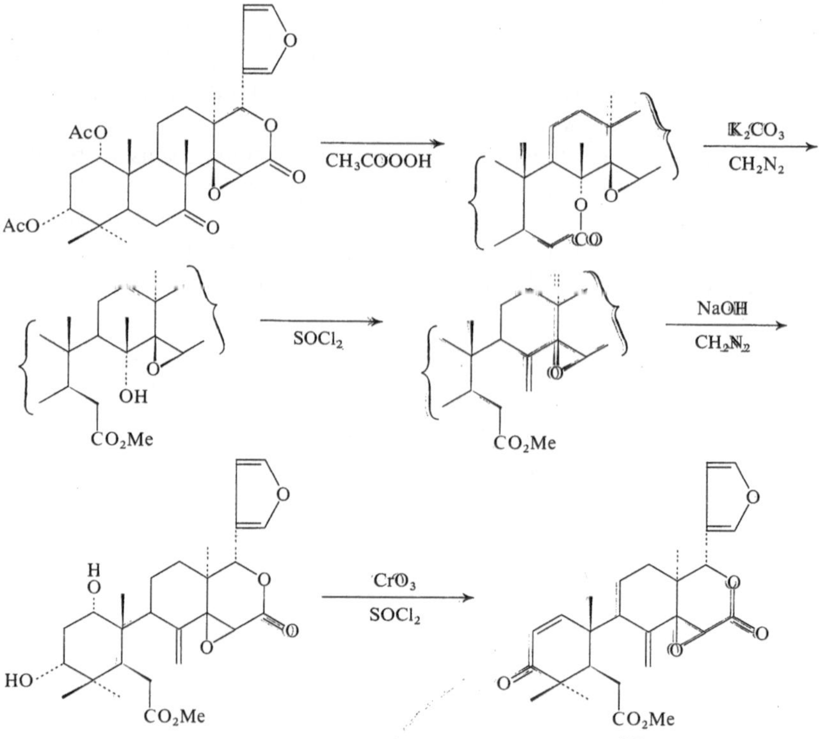

(37)

Scheme 6. Partial synthesis of andirobin

References, pp. 93—102

In compounds of this group, the B ring of khivorin derivatives is opened to a lactone or ester, apparently by oxidation of a 7-keto compound, without the formation of any subsequent carbon ring. This has been done in the laboratory by Baeyer-Villiger oxidation and andirobin (37) has been partially synthesised (74, 76) (Scheme 6). When a suitable derivative is prepared, cyclisation occurs spontaneously to give a 1,14-oxide; this can be formed from either a $\Delta^{14,15}$-1-hydroxy or a $\Delta^{1,2}$-14-hydroxy compound (90). The simplest example of this is methyl angolensate (38) which is one of the commonest limonoids. Cyclisation (Scheme 7) of the $1\alpha,3\alpha$-diol related to andirobin gave the 3α-alcohol, which was oxidised to methyl angolensate

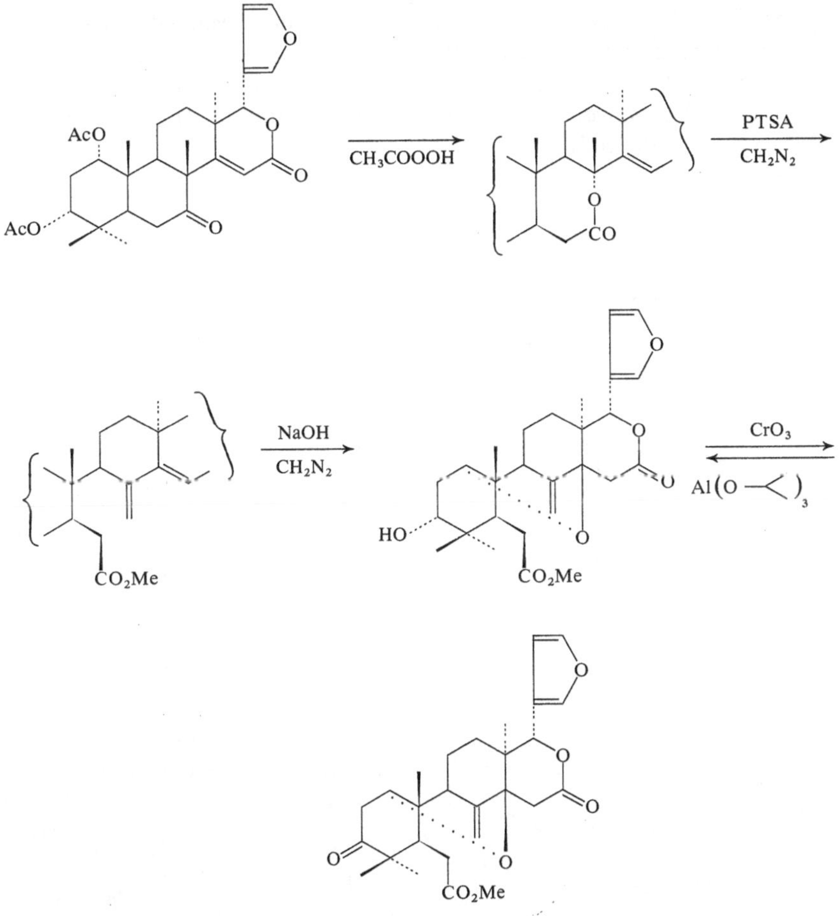

(38)
Methyl Angolensate

Scheme 7. Partial synthesis of methyl angolensate

2*

(74, 76). This alcohol can be obtained from methyl angolensate by borohydride reduction *(13)* or by rapid reduction with aluminium isopropoxide. Prolonged reduction with isopropoxide leads to equilibration to the equatorial 3β-isomer *(76)*.

An interesting member of this group is methyl 6,12-α-diacetoxy-angolensate **(39)** which is found in *Guarea thompsonii (77)*. This caused some problems at first, since the methyl group of the 12α-acetate is strongly shielded by the furan ring and resonates at δ 1.5. This effect has also been noticed in related compounds *(101)*, but does not occur in compounds of the havanensin group as mentioned earlier (P. 15). The configuration at C-6 in this compound is not known. In the related swietenine **(40)** *(61)* it is known to be *R* from x-ray crystallography, and it has been considered on spectroscopic grounds that methyl 6-hydroxyangolensate has the same configuration *(13)*. This corresponds to a 6-α-hydroxy group in the intact ring B, which is universal in all known 6-substituted limonoids except 6-hydroxyobacunol *(201)*.

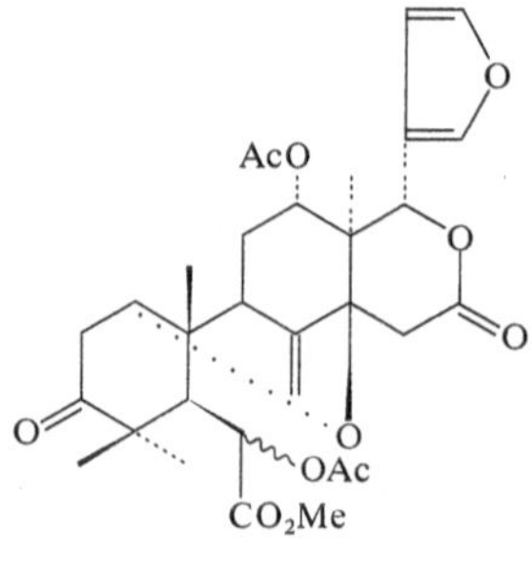

(39)

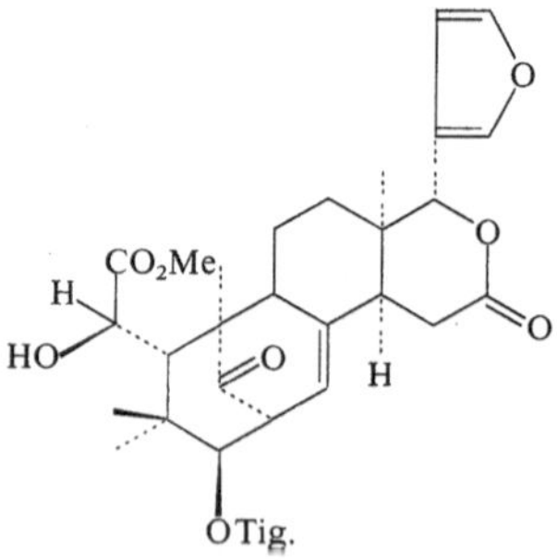

(40)
Swietenine

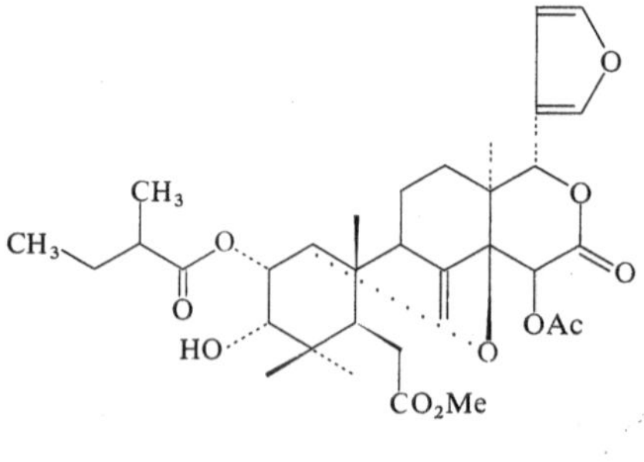

(41)
Ekebergin

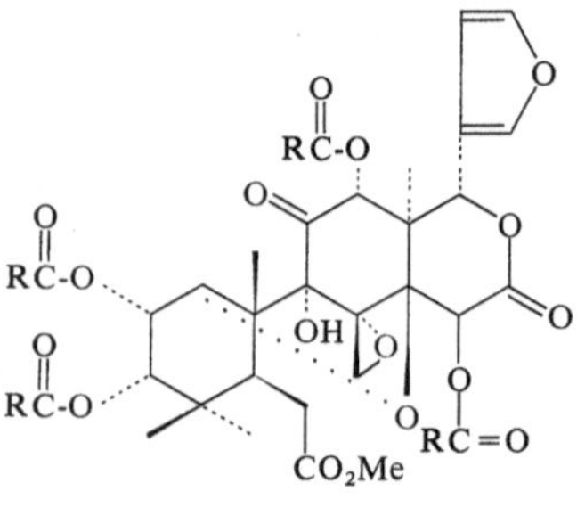

(42)
Ekebergolactone

Recently, more complex members of this group have been found in the genus *Ekebergia. E. capensis* (*197, 200*) contains ekebergin (**41**) and some related compounds, while *E. senegalensis* (*28, 57*) contains the more highly oxidised ekebergolactone (**42**). These compounds are unique in this group in containing a 15β-acetoxy substituent, although this also occurs in the prieurianin group.

A similar compound, deacetyl-2-deoxyekebergin (**42a**), is produced synthetically (*76*) as a by-product in the alkaline hydrolysis of the 1α,3α-diacetate corresponding to andirobin (Scheme 8). The mechanism of this change is obscure, since it takes place without inversion at C-14; possibly stabilisation of an intermediate by the methylene group at C-8 is involved. Whatever it may be, it seems reasonable that the mechanisms of the natural and the synthetic processes are similar and it may be significant that the 14β,15β-diols in the prieurianin series also have a methylene group at C-8.

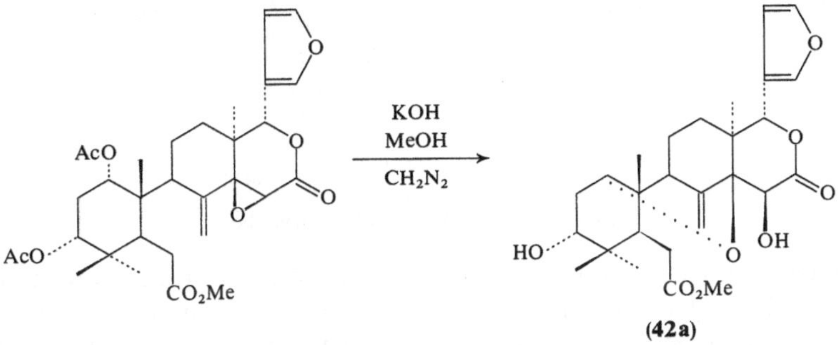

Scheme 8. Partial synthesis of ekebergin derivatives

5. Group IVb. Mexicanolide Group

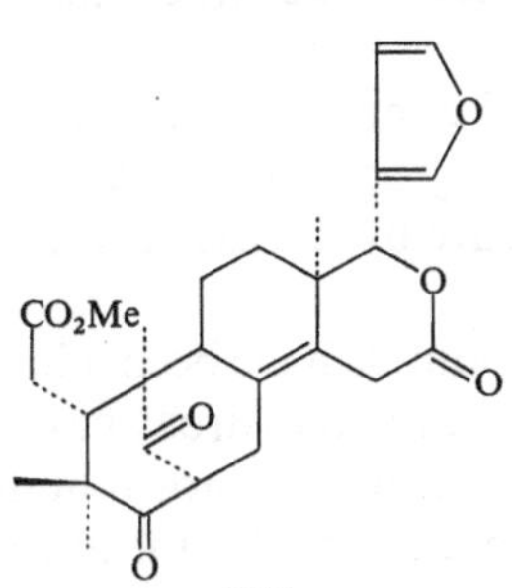

(**43**)
Mexicanolide

Compounds of this group are derived from 1,3-diketodiene lactones of the andirobin group by spontaneous Michael cyclisation, producing mexicanolide in the simplest case (*75, 76, 144*). In the synthesis (Scheme 9)

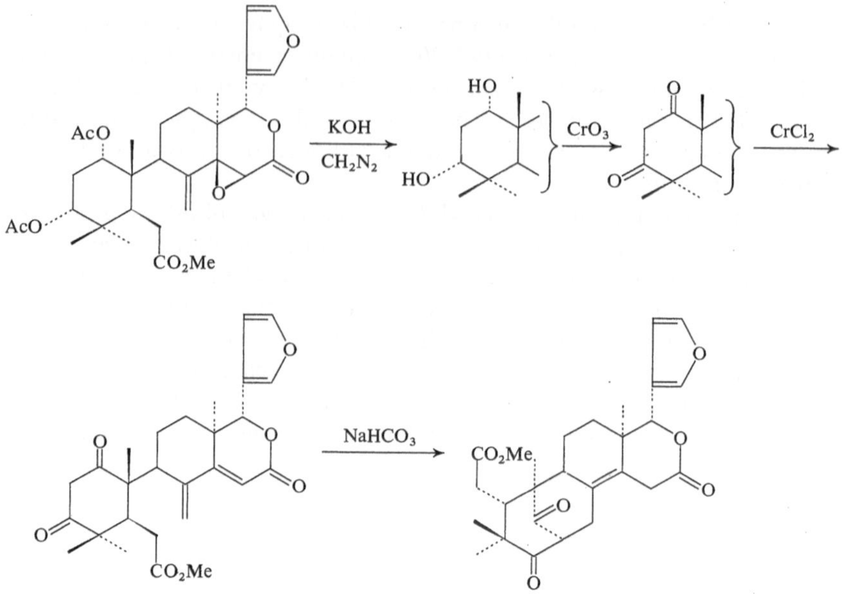

(43)

Scheme 9. Partial synthesis of mexicanolide

only mexicanolide (**43**), with the 8,14 double bond, is produced with no trace of the 8,30- or 14,15-isomers. These are common in natural products, and it is not known how they are produced. Synthetically, the only way they have been formed is by hydrogenation of the $\Delta^{8,30;\,14,15}$-diene (*130, 193*). This gives the $\Delta^{8,14}$- and $\Delta^{8,30}$-dihydro derivatives, with possibly a trace of the $\Delta^{14,15}$ compound. An earlier report (*21*) that carapin isomerised to mexicanolide on chromatography is untrue, and was due to the separation of an unrecognised mixture.

The $\Delta^{8,14}$, $\Delta^{8,30}$ and $\Delta^{14,15}$ compounds, as well as the $\Delta^{8,9;\,14,15}$-diene are all resistant to catalytic hydrogenation. The $\Delta^{14,15}$-16-oate system has been reduced to a dihydro derivative by the use of Raney nickel alloy in alkaline solution (*9*).

A number of a new compounds have been added to this group since the last review, which contain additional hydroxy or oxide groups. The simplest of these are the 2-hydroxy derivatives, known in the $\Delta^{8,30}$- and $\Delta^{8,14}$-series (*15, 159*). The biosynthetic mechanism for the introduction of this hydroxyl is not known and they have not been synthesised. They are readily recognised by the fact that H-3 resonates as a singlet instead of the usual rather wide doublet. Introduction of the 2-hydroxy has little effect on the resonance position of H-3, but a 2-acetoxy group (*15*) produces a downfield shift of $\delta\,0.4$. Next are the 8α-hydroxy compounds such as xyloccensin A

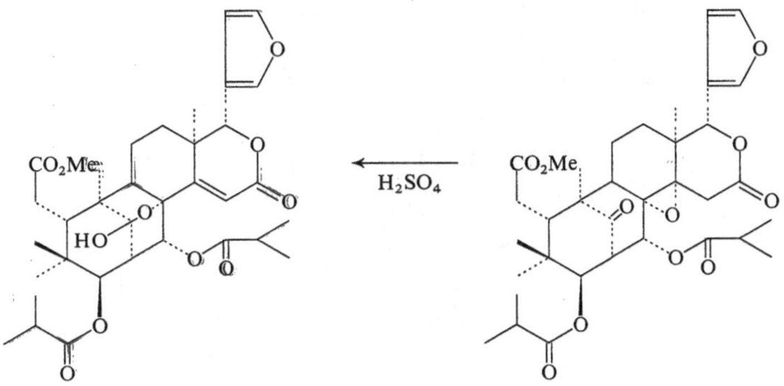

(44)
Xyloccensin A

(45)

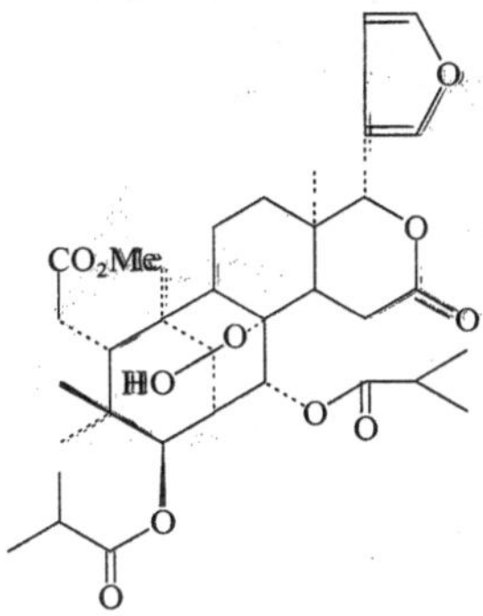

(46)
Xyloccensin F

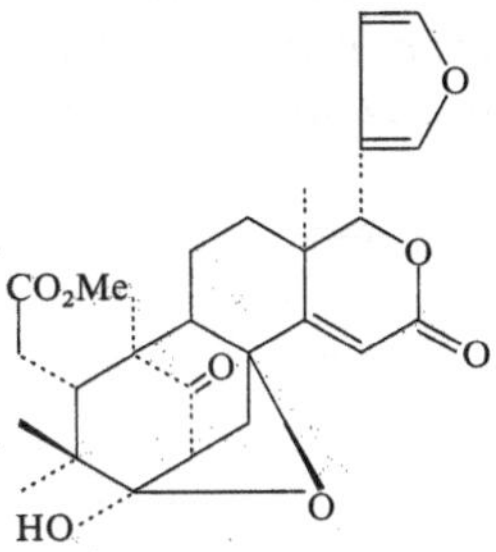

(47)

(44), which occur in *Xylocarpus moluccensis* (72). These spontaneously form 1,8-ketals, which are identified by the characteristic ketal carbon resonance at δ 109. It is probable that these are formed from 8,14α-oxides (45), which are also naturally occurring in *X. moluccensis* (200) and open with acid to give 8α-hydroxy-$\Delta^{14,15}$ compounds. The naturally occurring compounds such as xyloccensin F (46) which lack the 14,15-double bond appear to be produced by a reduction process, as must be the rare mexicanolide derivatives containing no double bond (2, 9, 191). Synthesis of 8α-alcohols or of 8,14-oxides has not been possible yet, as the 8,14 double bond is very resistant to oxidation in the laboratory. Allylic oxidation of carapin or mexicanolide with selenium dioxide gives an 8β-hydroxy-$\Delta^{14,15}$ compound, which forms a 3,8-ketal (47). This has also been found as a natural product (72).

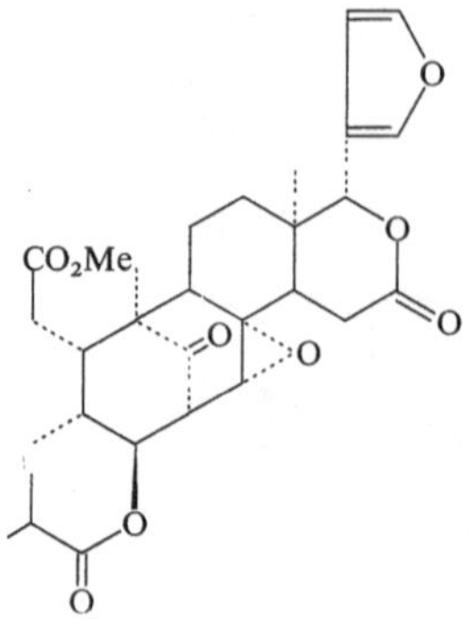

(48)   **(49)** 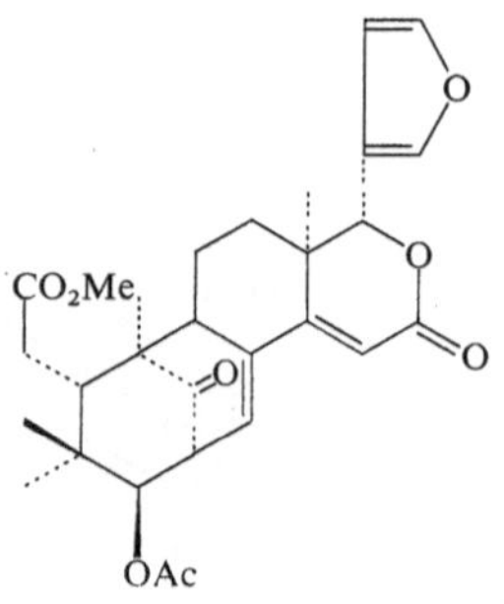   **(50)**
Xylocarpin   Angustidienolide

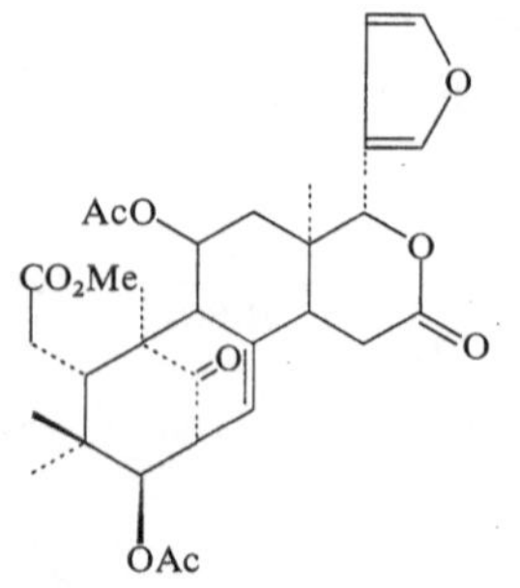

8,30-Oxides such as xylocarpin (48) are found in *Xylocarpus granatum* (*158*), *Swietenia humilis* (*159*) and *Swietenia macrophylla* (*200*), it is probable that they open to give 30α-hydroxy-$\Delta^{8,14}$ compounds (Scheme 10).

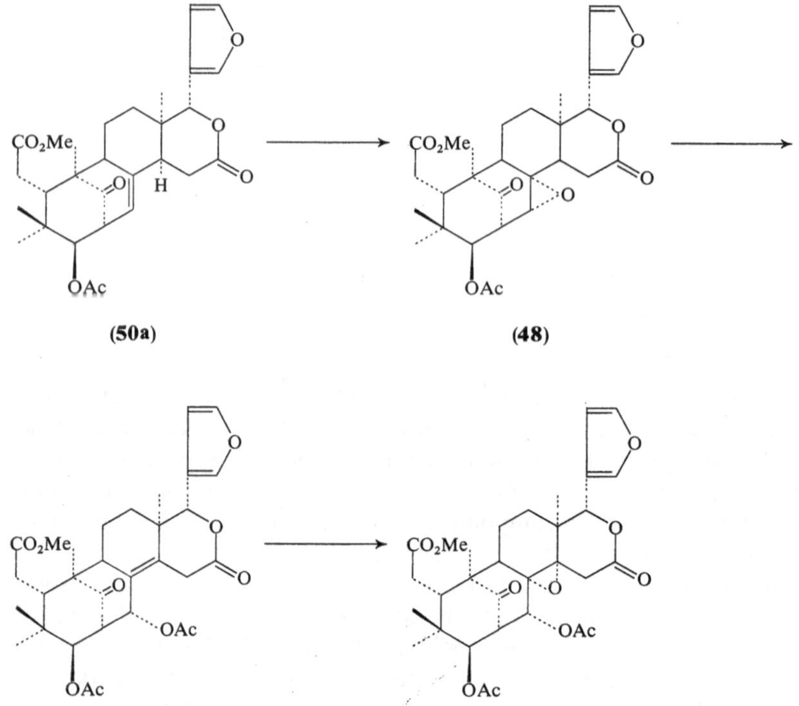

(50a)   **(48)**

unknown   **(45)**

Scheme 10. Hypothetical production of *Xylocarpus* compounds

Although these are not yet known, 30α-acyloxy-8,14-oxides are. The 8,30-double bond is also resistant to oxidation in the laboratory, and compounds of this type have not been prepared synthetically. Structural elucidation of all these compounds follows readily from n.m.r. spectroscopy; the use of ^{13}C n.m.r. is particularly useful in detecting the various oxygen functions. Only one mexicanolide derivative is so far known which is substituted in ring C; this is an 11β-acetate (49) isolated from *Khaya senegalensis* (13, 151).

Finally, the $\Delta^{8,30;\,14,15}$-diene augustidienolide (50) and its 2-hydroxy derivative occur in *Cedrela angustifolia* (130). Originally, the structure was misinterpreted as the H-30 resonance was not located in the ^1H n.m.r. spectra of angustidienolide, and the compounds were thought to be the $\Delta^{8,9;\,14,15}$-isomers. Both (50) and the $\Delta^{8,9;\,14,15}$-isomer have been synthesised from fissinolide (43a), by dehydrogenation with N-bromo-succinimide (193); a very much better yield of the diene is obtained from the $\Delta^{8,30}$ compound swietenine (65) (Scheme 11).

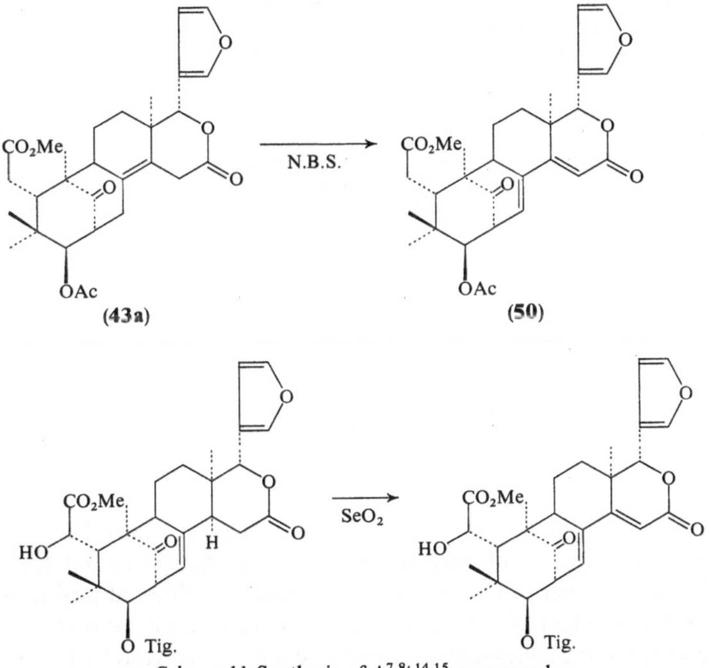

Scheme 11. Synthesis of $\Delta^{7,8;\,14,15}$ compounds

The nomenclature of these compounds is based upon the presumed biosynthesis of mexicanolide, in which the original methyl group C-30 of khivorin becomes the ring member joining C-8 and C-2. It has been proposed that the unsubstituted nucleus of mexicanolide should be called methyl meliacate (13).

6. Group IVc. Phragmalin Group

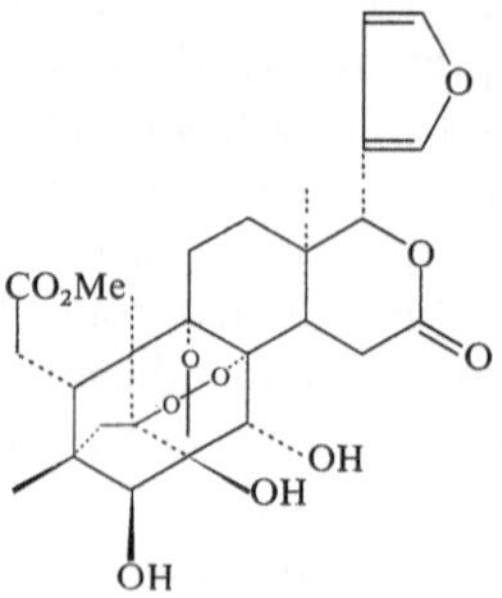

(52)
Phragmalin

Entandrophragmin, the longest known member of this group, was one of the first limonoids to be isolated from the Meliaceae (*19*) but the structure was unknown for a long time until it was deduced to be (**51b**) by x-ray crystallography of utilin (**51a**) (*102*).

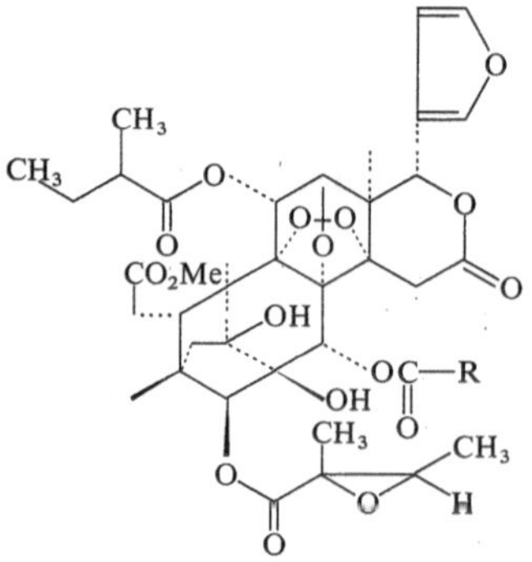

(51)
a R=Me Utilin
b R=iBu Entandrophragmin

Subsequently, several other members of this group have been isolated although they remain confined so far to *Entandrophragma* and a few other closely related genera. All contain an orthoacetate, either at positions 8, 9, 14 or at 1, 8, 9. The 1, 8, 9 orthoacetates are completely resistant to alkaline hydrolysis, but the 8,9,14-derivatives are readily hydrolysed with loss of the orthoacetate and β-furfuraldehyde (*19*), in a manner similar to the hydrolysis of gedunin. Presumably the reaction proceeds from the ring opened lactone. Loss of a proton from the C-17 hydroxyl then initiates a fragmentation reaction with loss of β-furfuraldehyde and the C-14 substituent and formation of a 13,14-double bond. With attack initiated in this way, complete hydrolysis of the orthoacetate follows.

Spectroscopic identification of members of the phragmalin group is fairly easy because of the presence of only three methyl groups, the characteristic singlets due to H-3 and H-30, and the orthoacetate carbon resonance at δ 119. The simplest member, phragmalin, (52) was found as an ester with nicotinic and acetic acids in *E. caudatum* (22), as a triacetate in *Xylocarpus moluccensis* (72) and as a series of esters in *Chukrasia* seed (62).

A frequent modification confined to this group is the presence of a C-acyl group at C-15. This gives a β-ketolactone structure, which was first identified in bussein (53) (100, 101). It has been suggested that this is produced by acyl transference from a C-30 ester. A second modification is the opening of the lactone ring to give a second methyl ester at C-16 and usually an acyloxy group at C-17, although in one case this is oxidised to a ketone (196). *Pseudocedrela kotschyii*, a small tree closely related to *Entandrophragma*, contains limonoids (54, 55, 56) of all three types (89, 137, 196).

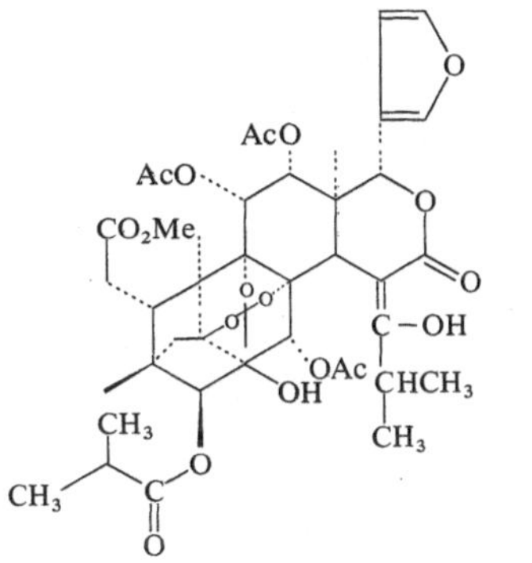

(53)
Bussein

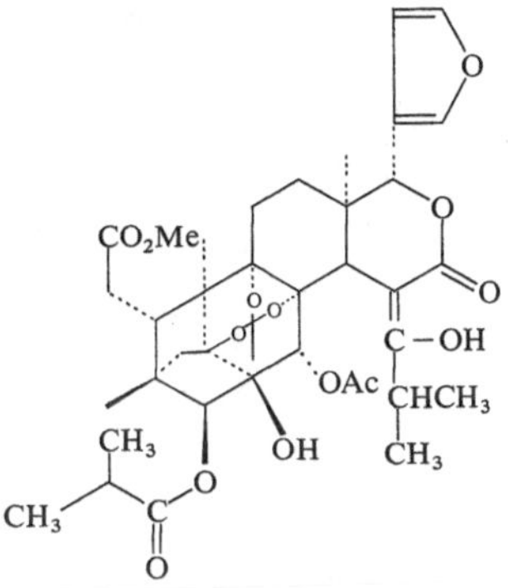

(54)
Pseudrelone A

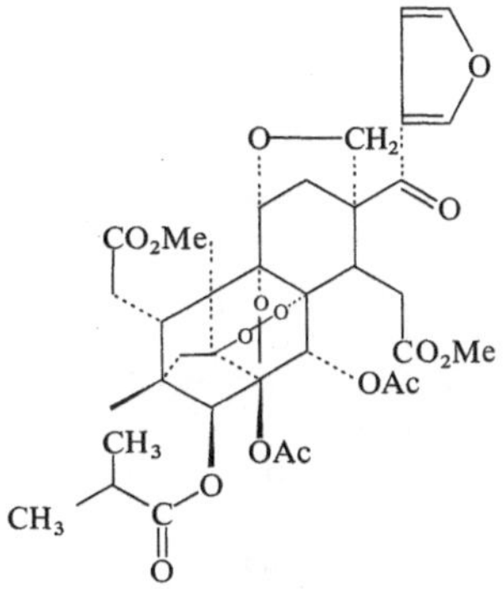

(55)
Pseudrelone B

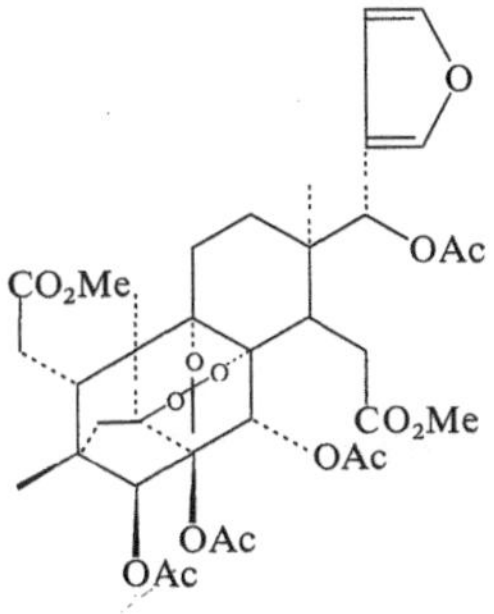

(56)
Pseudrelone C

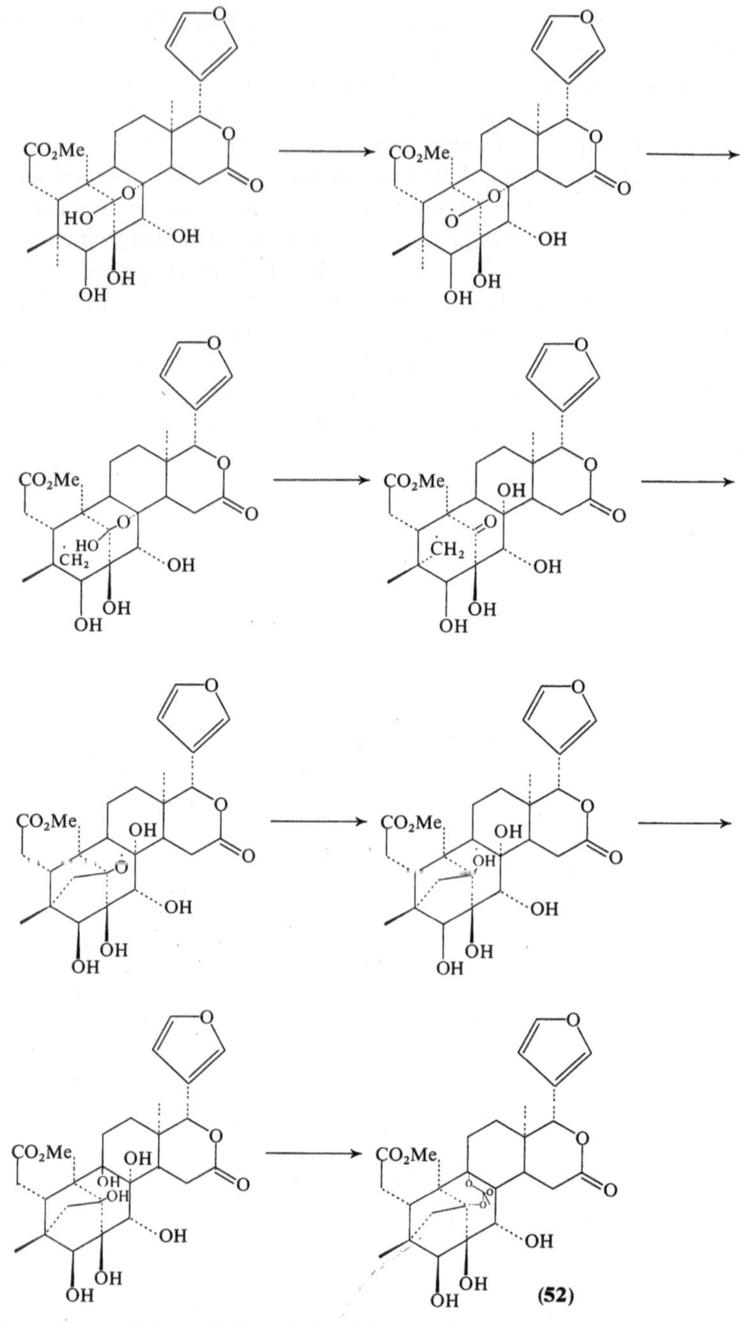

Scheme 12. Hypothetical biosynthesis of phragmalin

The most interesting question about phragmalin is how it is formed in the plant. Formally the C-29 methyl group is oxidised, but since the C-1 ketone is also reduced the transformation from mexicanolide is actually an isomerisation. It has been suggested that the precursor is a ketal of the *Xylocarpus* type, which yields an oxygen radical (*81*) (Scheme 12). This can then oxidise C-29 to a radical which can attack the ketonic form of the C-1 ketal, giving a second oxygen radical which finally oxidises C-9. Since C-8 oxidation is necessary to produce the original ketal, this scheme explains why phragmalin derivatives are always oxidised at C-8 and C-9.

An exception to this rule is tabularin, to which a ring A bridged structure unsubstituted at C-8 or C-9 has been ascribed (*34*). If this is correct, then the biosynthetic hypothesis is not correct in this case, and probably not in other cases. The structure assigned to tabularin was based on the discovery of a coupled chain of protons, considered to represent H-12, H-11, H-9, H-8, and H-15, and the similarity of the spectra assigned to atoms in ring A to those in similar positions in chukrasin. The diagnosis of the ring A bridge was largely based on the absence of a ketonic carbonyl resonance in the ^{13}C n.m.r. spectrum. As already mentioned, the introduction of the ring A bridge is an isomerisation and therefore cannot be detected analytically. In a weak spectrum such as the one obtained for tabularin carbonyl resonances can easily be missed, and if tabularin consists, as is so often the case, of a mixture of esters the count of C-methyl groups present may be unreliable because of confusion with ester groups. It would therefore be desirable for this work to be repeated.

7. Group V. Methyl Ivorensate Group (Rings A, B and D Opened)

(57)
Methyl ivorensate

This is still a very small group. The original member, methyl ivorensate (**57**), is found in small amount in *Khaya ivorensis* (*14*) and was synthesised by oxidation of methyl angolensate with perbenzoic acid (*15*). It has subsequently been found in *Soymida febrifuga* (*170*). *Khaya ivorensis* timber also yielded the ring A lactone (**58**) corresponding to mexicanolide (**43**) (*15*). This was not obtained by oxidation of mexicanolide and is presumably formed by cyclisation of the 1-ketone 3,4-lactone related to methyl ivorensate. Synthesis of (**58**) would present an interesting problem.

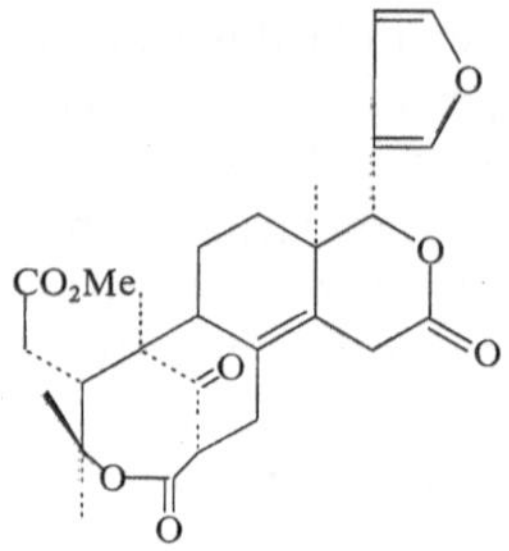

(**58**)

8. Group VI. Obacunol Group (Rings A and D Opened)

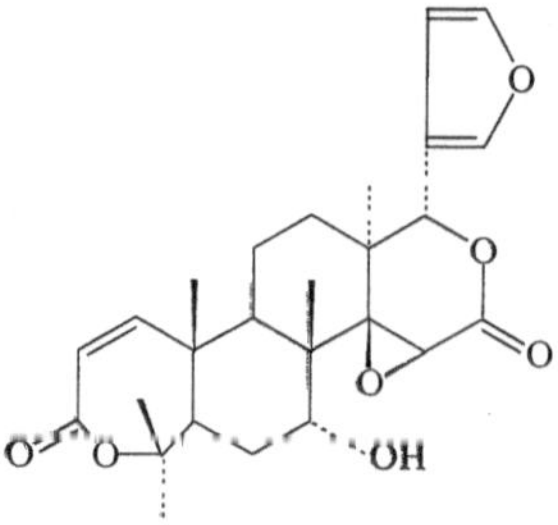

(**59**)
Obacunol

This group, to which limonin belongs, is characteristic of the Rutaceae and rare in Meliaceae. Obacunol (**59**) occurs in *Lovoa trichiliodes* (*7*) and in *Carapa procera* (*183*), dihydronomilin acetate (**60**) in *Xylocarpus granatum* (*16, 142*) and 6β-acetoxyobacunol acetate (**61**) in *Trichilia trifolia*. The stereochemistry of the 6-substituent in this compound is very unusual, but appears to be correct (*201*).

This group also includes some strange and much-altered compounds recently isolated from *Carapa procera* and *Carapa grandiflora* (*81*). The first of these, CP3 (**63**) was assigned the structure shown, partly on the basis of spectral data and partly on the basis of a presumed analogy to the spirolactone (**64**), of which the structure had previously been shown by x-ray crystallography. It was suggested that CP3 derived from a second spiro

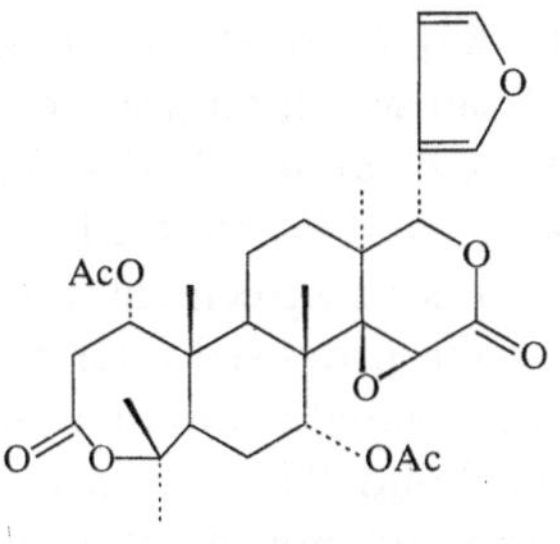

(60) 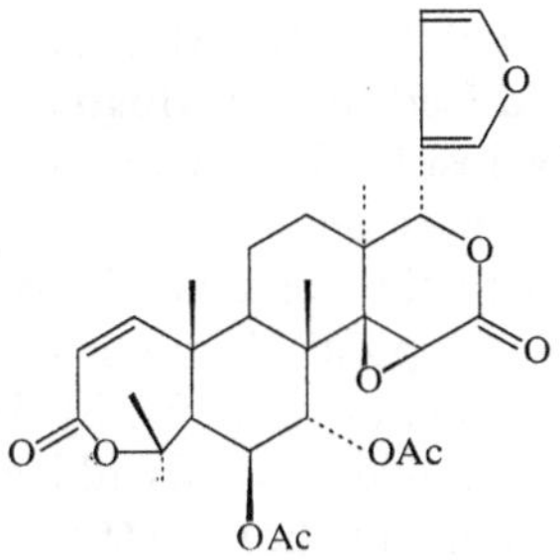   **(61)**

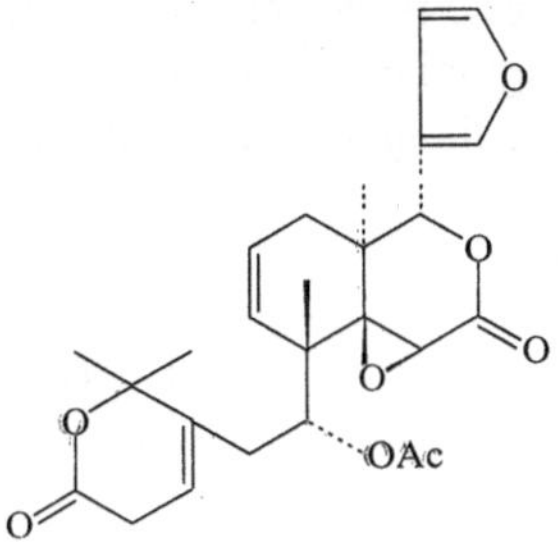

(63) 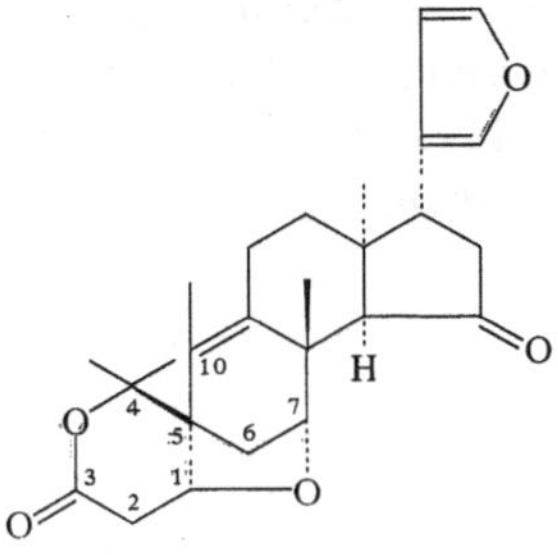   **(64)**

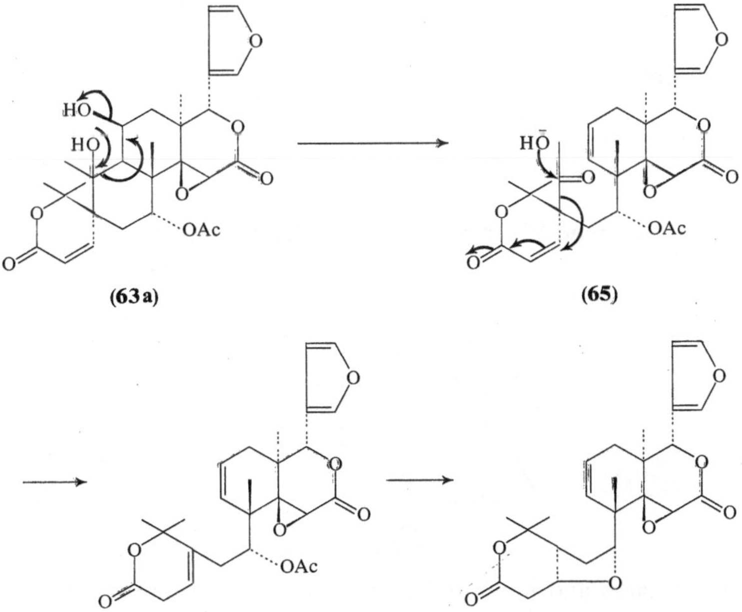

(63 a)   **(65)**

(63)   **(66)**

Scheme 13. Hypothetical biosynthesis of *Carapa* lactones

lactone (63 a) (Scheme 13) closely analogous to (64). *Retro*-Prins reaction of this would lead to a vinylogous β-diketone, from which an acetate could readily be lost to give the proposed structure. Acid catalysed esterification gave the diene ester, while basic hydrolysis caused addition of the hydroxyl produced to the conjugated system, both results in agreement with the structural proposal. A minor constituent was the proposed intermediate (65), still containing the methyl ketone, and the related compound (66) in which the 7α-hydroxyl has added to the double bond. The presence of the methyl ketone function in (65) supports the proposed biosynthetic scheme.

Examination of the related *Carapa grandiflora* gave two more related compounds (67, 68) in which the methyl ketone is present as an enol or ketal, cyclised to the hydroxyl group of the original compound. Opening of the enol ether of (67) with *p*-toluenesulphonic acid gave methyl ketone (65) obtained previously.

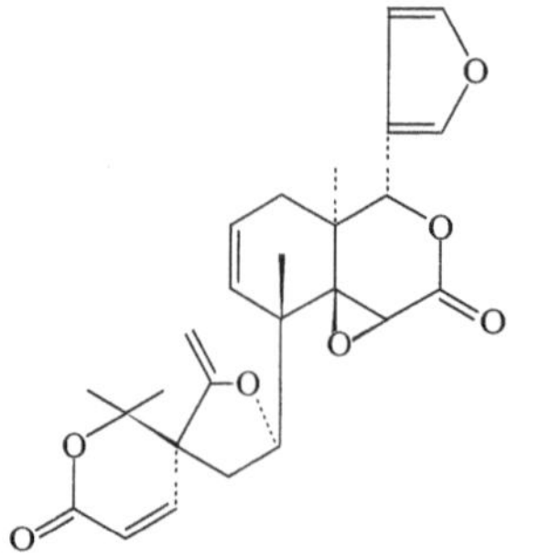

(67)

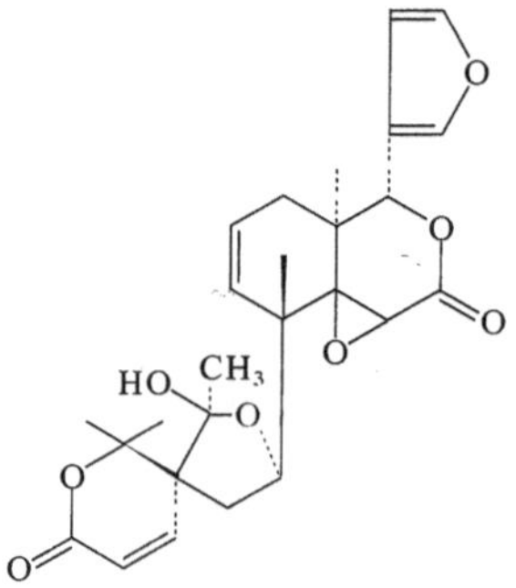

(68)

9. Group VII. Nimbin Group (Ring C Opened)

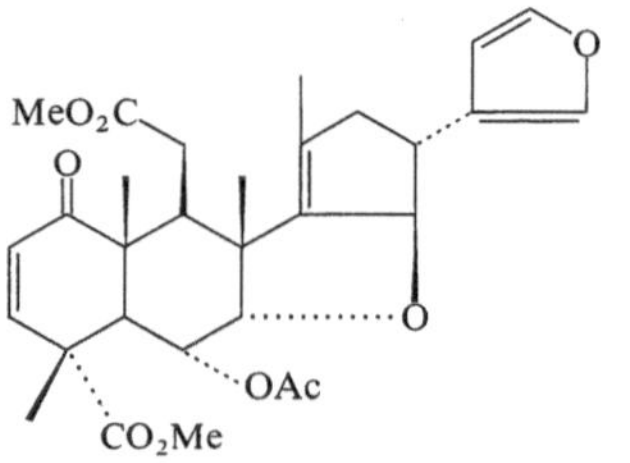

(69)
Nimbin

This is a large and important group containing some very complex compounds. At one time it was thought to be confined to the closely related genera *Melia* and *Azadirachta*, but heudebolin (70) a typical member of the group, was found in the bark of *Trichilia heudelottii* (*4*). This result is very curious. Although there is no doubt of the identity of the tree, there is always the possibility that extracts are confused during processing and it would be advisable to repeat this isolation. There are no very simple members of the group; the commonest feature in addition to the ring C opening is formation of a tetrahydrofuran ring between C-28 and C-6. Given that biosynthesis depends on the oxidative opening of ring C; similar to ring opening in other limonoids, the simplest derivative is the long-known salannin (**71**) (*103*) in which ring C is opened to a methyl ester. The subsequent changes can be rationalised as allylic rearrangement of a $\Delta^{14,15}$-13-ol, with addition of the 7α-hydroxyl to the allylic cation at C-15. This cyclisation probably depends on the extra flexibility bestowed on the molecule by opening of ring C, as 7,15-oxides do not occur in other groups.

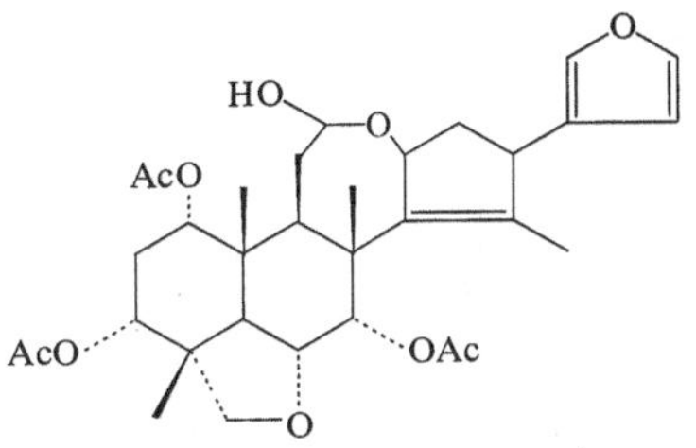

(**70**)
Heudebolin

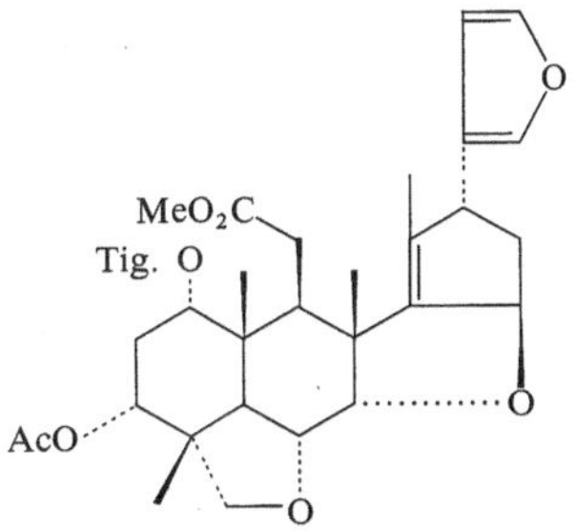

(**71**)
Salannin

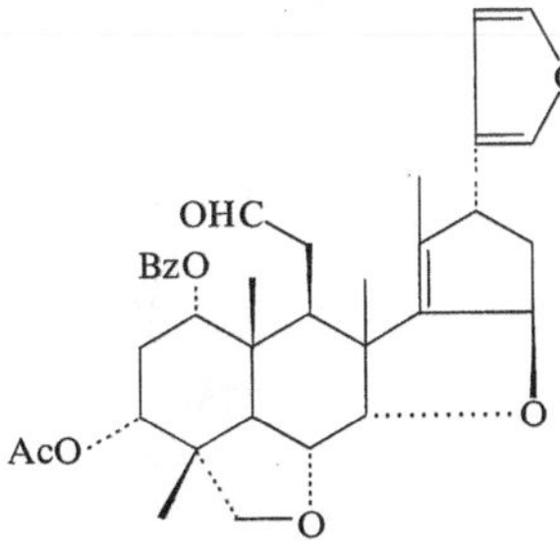

(**72**)
Ohchinal

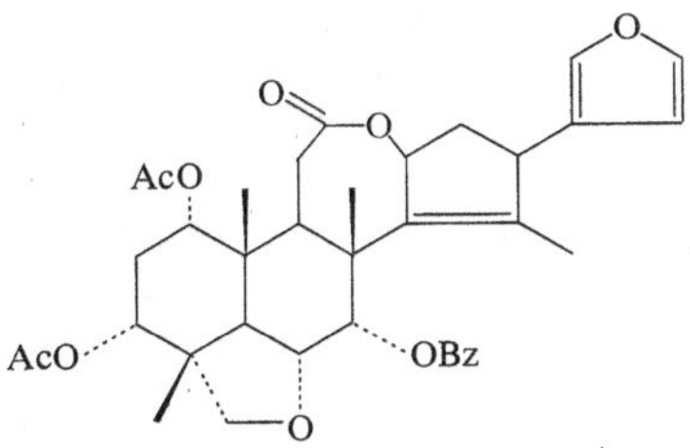

(**73**)
Ohchinolide

Ohchinal (**72**) (*148*) is similar except that C-12 is present as an aldehyde. The very common occurrence of C-12-aldehydes in this group suggests that the rationalisation presented in the previous paragraph is wrong and that the opening of ring C is not oxidative but hydrolytic. In this case the intermediate will be a 12β-hydroxy-14,15β-oxide (Scheme 14) which will

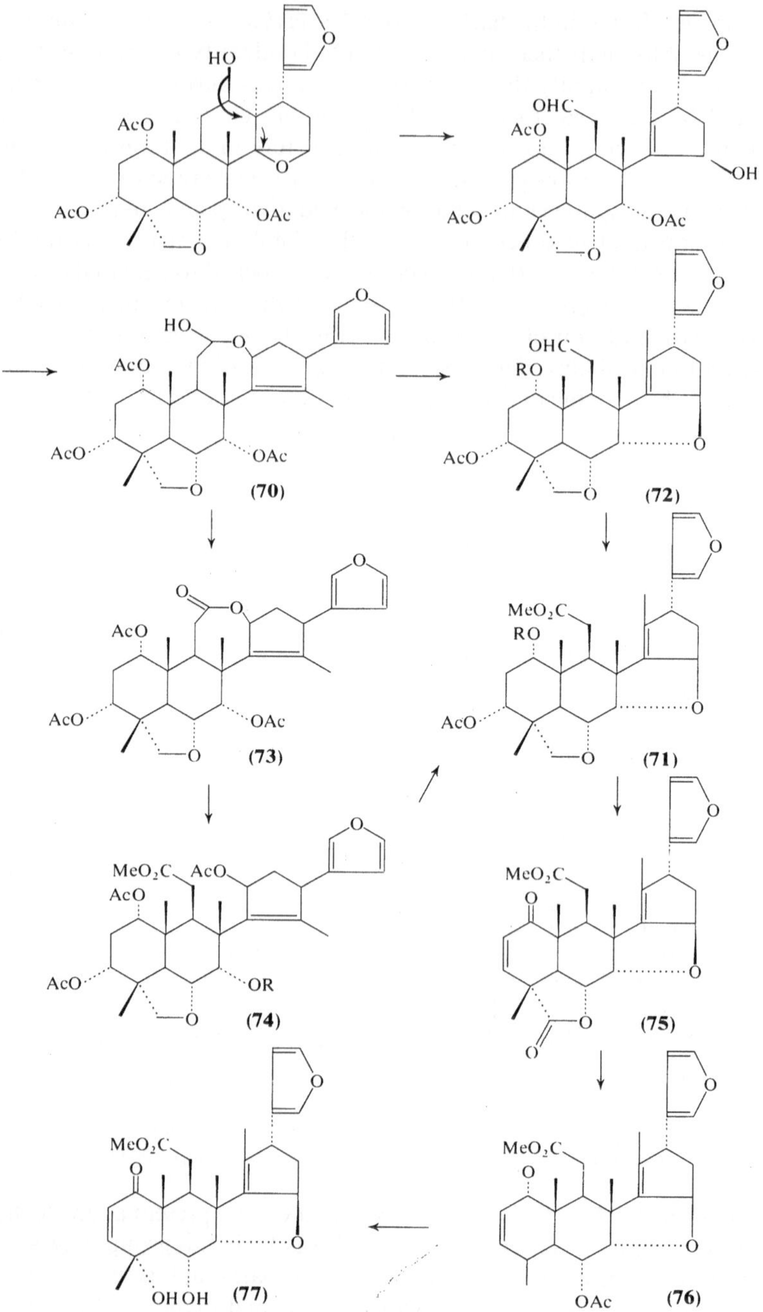

Scheme 14. Hypothetical biosynthesis of nimbin derivatives

open to give a 12-aldehydo$\Delta^{13,14}$-15β-ol. However, this proposal does not give such a simple explanation of the formation of the 7α,15β-oxide ring.

In heudebolin (**70**), which represents a common type in this group, the C-12-aldehyde has cyclised in a hemi-acetal with the 15-hydroxyl group. Related to this again is ohchinolide (**73**) in which the hemiacetal has been replaced by a lactone. Ochi has shown by x-ray crystallography (*146*) that the 15-oxygen of ohchinolide is β which, since ring D has rotated about the C-8, C-14-bond corresponds to a 15α-OH configuration in the original intact limonoid. ^{1}H n.m.r. measurements by KRAUS (*110*) indicate that the hemiacetals have the same configuration. This provides another biogenetic problem. A possible unifying hypothesis (Scheme 14) for these diverse products would be that ring opening proceeds originally from a 12β-hydroxy-14,15β-oxide by hydrolysis, as outlined above, and that the hydrate of the aldehyde attacks the allylic C-15 with inversion to give compounds similar to heudebolin which may then be oxidised to lactones like (**73**). Opening of the lactone or hemi-acetal ring C is accompanied by a second allylic substitution at C-15 by the 7α-hydroxyl, giving rise to compounds similar to salannin (**71**) and ohchinal (**72**).

Nimbolidin (**74**) (*110*) represents then an intermediate stage between ohchinolide and salannin. The next stage is oxidation of ring A to a 1-oxo compound, with elimination of the substituent on C-3, and oxidation of C-28 to a carboxyl group. This gives the lactone nimbolide (**75**) (*85*), the hydroxy ester form of which is nimbin (**69**). Decarboxylation at C-4 then gives nimbinene (**76**) (*113*), which is oxidised to a C-4-alcohol in nimbandiol (**77**). This type of oxidation giving rise to *nor*-derivatives is fairly common in terpenes and can occur during isolation of such compounds, especially when C-4 aldehydes are involved (*91*).

(78)
Azadirachtin

Azadirachtin (**78**) which is considerably more complex has been obtained from some seed specimens of *Azadirachta indica*. Thus NAKANISHI obtained 0.3% yield of (**78**) from a 300 g. sample from Mombasa (*202*), while Kraus working with 10 kg. from Maiduguri (*114*) did not obtain any. The structure was elucidated by advanced spectroscopic techniques. Preliminary investigation showed the presence of an acetate, a tiglate, two methoxy carbonyls, two other olefinic protons, one secondary and one tertiary hydroxyl, and two quaternary methyl groups, one of which resonated at the exceptionally low field of δ 2.06. The nature of the 35 carbon atoms was established by ^{13}C n.m.r. techniques, including partially relaxed Fourier transform spectroscopy combined with simultaneous continuous wave decoupling, which greatly simplified the assignment of peaks in congested regions. This showed, among others, the presence of four carbonyl carbons, eight other quaternary carbons including two ketals, ten saturated carbon atoms attached to oxygen, and also established that the olefin was an enol ether. ^{1}H n.m.r. spectroscopy aided by solvent shifts effected with benzene and Eu(fod)$_3$ showed two isolated AX pairs of 2H-30 and 2H-32, the systems H-1, 2H-2, H-3; H-5, H-6, H-7; and H-15, 2H-16, H-17, and a one proton singlet at H-9. Combination of these data led to the structure of azadirachtin. Acetylation esterified only the tertiary hydroxyl at C-14, the secondary one at C-7 being sterically hindered, and produced an anisotropic shift of the resonance of the proton vicinal to the tiglate. This demonstrated that the tiglate was at C-1 as in salannin, which shows a similar shift.

The configuration of the C-20 hydroxyl is α, *cis* to the 13-methyl group and responsible for its extreme deshielding to δ 2.06. A long range coupling was observed between the methyl resonance at δ 1.72 and H-9, thus locating the methyl at C-10 and showing their *trans*-periplanar relation. The methyl resonance at δ 2.06 showed a n.o.e. with H-9, thus further demonstrating the stereochemistry.

Azadirachtin probably arises from a compound similar to heudebolin by further oxidation *inter alia* at C-29, C-30 and by hydroxylation of the double bond.

No compounds with rings C and D opened, or with a 12-keto group and a lactone ring D, are known. It has been pointed out (*81*) that a 12-keto ring D lactone would be susceptible to reversed aldol reaction with loss of β-furfuraldehyde to give a quassinoid, and that this may well be the origin of quassinoids in nature. The similarity in ring A oxidation levels between nimbin and many quassinoids lends support to this view.

10. Group VIII. Toonafolin Group (Ring B Opened)

This group has only recently been discovered by Kraus (*119*) in further work on *Toona ciliata*. Toonafolin (**79**) (*118*) is the ring B lactone corresponding to cedrelone with the addition of a 1,11α-ether bridge. The

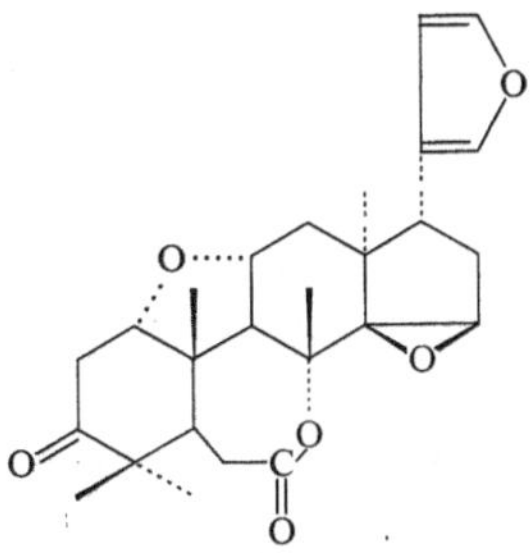

(79)
Toonafolin

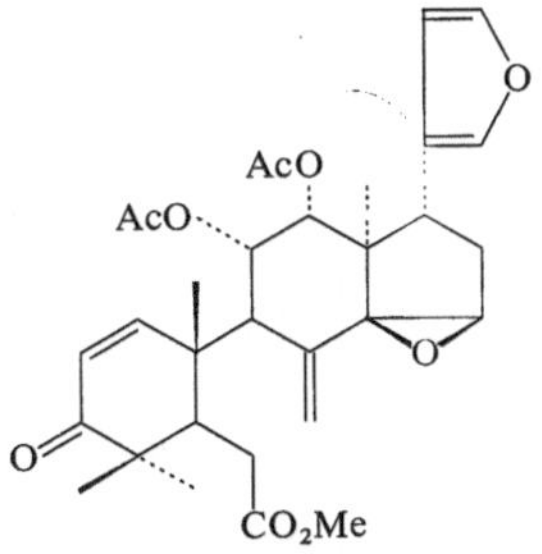

(80)
Toonacilin

structure was determined straightforwardly by n.m.r. techniques. Alkaline hydrolysis opened the lactone, thionyl chloride then gave the expected exomethylene compound. Toonacilin (**80**) (*119*) already has the lactone opened and has in addition an 11α,12α-diacetoxy system. The structure of this substance was determined by x-ray crystallography.

11. Group IX. Evodulone Group (Ring A Opened)

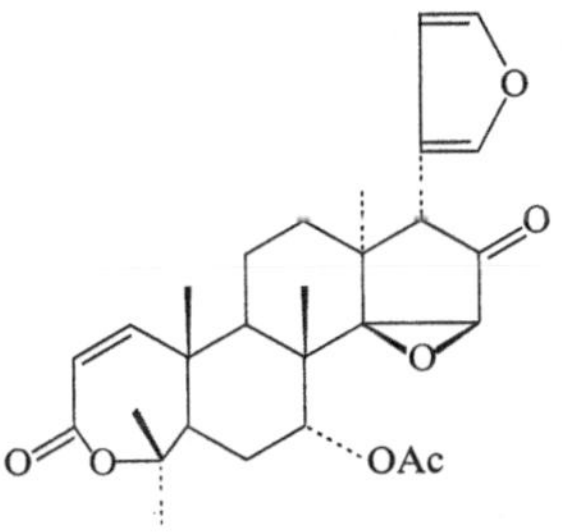

(81)
Evodulone

This group has also been discovered only recently, although representatives of it have for some time been suspected to be precursors of the prieurianin group of limonoids. Members have now been discovered in several taxa, as minor companions of prieurianin limonoids, in *Carapa* species, and in *Toona sureni*. *T. sureni* contains surenin (**82**) a simple ring A lactone corresponding to 6-acetoxyepoxyazadirone, and surenone (**83**), the corresponding 7-ketone (*115*), the structures followed readily from n.m.r. data. Proceranone (**84**), gumulin (**85**), and evodulone (**81**) are corresponding 6-deoxy compounds from *Carapa procera* (*182, 183, 184*) where they are

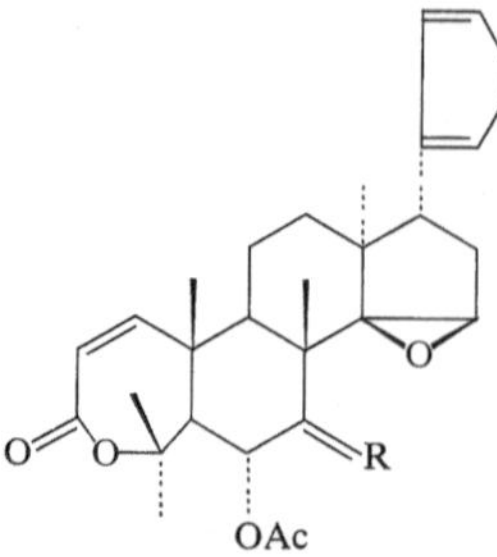

(82) R=H, a-OAc
Surenin
(83) R=O
Surenone

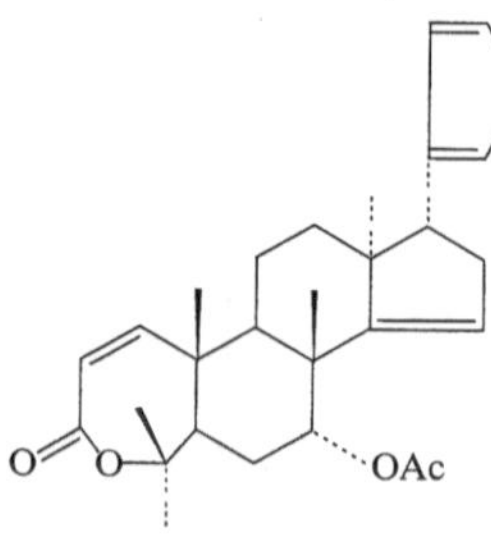

(84)
Proceranone
(85)
Gumulin
14,15β-epoxide

accompanied by a complex spirolactone (64) of which the structure was determined by x-ray crystallography (41). A study of *Trichilia dregeana* (136), gave three compounds of this group. The first, dregeana 3 (86), is a simple ring A lactone, while the other two (87, 88) also have a C-29 hydroxy group. These last compounds contain the characteristic 3-ethyl- and 3-methyl-2-hydroxybutyric acids of the prieurianin group and most probably represent biosynthetic intermediates. Finally, *Nymania capensis* has given a product (89) (132) in which ring A is opened to an isopropyl derivative similar to canaric acid. This has also a modified furan ring.

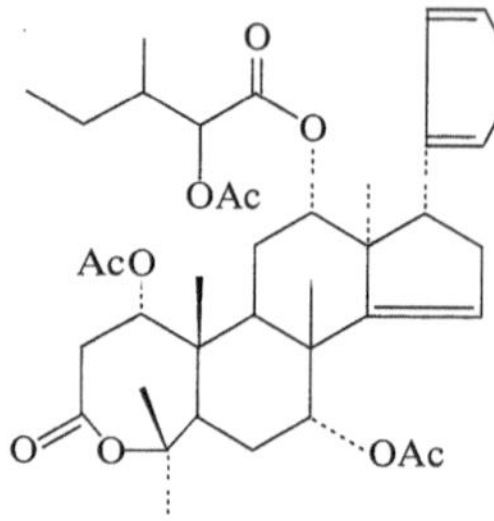

(86)
Dregeana 3

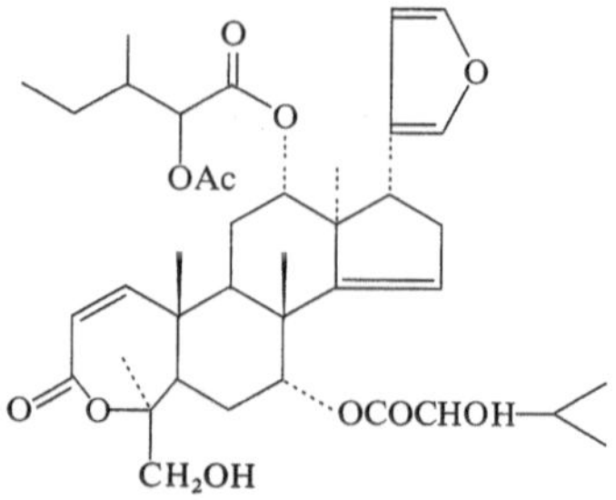

(87), (88) 1,2-Dihydro, 1 a-OAc

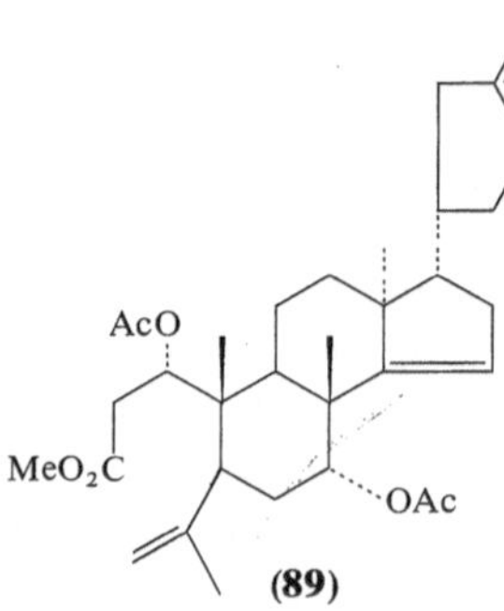

(89)

12. Group X. Prieurianin Group (Rings A and B Opened)

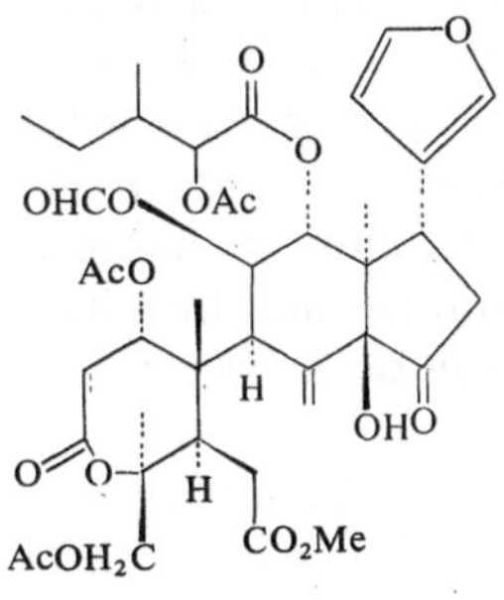

(90)
Prieuranin

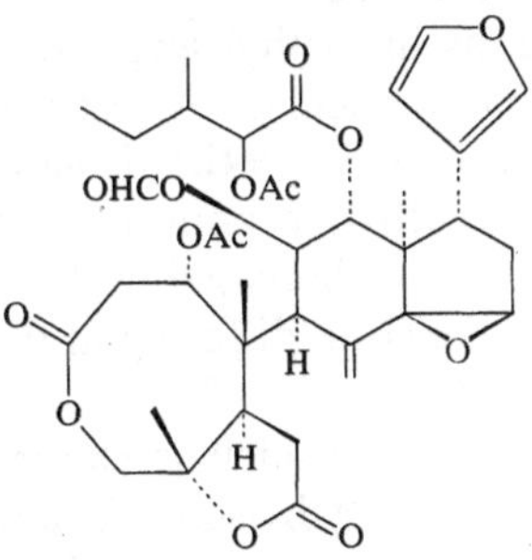

(91)
Dregeanin

Prieurianin (**90**) (*28*) and dregeanin (**91**) (*186*) were the first two members of this group to be described in 1965. Similar compounds have been found in the closely related genera *Trichilia* and *Guarea,* in *Aphanamixis,* and in *Nymania.* Most have the 11β-formyl,12α-(3′-methyl-2′-hydroxy)valerate system which also occurs in simpler *Trichilia* compounds such as heudelottin E (**18**).

A major problem in the structural work was that prieurianin and dregeanin gave badly resolved n.m.r. spectra, and it was only in 1975 that it was discovered, following a suggestion by NAKANISHI, that this was due to restricted internal rotation in the molecule, the effects of which can be overcome by warming the solution. At 60° well resolved spectra are obtained. After this, the structure of prieurianin was soon elucidated and at the same time independently confirmed by x-ray crystallography.

The rotation involved is presumably about the 9-10 bond which joins two halves of the molecule, but the details are unknown. Rohituka-7 (**92**) (*35*) of which the structure has been confirmed by x-ray crystallography (*109*), gives sharp spectra at room temperature.

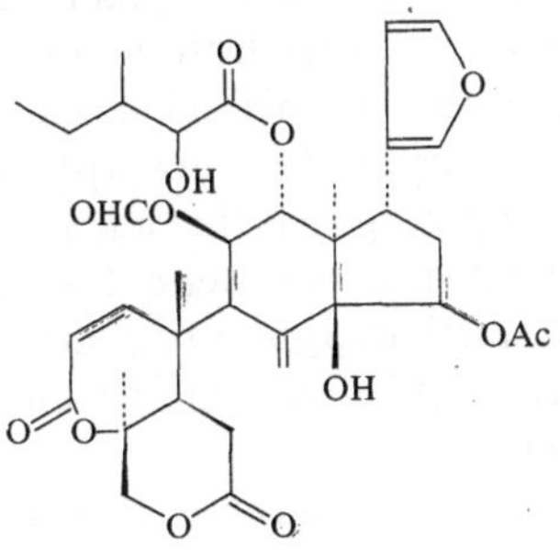

(92)

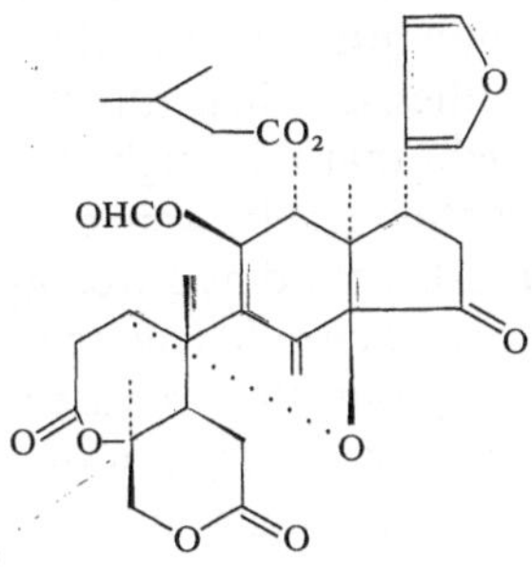

(93)

The identification of a number of compounds related to prieurianin was straightforward. One of the variants discovered contained a 1,14-oxide bridge, as in methyl angolensate (**38**). This was named polystachin (**93**) (*135*). Of particular interest for subsequent developments is the dilactone epoxide G.T.B., (**94b**), isolated from *Guarea thompsonii* (*63*). This is isomeric with dregeanin, and the two give a common methanolysis product, which will be discussed in detail later. It follows that they have the same carbon skeleton, differing in the arrangement of the lactone rings or ester groupings.

Dregeanin, which is stable to refluxing methanol, gives an unusual reaction with hydrogen over a platinum catalyst in methanol. The product has added one molecule of hydrogen and one of methanol, forming a dihydro derivative methyl ester. This no longer has the i.r. maximum at $1787 \, \text{cm}^{-1}$ shown by dregeanin and on oxidation gives a product with a u.v. maximum at 253 nm. ($\varepsilon \, 8 \times 10^3$), shifting in alkali to 301 nm. From the fact that hydrolysis gives the same product as does G.T.B., two structures are possible for dregeanin, (**91**) and (**95**).

(**94a**)
(**94b**) 1,2-Dihydro,1α-OAc   (**95**)

Structure (**91**) contains a γ-lactone ring which explains the i.r. maximum, but it was not originally seen how it could explain the hydrogenation results or how the unusual eight membered lactone ring would be biosynthesised. Structure (**95**) contains a very strained δ lactone, and it was thought that this might account for the observed i.r. absorption. The hydrogenation results were explained by supposing that the new saturated centre at C-8 in the dihydro compound imposed a greater steric strain on the molecule which would lead to spontaneous methanolysis. Oxidation of the 1-hydroxy group thus released would give a β-ketolactone which could explain the u.v. absorption of the product.

Further investigation of the hydrolysis of dregeanin raised more problems than it solved, and progress was only made after the discovery of a new type of compound, isolated independently by TEMPESTA *et al.* (*107*) in

Trichilia hispida (**96**) and by MacLachlan (*132*) in *Nymania capensis* (**97**). A similar compound (**98**) also occurs in *Guarea thompsonii* (*200*).

Nymania 1 is a methyl ester, analysing for a methanolysis product of prieurianin, with which it co-occurs. The ^{13}C-n.m.r. spectrum contains a singlet at δ 119.5, indicating an orthoester, and this led to the structure (**97**). A similar orthoester had previously been found in the hydrolysis products of dregeanin (*63, 133*).

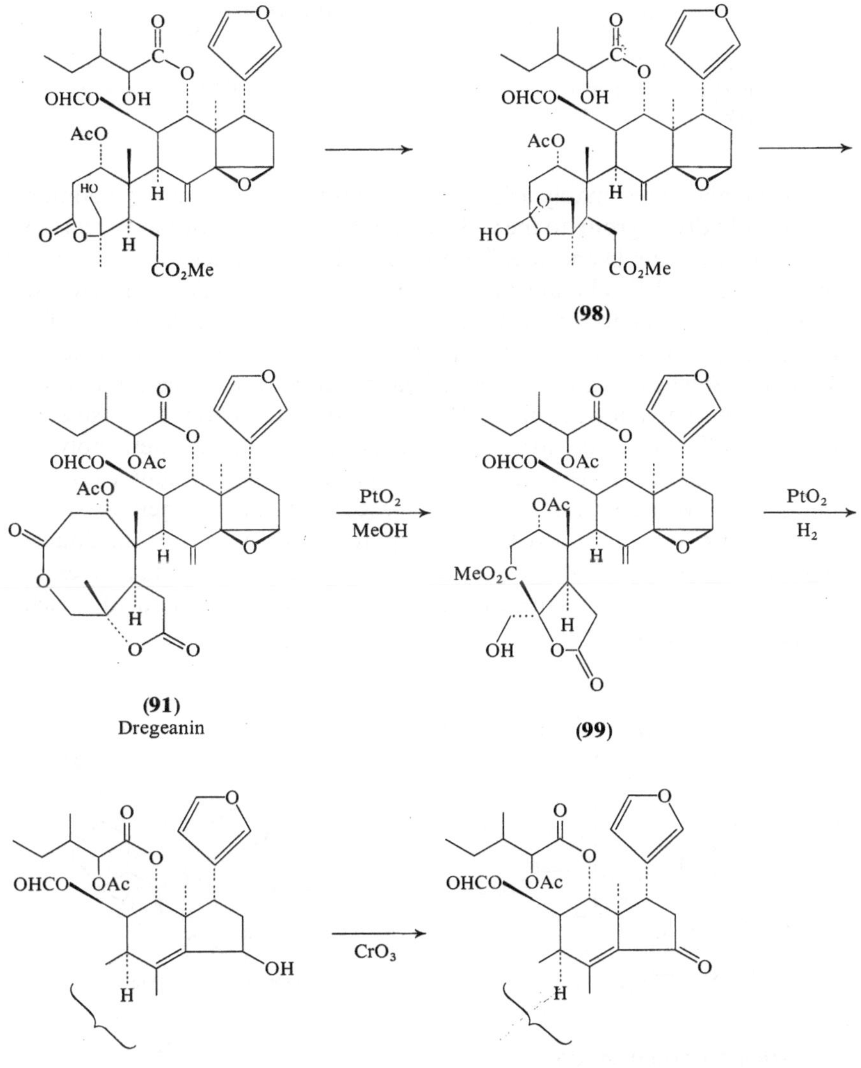

(98)

(91)
Dregeanin

(99)

Scheme 15. Hypothetical biosynthesis and reactions of dregeanin

(96) 14β-OH,15β-tiglyloxy
(97) 14β-OH,15-oxo
(98) 14,15β-epoxide

(99)

Nymania 1 presumably arises from 29-deacetylprieurianin by addition of the 29-hydroxy group to the C-3 lactone carbonyl as shown in Scheme 15 for the related ring D epoxide. Reversion of this orthoester to a lactone by breaking the C-3/O-4 bond will give a tertiary alcohol, which can lactonise with the C-7 ester to give (**91**), the alternative structure considered for dregeanin (*198*). A reinvestigation of the hydrogenation of dregeanin (*200*) showed that when dregeanin was shaken in methanol with platinum oxide in air, it was methanolysed giving D_5, (**99**), a compound previously isolated from *Trichilia prieuriana* (*63, 133*). The mechanism of this methanolysis, which is specifically dependent on the presence of platinum oxide, is unknown. Hydrogenation of D_5 then proceeds by allylic opening of the epoxide ring to give a 15-hydroxy group, followed by shift of the double bond to the more substituted position. Oxidation of the product gives an α,β-unsaturated ketone, the calculated absorption maximum of which agrees with that found.

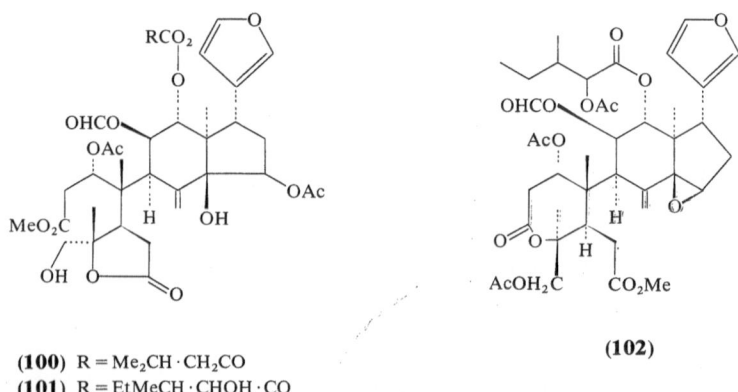

(100) R = Me$_2$CH·CH$_2$CO
(101) R = EtMeCH·CHOH·CO

(102)

References, pp. 93—102

Related to D_5 are the rohituka substances 1 and 2 (**100, 101**) (*35, 133*) which have analogous structures with a ring D glycol in place of the oxide present in D_5. The stereochemistry at C_{15} is assigned on the basis of x-ray crystallography of the related (**92**) (*109*). Originally, these compounds were thought to have structures analogous to prieurianin (**90**), but lacking the C-29 acetyl group. This was shown to be incorrect by the isolation of (**102**) from *Guarea guidona* (*131*), which was different from D_5 acetate (*133*).

Dregeanin and two γ-lactones (**103**) obtained from it on hydrolysis, to be described further below, show an i.r. maximum near $1790\,cm^{-1}$, while D_5 and the two rohituka substances show maxima at $1775\,cm^{-1}$. The reason for this is not known. It is possible that D_5 may have the isomeric structure (**104**) with a δ-lactone and a tertiary hydroxy group. The arguments against this alternative structure are that all six of these compounds show a C-4 resonance at about δ 90, distinct from any other dregeanin derivative, which has been explained on the basis that C-4 is included in a γ-lactone ring, (*133*). Also D_5 and the rohituka compounds show a shift in the 2H-29 resonance from ca. δ 3.8 (s) to δ 4.1, 4.3 (d, J = 12 Hz.) on acetylation, which is explained as due to the acetylation of a primary alcohol. Prieurianin has the C-4 resonance at δ 84.6, rohitukin, a 7,29 lactone, at δ 79.4 ppm.

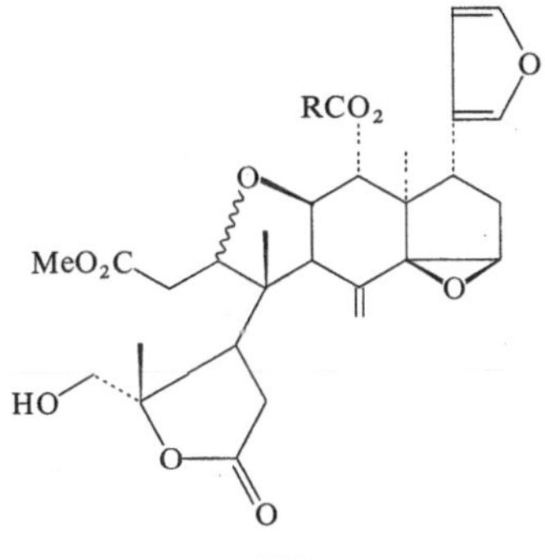

(**103**) 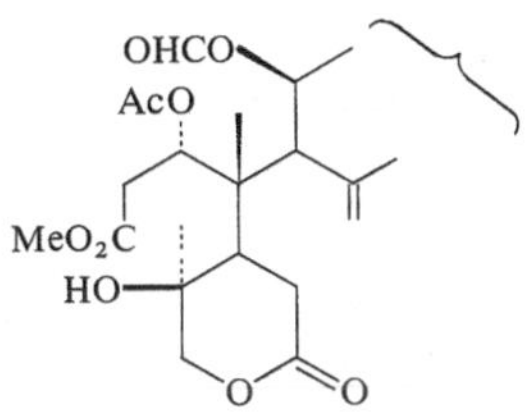   (**104**)

Investigation of the methanolysis of dregeanin and related compounds has given results which are still not fully understood. As shown above, dregeanin is methanolysed under mild conditions to give D_5. On more vigorous treatment, dregeanin and D_5 both give four products, a hemi-orthoester, a dilactone, and two γ-lactones. All these contain a 1,11-oxide link, detected by ^1H n.m.r. spectroscopy and also found in some natural products (*35*). The first three give the same equilibrium mixture under methanolysis conditions, the fourth is stable and is not formed from the other three. The only possible structures for the two γ-lactones are the two C-1 isomers of structure (**103**). Since they are not interconverted, it follows that the ether link is stable once formed.

Methanolysis of D_4 or of G.T.B. (94b) under similar conditions gives only the stable γ-lactone, with none of the other products. These results have been rationalised on the basis that in the methanolysis of D_4 and G.T.B. the oxide is formed by addition of an 11-alkoxy ion to the 1,2-double bond before opening of the lactone ring A. In the lactone, the β-face is hindered and so only the 1α-ether will be formed. In D_5 ring A is already open and there is no similar hindrance. No explanation has been given why the α-oxide, but not the β-oxide, should be stable to further methanolysis.

The dilactone methanolysis product has no i.r. absorption above $1760\,cm^{-1}$ and therefore is not a γ-lactone. This leaves only one structure for it, namely (105). It is very strange that this is the compound which, on the above hypothesis, cannot be formed from D_4 for steric reasons. The analogous compound with a 1α-ether linkage and a ring D ketol in place of the epoxide, is a natural product, rohituka-3 (35).

The non-crystalline hemi-orthoester formed on methanolysis of dregeanin is not stable, and on standing changes to the dilactone above. It could be either (106) or (107). It is not known which of these is correct, but it has been suggested (133) that the ^{13}C-n.m.r. spectrum is more consistent with (107), since the C-4 resonance at δ 84.6 is distinct from that in hispidin (96) or nymania 1 (97) in both of which it occurs at δ 82.6.

(105)   (106)   (107)

Similar methanolysis of prieurianin gives three products. The first two appear to be analogous to the two γ-lactones obtained from G.T.B. This can be rationalised on the same basis as before, either by supposing that the A ring in prieurianin is more readily opened than that in G.T.B., or that prieurianin is less hindered because it does not contain the C-4/C-7 lactone. Both these compounds appear to be stable to further methanolysis, unlike dregeanin.

The third product, isolated as its acetate, is the only methanolysis product known to have lost the 12-substituent. This was originally assigned structure (108a), but this formula does not explain the observed C-14 resonance at δ 87.0, whereas in all the other 14,15-ketols it is near δ 81. The C-14 signal is also found at δ 87 in polystachin (93) which has a 1,14-ether

(108a)   **(108b)**

link. There seems little doubt that the compound is really **(108b)**, the 11-hydroxyl being sterically hindered and therefore resistant to acetylation.

Investigation of the methanolysis of rohitukin was hindered by the shortage of material; the only product isolated appeared to be analogous to the 1β γ-lactone obtained from dregeanin.

Natural products of intermediate oxidation stages are also known. Dregeana 2 **(109)** *(136)* and nymania 3 **(110)** *(132)* have C-29 unoxidised and acetates at C-11 and C-12 while surenolactone **(111)** *(116)* has ring C unsubstituted and lactones in rings A and B. It is not likely that this compound is phytochemically related, though it formally belongs in this group. Biologically it appears to be related to toonafolin in group VIII.

(109)   **(110)**

(111)

The stereochemistry of this group still presents problems. As already mentioned it is not understood why some compounds but not others exhibit restricted rotation. Equally it is not known why the 15-ketone polystachin (**93**) exists as the 1,14 ether, while the corresponding 15β-acetate, rohituka-7 (**92**), exists in the open form and shows no sign of cyclising.

The biosynthesis of the ring D functionalities is interesting. The 14,15β-oxides represent the normal limonoid structure. How the 14β,15β-diol system is formed is not known; the mechanism may be similar to that operating in the case of the partially synthetic ekebergin derivative referred to on p. 21 (*76*). The mechanism of formation of the ketols is more interesting. The first hypothesis is the obvious one, namely that they are synthesised by oxidation of the corresponding diols, but this need not necessarily be so. Compounds containing 15-keto groups are quite common [cf. neohavanensin (*44*) or amoorastatone (*168*)], and are produced by the very ready acid-catalysed opening of the 14,15-epoxides. A second possibility for the genesis of the ketols is therefore hydroxylation at C-14 of a C-15 ketone.

III. Outlook

The limonoids of Madagascar species of the Meliaceae remain uninvestigated, and since these form a distinct biological group, they may easily contain limonoids of new types. Otherwise, all the main biological groupings have now been investigated and it seems probable that further progress will depend on the application of better separation techniques. These may yield many new compounds and types of compound in the future.

Many South American species and even genera remain uninvestigated, but since most of these are rather closely related to the *Trichilia-Guarea* group, it is likely that they will contain similar extractives.

Further work on the biosynthesis of limonoids is required. Experiments aimed at laboratory simulation of production of phragmalin and prieurianin are in hand, but experiments using labelled precursors in biological systems are badly needed. Research on the probable relationship of limonoids and quassinoids is also required.

IV. Tables

Table 1. *Protolimonoids*

(a) Side chain uncyclised. All these compounds belong to the euphane/tirucallol group with $\Delta^{7,8}$-14β methyl

Name	Structure	Origin	References
A-ring *seco*			
Entandrolide	A-ring, lactone, $\Delta^{24,25}$, 20 ξ H	*Entandrophragma angolense*	(161)
—	5-isopropenyl-3,21-dicarboxylic acid, $\Delta^{24,25}$, 20 ξ H	*Guarea cedrata*	(161)
—	3-monomethyl ester of above	*G. cedrata*	(17)
A-ring intact			

Table 1 (continued)

Name	Structure	Origin	References
Methyl kulonate	3-oxo, 16β-OH, 21-carboxylic acid methyl ester, $\Delta^{24,25}$, 20βH	Melia azedarach	(56)
Kulactone	16→21-lactone of methyl kulonate	M. azedarach	(53)
Kulolactone	3α-OH-dihydro kulactone	M. azedarach	(53)
Sendanolactone	6-oxokulactone	M. azedarach	(52)
Kulinone	3-oxo-16β-hydroxy, $\Delta^{24,25}$, 20βH	M. azedarach	(52)
Sapelin F	3α,21,23R,24S,25-pentahydroxy, 20αH	E. cylindricum	(49)
	3α,21,23R-trihydroxy-24S,25-epoxy, 20αH	E. utile	(161)

(b) Side chain containing a C-21, C-23R five membered acetal ring

Name	Structure	Origin	References

(i) Euphol-tirucallol group with $\Delta^{7,8}$-14β-methyl

3-Deoxymelianone	24R, 25-epoxy, 20αH	Melia azedarach	(178)
Melianone	3-oxo, 24R, 25-epoxy, 20αH	M. azedarach	(122, 123)
Melianol	3α-OH-dihydromelianone, 20αH	M.azedarach	(122)
epi-Turreanthin*	melianol acetate	Turreanthus africanus / Aphanamixis polystacha	(36) / (36, 55)
Turreanthin*	3β-OAc-dihydromelianone	T. africanus	(26, 27)
Meliantriol	3β,24R, 25-trihydroxy, 20αH	M. azedarach / Azadirachta indica	(123)

* In chloroform at room temperature this compound consists of an equilibrating mixture of 21-epimers (32).

(ii) *Apo* series. 7α-OH-$\Delta^{14,15}$-8β-methyl

Chisocheton A	3-oxo-24R,25-epoxy-21α-acetate, 20αH	Chisocheton paniculatus	(64)

D. A. H. TAYLOR:

Table 1 (continued)

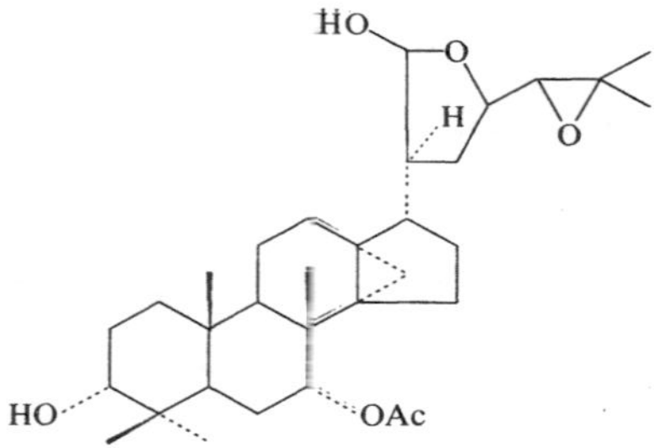

Name	Structure	Origin	References
Chisocheton D	3α-OH-dihydro-A	C. paniculatus	(64)
Chisocheton C	3α-acetoxy,24R,25-dihydroxy, 20αH	C. paniculatus	(64)
—	3β-acetoxy-21,23-lactone, $\Delta^{24,25}$, 20αH	M. azedarach	(125)

(iii) Other

| Glabretal | 3α-OH, 7α-OAc, 24R, 25-epoxy-14α, 18-cyclo, 20αH | Guarea glabra | (92) |
| — | Also the following: 3-(2′-hydroxyvaleryl), 3α-angelyl,3α-tiglyl,3α-methacryl and 3-oxo | G. glabra | (92) |

(c) Side chair containing a C-21,24R six membered oxide ring

Name	Structure	Origin	References
	(i) Euphol-tirucallol group with $\Delta^{7,8}$-14β-methyl		
Bourjotinolone A	3-oxo, 23R, 25-dihydroxy, 20αH	Trichilia hispida	(33, 108)
Sapelin A	3α,23R,25-trihydroxy, 20αH	Entandrophragma cylindricum	(48)
		E. utile	(161)
		T. hispida	(108)
	(ii) Apo series. 7α-OH, $\Delta^{14,15}$-8β-methyl		

Table 1 (*continued*)

Name	Structure	Origin	References
Sapelin C	3-oxo, 23R, 25-dihydroxy, 20αH	*E. cylindricum*	(49)
Sapelin D	3α, 23R, 25-trihydroxy, 20αH	*E. cylindricum*	(49)
Grandifoliolenone	$\Delta^{1,2}$-sapelin-C-7-acetate	*Khaya grandifolia*	(66)
16-Oxo-grandifoliolenone		*K. grandifoliola*	(67)
Chisocheton B	Sapelin-D-3-acetate	*Chisocheton paniculatus*	(64)
Melianin A	1α,OAc, sapelin-D-3-benzoate	*Melia azedarach*	(153)
Substance A	1,3-di-OAc, 16-oxo, $\Delta^{24,25}$-7-acetate, 20αH	*K. ivorensis*	(15)
24 -Hydroxy-grandifoliolenone		*E. spicatum*	(79)

(d) Side chain containing a C-21 to C-25 seven membered oxygen ring

Name	Structure	Origin	References

(i) Euprol-tirucallol group with $\Delta^{7,8}$-14β-methyl structure

Hispidone	3-oxo, 23R,24S-dihydroxy, 20αH	Trichilia hispida	(106)
Sapelin B	3α-hydroxy, 20αH	E. cylindricum T. hispida	(48) (108)

(ii) Apo series with 7α-OH, $\Delta^{14,15}$-8β-methyl

Sapelin E	3-oxo, 23R,24S-dihydroxy	E. cylindricum	(49)
Melianin B	1α-AcO, 3α-benzoyloxy, 23R,24S-dihydroxy, 7-acetate	M. azedarach	(153)

(e) Side chain containing a C-22 to C-25 five membered oxygen ring

(i) Euphol-tirucallol series with $\Delta^{7,8}$-14β-methyl

D. A. H. Taylor:

Table 1 (continued)

Name	Structure	Origin	References
Odoratone	3-oxo, 23ξ,24ξ-dihydroxy	Cedrela glaziovii C. odorata	(60) (46)
Odoratol	3α,23ξ,24ξ-trihydroxy	C. glaziovii C. odorata	(60) (46)
24-Isoodoratol		C. glaziovii C. odorata	(60) (46)

(ii) Other

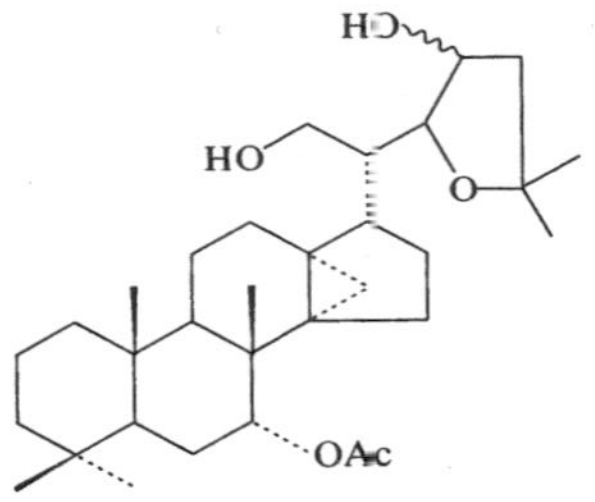 Name	Structure	Origin	References
Compound 26	3-oxo, 7α-OAc, 21,23ξ-dihydroxy, 14α,18-cyclo	G. glabra	(92)

References, pp. 93—102

Table 2. *Havanensin Group*

Name	Structure	Origin	References
(1) Vepinin	3-oxo, Δ^{1,2}, 7α,15β-oxide	*Azadirachta indica*	(141)
(2) Δ^{14,15} Azadirone	3-oxo-Δ^{1,2}, 7α-OAc	*A. indica* *Khaya anthotheca*	(121) (98)
6α-Acetoxyazadirone		*Chisocheton paniculatus* *Dysoxylon binectariferum*	(176) (181)
Chisocheton E	20,21,22,23-tetrahydro, 23-oxo, 6α-acetoxyazadirone	*C. paniculatus*	(64)
Chisocheton F	23-dihydrochisocheton E	*C. paniculatus*	(64)
Chisocheton G	21,23-dihydro, 21-hydroxy, 23-oxo, 6α-acetoxyazadirone	*C. paniculatus*	(64)
Compound 31	21,23-dihydro, 23-hydroxy, 21-oxo, 6α-acetoxyazadirone	*C. paniculatus*	(64)
Meldenin	3-oxo, 6α-OAc, 7α-hydroxy	*A. indica*	(58)
11α-Acetoxyazadirone		*K. anthotheca*	(98)
11β-Acetoxyazadirone		*K. anthotheca*	(88, 98)

Table 2 (continued)

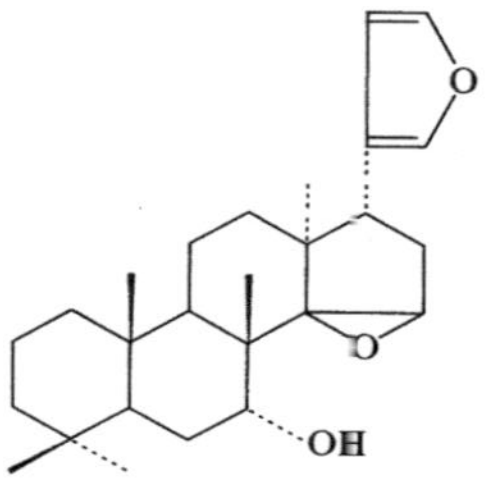

Name	Structure	Origin	References
Deoxyhavanensin 3,7-diacetate	1α,3α,7α-trihydroxy	K. anthotheca C. paniculatus	(9, 155) (64)
Vilasinin	1α,3α,7α-trihydroxy, 28, 6α-oxide	A. indica	(164)
Vilasinin 1,3-diacetate		C. paniculatus	(64)
Vilasinin triacetate		A. indica	(164)
Nimbolin A	Vilasinin-1,3-diacetate-7-cinnamate	A. indica Melia azedarach M. birmannica	(86) (86) (25)
Nimbidinin	12-oxovilasinin triacetate	A. indica	(134)
Sendanal	1α,3α,6α,7α,12α-pentahydroxy, 28-oxo, $\Delta^{14,15}$-diacetate	M. azedarach	(149)

(3) 14,15-epoxy,α-hydroxy

Trichilenone	3-oxo, $\Delta^{1,2}$, 7-acetate	T. havanensin	(44)

Compound	Description	Source	Ref.
Havanensin 1,7-diacetate 3,7-diacetate 1,3,7-triacetate	1α,3α-dihydroxy	— K. anthotheca T. havanensis	(44) (9, 155) (44)
Trifolin	Havanensin, 1-acetoxy, 7-(2'-OH-3'-methylbutyryloxy)	T. trifolia	(201)
Heudelottin C	3-oxo, Δ^{1,2}, 7α-(2'-hydroxy,3'-methylbutyryl, 11β-hydroxy,12α-(2'-hydroxy,3'-ethylbutyryloxy)	T. heudelottii	(9, 154, 156)
Heudelottin E	11-formoxy Heudelottin C	T. heudelottii	(9, 154, 156)
Heudelottin F	Heudelottin E-12-(2'-acetate)	T. heudelottii	(9, 154, 156)
Turrea A	1α,3α,7α,11β,12α-pentahydroxy-28-oic acid methyl ester 1,7,12-triacetate	Turrea floribunda	(18)
Turrea B	1α,3α,7α,11β,12α-pentahydroxy-28-oic acid methyl ester 1,7,11-triacetate, 12-isobutyrate	Turrea floribunda	(18)
(3a) 14,15-epoxy-19-hydroxy, 29-hemiacetal			
Amoorastatin	1α,7α-dihydroxy-3α-OAc, 11-oxo	Aphanamixis grandifolia	(167)
12α-Hydroxyamoorastatin		A. grandifolia	(168)
Sendanin	12α-acetoxyamoorastatin-29-acetate	M. azedarach	(145)

Table 2 (continued)

Name	Structure	Origin	References
29-Deacetyl sendanin	12α-acetoxyamoorastatin	M. azedarach	(147)
Aphanastatin	2α,12α-dihydroxyamoorastatin-1-acetate, 29-isobutyrate	A. grandifolia	(166)
Trichilin A	2α,12β-dihydroxyamoorastatin-2-acetate, 29-isobutyrate	Trichilia emetica	(138)
Trichilin B	2α,12α-dihydroxyamoorastatin-2-acetate, 29-isobutyrate	T. emetica	(138)
Trichilin E	2α,12α-dihydroxyamoorastatin-1-acetate, 29-isobutyrate	T. emetica	(138)
Trichilin D	12-deoxytrichilin A	T. emetica	(138)
Trichilin C	1α,2α,3α,7α,11β-pertahydroxy, 12-oxo, 2,3-diacetate, 12-isobutyrate	T. emetica	(138)

(4) 15-oxo, 7α-hydroxy

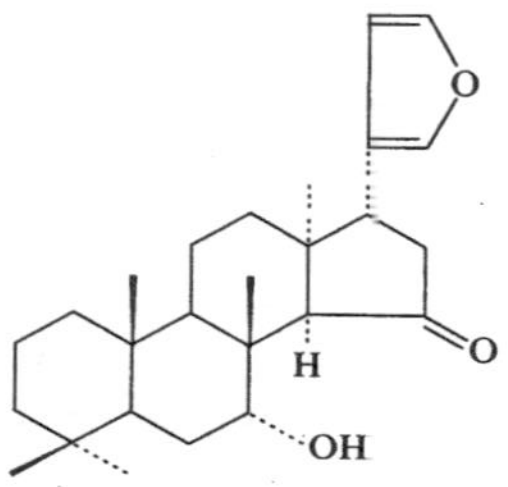

Neotrichilenone	3-oxo, Δ^{1,2}, 7-acetate	A. indica	(114)
Neohavanensin	1α,3α-dihydroxy	T. havanensis	(44)

(4a) 19-hydroxy-29-hemiacetal-15-oxo

Amoorastatone
(= Neoamoorastatin)   3-acetate   *A. grandifolia*   (168)

(5) 16-oxo, $\Delta^{14,15}$, 7α-hydroxy

Azadiradione	3-oxo, $\Delta^{1,2}$, 7-acetate	*A. indica*	(121)
17-Epiazadiradione		*A. indica*	(111)
Azadiradione benzoate	3-oxo, $\Delta^{1,2}$, 7-benzoate	*A. indica*	(114)
6α-Acetoxyazadiradione Dideacetyl-6α-acetoxy-azadiradione		*C. paniculatus*	(176)
17β-Hydroxyazadiradione		*A. indica*	(111)

D. A. H. TAYLOR:

Table 2 (continued)

Name	Structure	Origin	References
17β-Hydroxy-6α-acetoxy-azadiradione		C. paniculatus	(64)
Deoxykhayanthone	1α,3α-diacetoxy-7-acetate	K. nyasica	(190)
(6) 16-oxo-14,15-epoxy, 7α-hydroxy			
Epoxyazadiradione (= Nimbinin)	3-oxo, $\Delta^{1,2}$, 7-acetate	A. indica	(121)
Epoxyazadiradione benzoate	3-oxo, $\Delta^{1,2}$, 7-benzoate	A. indica	(114)
6α-Acetoxy-epoxyazadiradione		C. paniculatus Carapa guyanensis	(64) (129)
Diepoxyazadiradione	3-oxo, 1,2-epoxy, 7-acetate	A. indica	(114)
—	3-oxo, 1-methoxy, 7-acetate	A. indica	(114)
Grandifolione Khayanthone	1α,3α-diacetoxy 1α,3α-diacetoxy, 7-acetate	Khaya grandifolia K. anthotheca K. nyasica	(59) (9, 12) (190)

(7) 14,15-Epoxide, ring B diosphenol

Cedrelone	3-oxo, $\Delta^{1,2}$	Toona ciliata T. sureni K. anthotheca	(93, 94) (116) (2, 9)
Dihydrocedrelone	3-oxo	T. ciliata	(54)
Anthothecol	11α-acetoxycedrelone	K. anthotheca	(2, 30)
Deacetylanthothecol	11α-hydroxycedrelone	K. anthotheca	(9)
—	3-oxo-1α, 11-oxy, $\Delta^{9,11}$	K. anthotheca	(98)
Hirtin	3-oxo, $\Delta^{1,2}$, 11β-OAc, 12α-propionyloxy-29-oic acid methyl ester	Trichilia hirta	(47)

Table 3. *Gedunin Group*

Name	Structure	Origin	References
Gedunin		*Azadirachta indica*	(114, 121)
		Cabralea eichleriana	(175, 203)
		Chisocheton paniculatum	(176)
		Entandophragma angolense	(19)
		E. delevoyi	(186)
		Trichilia trifolia	(201)
		Xylocarpus granatum	(186)
Dihydrogedunin	1,2-dihydrogedunin	*Guarea thompsonii*	(30, 104)
Photogedunin	21,23-dihydro, 21-oxo, 23-hydroxygedunin	*Cedrela odorata*	(40)
7-Deacetylgedunin		*A. indica*	(121)
		Cabralea eichleriana	(175, 203)
		Chukrasia tabularis	(62)
		Khaya ivorensis	(15)
		K. grandifoliola	(2)
		Pseudocedrela kotschyii	(89)
		Trichilia trifolia	(201)
7-Deacetylphotogedunin		*C. eichleriana*	(175, 203)
7-Deacetylgedunin benzoate		*A. indica*	(114)

Compound	Modification	Species	Reference
7-Oxo-7-deacetoxygedunin		Cabralea eichleriana	(175, 203)
		Carapa guyanensis	(162, 163)
		Cedrela odorata	(30)
		Guarea guidona	(131)
		K. ivorensis	(15)
		K. senegalensis	(13, 30)
		P. kotschyii	(28, 89, 196)
3α-Gedunol	3α-hydroxy	G. thompsonii	(104)
7-Oxogedunol	3α-hydroxy, 7-oxo	G. thompsonii	(104)
6α-Acetoxygedunin		Carapa guyenensis	(129)
		Chisocheton paniculatum	(64)
6α-Hydroxygedunin		Carapa guyanensis	(129)
11β-Acetoxygedunin		C. guyanensis	(70)
6α,11β-Diacetoxygedunin		C. guyanensis	(70)
Khivorin		Khaya anthotheca	(12)
		K. grandifoliola	(30)
		K. ivorensis	(29)
		K. madagascariensis	(2)
		K. nyasica	(190)
		K. senegalensis	(10)

Table 3 (continued)

Name	Structure	Origin	References
3-Deacetylkhivorin		K. anthotheca	(12)
		K. ivorensis	(15)
		K. madagascariensis	(191)
		K. nyasica	(190)
		K. senegalensis	(10)
7-Deacetylkhivorin		K. ivorensis	(15)
		K. senegalensis	(2)
3,7-Dideacetylkhivorin		K. senegalensis	(2)
7-Oxo-7-deacetoxykhivorin		K. grandifoliola	(2)
		K. senegalensis	(30)
3-Deacetyl-7-oxo-7-deacetoxykhivorin		K. senegalensis	(10)
11β-Hydroxykhivorin		K. nyasica	(80, 187)
11β-Acetoxykhivorin		K. madagascariensis	(188, 191)
		K. nyasica	(2, 190)

Table 4. *Rings B and D Opened*

(a) Andirobin group

Name	Structure	Origin	References
Andirobin		Carapa guyanensis Cedrela odorata	(162, 163) (157)
Deoxyandirobin	$\Delta^{14,15}$	K. grandifoliola Soymida febrifuga	(2) (169)
—	21,23-dihydro-21-methoxy-23-oxo-deoxyandirobin	S. febrifuga	(169)
Methyl angolensate		Carapa procera Cedrela odorata E. angolense E. utile Guarea thompsonii K. grandifoliola K. ivorensis K. senegalensis Soymida febrifuga Swietenia mahogani	(183) (45) (19) (19) (77, 104) (28) (15) (3) (6) (189)

Table 4 (continued)

Name	Structure	Origin	References
Methyl-6-hydroxyangolensate		K. grandifoliola	(59)
		K. ivorensis	(15)
		K. senegalensis	(13)
		Swietenia mahogani	(189)
Methyl-6-acetoxyangolensate		K. grandifoliola	(2)
Methyl-6,12α-diacetoxy-angolensate		G. thompsonii	(77)
Ekebergin	2α-(2'-methylbutyryloxy),3α-hydroxy,15β-OAc-methyl angolensate	Ekebergia capensis	(197)

Also: 3-oxo, 3-acetate, and 16-deacetyl ekebergin			(200)
Ekebergolactone A	2α,3α,9α,12α,15β-pentahydroxy, 11-oxo, 8, 30-oxide diacetate, isobutyrate, 2-methylbutyrate	E. senegalensis	(28, 57)

Ekebergolactone B

2α,3α,9α,12α,16β-pentahydroxy, 11-oxo, 8, 30-oxide
triacetate, isobutyrate
Location of esters undertermined.

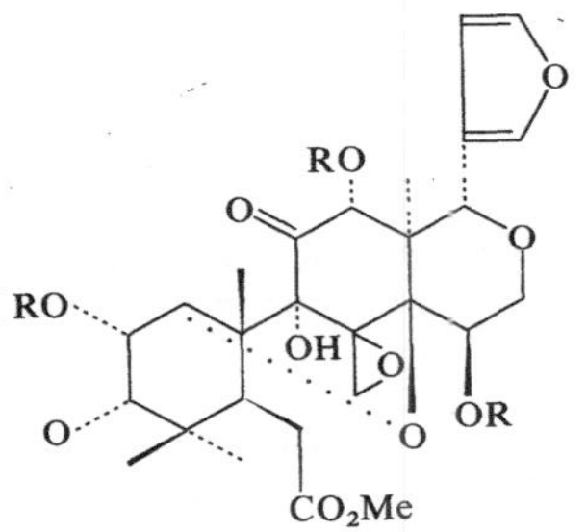

(b) Cyclisation forming a new ring B. Mexicanolide group

Name	Structure	Origin	References
	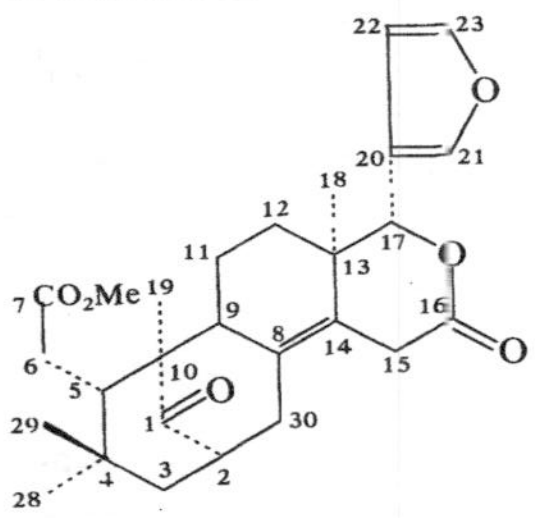		
Dihydrokhayasin	3β-isobutyryloxy, 8,14-dihydroxy	K. anthotheca	(2, 9)
2-Hydroxydihydrofissinolide	2β-hydroxy, 3β-OAc	K. madagascariensis	(2, 191)

5*

Table 4 *(continued)*

Name	Structure	Origin	References
Mexicanolide	3-oxo	Carapa procera	(30)
		Cedrela odorata	(1, 30, 68, 69)
		K. grandifoliola	(2, 28)
		K. ivorensis	(2, 29)
		K. senegalensis	(2, 13)
Carapin	3-oxo, $\Delta^{14,15}$ instead of $\Delta^{8,14}$	Carapa procera	(21, 30)
Fissinolide	3β-OAc	Structure	(130, 192, 193)
		Cabralea eichleriana	(203)
		Cedrela fissilis	(204)
		K. grandifoliola	(11, 192)
		K. ivorensis	(15)
		K. madagascariensis	(191)
		K. nyasica	(190)
		K. senegalensis	(2)
Khayasin	3β-isobutyryloxy	K. grandifoliola	(2)
		K. madagascariensis	(2)
		K. senegalensis	(30)
Cabralin	21,23-dihydro-21-oxo-23-hydroxyfissinolide	Cabralea eichleriana	(175, 203)
iso-Cabralin	21,23-dihydro-23-oxo-21-hydroxyfissinolide	C. eichleriana	(175, 203)
—	3β-hydroxy	C. eichleriana	(175, 203)
		Cedrela odorata	(157)
		K. ivorensis	(15)
		K. madagascariensis	(2)

Compound	Modification	Species	Ref.
—	3β-OAc, $\Delta^{8,30}$ instead of $\Delta^{8,14}$	K. senegalensis K. ivorensis K. nyasica Xylocarpus granatum	(2, 3) (15) (190) (158)
—	3-isobutyryloxy, $\Delta^{8,30}$ instead of $\Delta^{8,14}$	K. senegalensis K. nyasica	(2) (190)
Febrifugin	3β-tigloyloxy, $\Delta^{8,30}$ instead of $\Delta^{8,14}$	Soymida febrifuga	(173)
—	3β-OAc, $\Delta^{14,15}$ instead of $\Delta^{8,14}$	K. nyasica	(190)
—	3β-hydroxy, $\Delta^{14,15}$ instead of $\Delta^{8,14}$	K. grandifoliola	(2)
2-Hydroxyfissinolide Also acetate	3β-OAc, 2-hydroxy	K. ivorensis	(15)
2-Hydroxykhayasin	3β-isobutyryloxy, 2-hydroxy	K. madagascariensis K. senegalensis	(2) (200)
—	3β-OAc, 2-hydroxy, $\Delta^{8,30}$ instead of $\Delta^{8,14}$	K. nyasica	(190, 200)
—	3β-tigloyloxy, 2-hydroxy, $\Delta^{8,30}$ instead of $\Delta^{8,14}$	Swietenia humilis	(159)
—	3β-isobutyryloxy, 2-hydroxy, $\Delta^{8,30}$ instead of $\Delta^{8,14}$	S. humilis	(159)
Swietenolide	3β,6-dihydroxy	S. macrophylla	(50)
Swietenolide 3-tiglate		K. ivorensis S. macrophylla	(158) (200)
Swietenolide 3-isobutyrate		K. ivorensis	(158)
Swietenolide diacetate		K. ivorensis S. macrophylla	(15) (200)

Table 4. (continued)

Name	Structure	Origin	References
Swietenine	3β-tigloyloxy, 6-hydroxy, $\Delta^{8,30}$ instead of $\Delta^{8,14}$	K. nyasica S. macrophylla	(190) (61)
Swietenine acetate		S. macrophylla	(200)
—	3β-isobutyryloxy, 6-hydroxy, $\Delta^{8,30}$ instead of $\Delta^{8,14}$	K. nyasica	(190)
—	3β-OAc, 6-hydroxy, $\Delta^{14,15}$ instead of $\Delta^{8,14}$	K. senegalensis	(200)
6-Hydroxymexicanolide		Cedrela odorata K. senegalensis	(157) (2)
—	3β,11β-di-OAc, $\Delta^{8,30}$ instead of $\Delta^{8,14}$	K. senegalensis	(2, 3)
Angustidienolide	3β-OAc, $\Delta^{8,30;14,15}$	C. angustifolia	(130, 193)
2-Hydroxyangustidienolide		C. angustifolia	(130, 193)

Name	Structure	Origin	References
Xylocarpin	3β-hydroxy, 8α,30α-oxide	X. granatum	(158)
Humilin B	3β-OAc, 2-hydroxy, 8α,30α-oxide	S. humilis	(159)

Name	Structure	Origin	References
Swietenine acetate oxide	3β-tigloyloxy, 6-OAc, 8α,30α-oxide	S. macrophylla	(200)
Xyloccensin H*	3β-isobutyryloxy, 8α,14α-oxide	X. moluccensis	(200)
Xyloccensin G*	3β,30α-diisobutyryloxy-8α,14α-oxide	X. moluccensis	(200)
Xyloccensin C*	3β-isobutyryloxy, 2,8α-dihydroxy, 1,8 ketal	X. moluccensis	(72, 200)
Xyloccensin B*	3β,30α-diisobutyryloxy-8α-hydroxy, 1,8-ketal	X. moluccensis	(72)
Xyloccensin F*	3β,30α-diisobutyryloxy, 2,8α-di-hydroxy, 1,8-ketal	X. moluccensis	(72)
Xyloccensin A*	3β,30α-diisobutyryloxy-8α-hydroxy, $\Delta^{14,15}$, 1,8-ketal	X. moluccensis	(72)
Xyloccensin D*	3β,30α-diisobutyryloxy, 2,8α-di-hydroxy, $\Delta^{14,15}$, 1,8-ketal	X. moluccensis	(72)
—	3-oxo, 8β-hydroxy, $\Delta^{14,15}$, 3,8-ketal	Cedrela glaziovii	(72)

* Also contains 2-methylbutyryl esters.

(c) Ring A bridged. Phragmalin group

Table 4 (continued)

Name	Structure	Origin	References
—	2-hydroxy-3-OAc, 30-nicotinyloxy, 1,8,9-orthoacetate	E. caudatum	(22)
—	2-hydroxy-3-nicotinyloxy, 30-OAc, 1,8,9-orthoacetate	E. caudatum	(22)
Phragmalin triacetate	2-hydroxy-1,2,30-triacetoxy, 1,8,9-orthoacetate	X. moluccensis	(72)
—	2-hydroxy-3,30-diisobutyryloxy, 1,8,9-orthoacetate (and photoderivative)	Chukrasia tabularis	(62)
—	2-hydroxy-3-isobutyryloxy, 30-propionyloxy, 1,8,9-orthoacetate (and photoderivative)	C. tabularis	(62)
—	2-hydroxy-3,30-diisobutyryloxy, 12α-OAc, 1,8,9-orthoacetate	C. tabularis	(62)
—	2-hydroxy-3-isobutyryloxy, 30,12α-diacetoxy, 30-propionyloxy 1,8,9-orthoacetate	C. tabularis	(62)
Febrinin A and B	3-tigloyloxy, 2-OAc, 30-isobutyryloxy, 1,8,9-orthoacetate	Soymida febrifuga	(172)
Utilin	1,2-dihydroxy, 3-[2′-(S),3′(R)-epoxy-2′-methylbutyryloxy], 30α-OAc, 11β-(2′-methylbutyryloxy)-14α-hydroxy-8,9,14-orthoacetate	E. utile	(19, 102)
Entandrophragmin	30-isobutyrate corresponding to utilin	E. cylindricum E. utile E. spicatum	(19, 102, 186) (19) (5, 79)
Candollein	3-(2′-methylbutyrate) corresponding to entandrophragmin	E. candollei	(5, 99)
E₃	3-(2′-hydroxy) derivative of candollein	E. cylindricum	(99)

Pseudrelone A$_1$	2-hydroxy-3-isobutyryloxy, 15-acetyl, 30-OAc, 1,8,9-orthoacetate	Pseudocedrela kotschyii	(89)
Pseudrelone A$_2$	15-isobutyryl analog of above	P. kotschyii	(89)
Bussein	Structure	E. bussei E. caudatum	(5, 186) (100, 101)
—	11-isobutyryl analog of bussein	E. spicatum	(79)
Chukrasin A		Chukrasia tabularis	(171)

R^1=H, R^2, R^3=Ac and i-Bu, R^4=OH

N. B. Also unisolated isobutyric acid esters.

Table 4 (continued)

Name	Structure	Origin	References
Chukrasin B	Above entry, $R^1 = H$, R^2, $R^3 = i$-Bu, $R^4 = H$	Chukrasia tabularis	(171)
Chukrasin C	$R^1 = H$, R^2, $R^3 = Ac$ and i-Bu, $R^4 = H$	Chukrasia tabularis	(171)
Chukrasin D	$R^1 = Ac$, R^2, $R^3 = Ac$ and i-Bu, $R^4 = H$	Chukrasia tabularis	(171)
Chukrasin F	$R^1 = Ac$, R^2, $R^3 = i$-Bu, $R^4 = H$	Chukrasia tabularis	(171)
Pseudrelone C		P. kotschyii	(137)
Procerin	3β-propionyl analog of pseudrelone C	Carapa procera	(194)

Pseudrelone B

Pseudocedrela kotschyii   (196)

Table 5. *Methyl Ivorensate Group*

Name	Structure	Origin	References
Methyl ivorensate		*Khaya ivorensis* *Soymida febrifuga*	(14, 15) (170)
Methyl-21,23-dihydro-21-methoxy-23-oxo-ivorensate		*S. febrifuga*	(170)

Table 5 *(continued)*

Name	Structure	Origin	References
Compound C	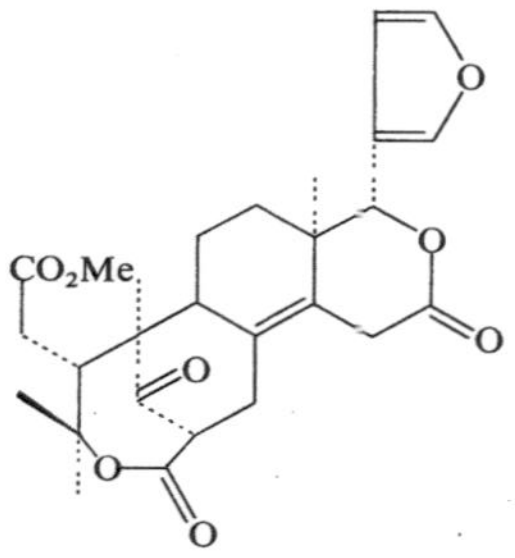	*K. ivorensis*	*(15)*

Table 6. *Obacunol Group*

Name	Structure	Origin	References
Obacunol	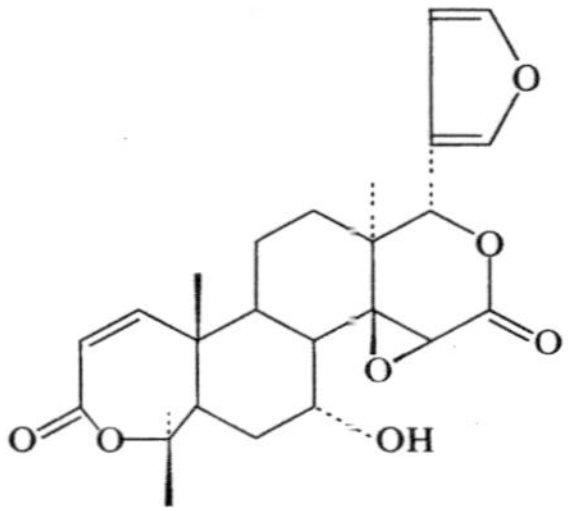	*Lovoa trichiliodes*	*(7)*

Obacunol acetate   *Carapa procera*   *(183)*

6β-Acetoxyobacunol   *Trichilia trifolia*   *(201)*

Dihydronomilin acetate   1α-acetoxy-1,2-dihydroobacunol acetate   *Xylocarpus granatum*   *(16, 142)*

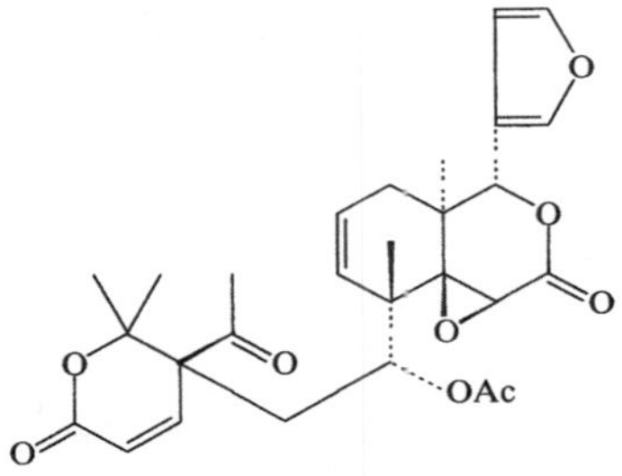

C. procera   *(81)*

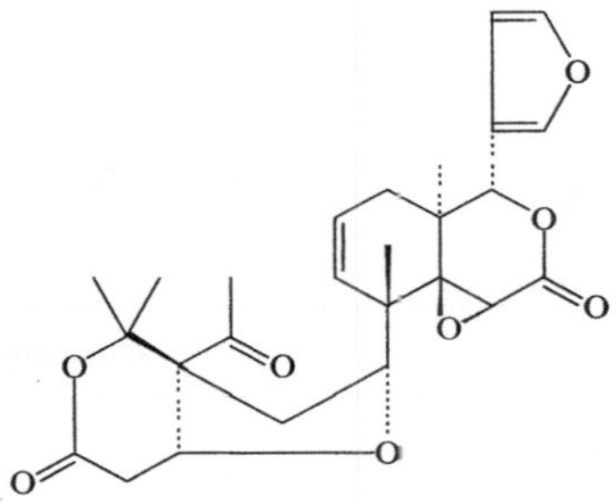

C. procera   *(81)*

D. A. H. TAYLOR:

Table 6 *(continued)*

Name	Structure	Origin	References
CP3		*C. procera*	*(81)*
CG3		*C. grandiflora*	*(81)*
CG2	Anhydro-CG3 ($\Delta^{10,19}$)	*C. grandiflora*	*(81)*

Table 7. *Nimbin Group*

Name	Structure	Origin	References
Heudebolin		*Trichilia heudelottii*	(4)
Nimbolin B	7-cinnamoyl analog of heudebolin	*M. azedarach* *A. indica*	(86)
Nimbolinin B	7-tigloyl analog of heudebolin and 1-deacetyl derivative	*M. azedarach*	(110)
Ohchinolide A		*M. azedarach*	(110, 146)
Ohchinolide B	7-tigloyl analog of ohchinolide A	*M. azedarach*	(110, 146)

Table 7 (*continued*)

Name	Structure	Origin	References
Nimbolidin A	MeO$_2$C, AcO, AcO, AcO, OBz (structure)	*M. azedarach*	(110)
Nimbolidin B	7-tigloyl analog of nimbolidin A	*M. azedarach*	(110)
Salannol	MeO$_2$C, CO$_2$, CH$_2$, HO (structure)	*A. indica*	(113)
Ohchinin acetate	3-acetyl, 1-cinnamoyl analog of salannol	*M. azedarach*	(148)
Salannin	3-acetyl, 1-tigloyl analog of salannol	*A. indica*	(103)
Ohchinal	3-acetyl, 1-benzoyl, 12-aldehydo analog of salannol	*M. azedarach*	(148)

Azadirachtin   *A. indica*   (202)

Nimbolide   *A. indica*   (85)

Nimbin   *A. indica*   (97, 139)

D. A. H. TAYLOR:

Table 7 (continued)

Name	Structure	Origin	References
Nimbinene		A. indica	(113)
Deacetylnimbinene		A. indica	(113)
Nimbandiol		A. indica	(113)
6-acetylnimbandiol		A. indica	(113)

Table 8. *Toonafolin Group*

Name	Structure	Origin	References
Toonafolin	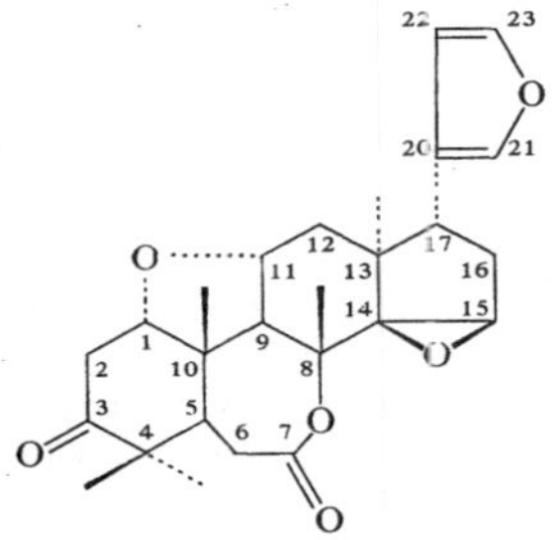	*Toona ciliata*	(118)
Toonacilin	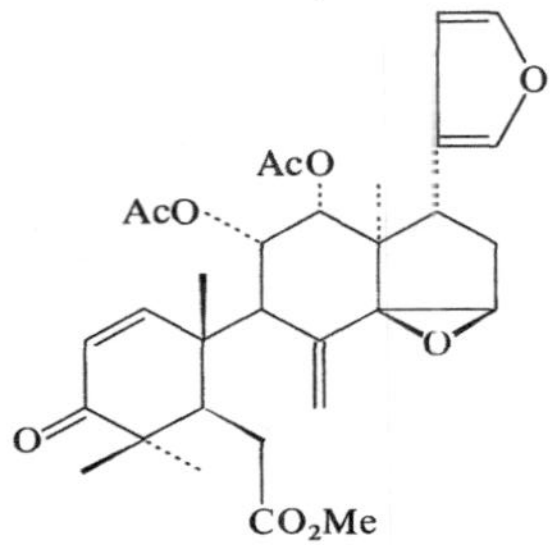	*T. ciliata*	(119)
6-Acetoxytoonacilin		*T. ciliata*	(119)
21-hydroxytoonacilide	21,23-dihydro-21-hydroxy-23-oxotoonacilin	*T. ciliata*	(117)
23-hydroxytoonacilide	21,23-dihydro-23-hydroxy-21-oxotoonacilin	*T. ciliata*	(117)

6*

D. A. H. TAYLOR:

Table 9. *Evodulone Group*

Name	Structure	Origin	References
Proceranone	$\Delta^{1,2}$, 7-Ac	*Carapa procera*	(184)
Dregeana-3	1α-OAc, 12β-(2'-acetoxy-3'-methylvaleryloxy)-7-Ac	*Trichilia dregeana*	(136)
Dregeana 4	1α-OAc, 29-hydroxy, 12β-(2'-acetoxy-3'-methylvaleryloxy), 7-(2'-hydroxy-3'-methylbutyryl)	*T. dregeana*	(136)
Dregeana 5	$\Delta^{1,2}$, 29-hydroxy, 12β-(2'-acetoxy-3'-methylvaleryloxy), 7-(2'-hydroxy-3'-methylbutyryl	*T. dregeana*	(136)
Gumulin	$\Delta^{1,2}$, 14β,15β-epoxy, 7-Ac	*C. procera*	(183)
Surenin	$\Delta^{1,2}$, 14β,15β-epoxy, 6-OAc, 7-Ac	*Toona sureni*	(115)
Surenone	$\Delta^{1,2}$, 14β,15β-epoxy, 6-OH, 7-dehydro	*T. sureni*	(115)
Evodulone	$\Delta^{1,2}$, 14β,15β-epoxy, 16-oxo, 7-Ac	*C. procera*	(182)

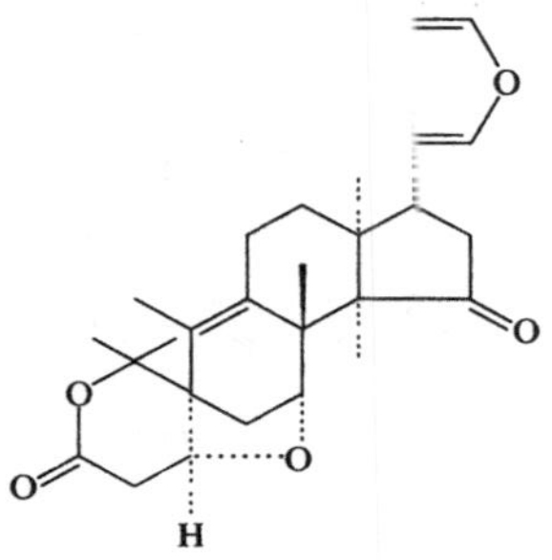

C. procera (41)

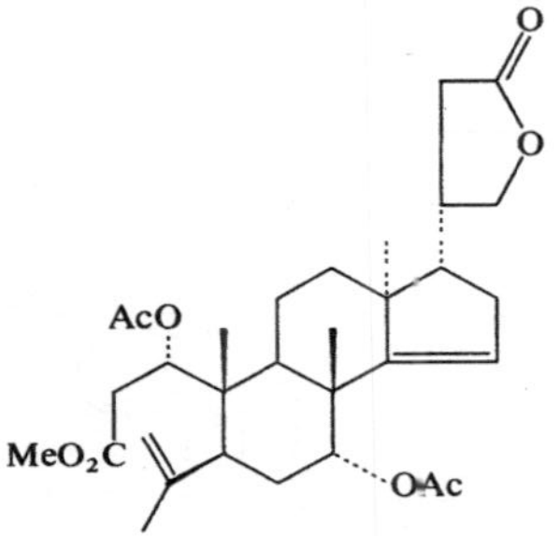

Nymania capensis (132)

Nymania 2

D. A. H. TAYLOR:

Table 10. *Prieurianin Group*

Name		Structure	Origin	References
Surenolactone	1		*Toona sureni*	(*116*)
Nymania 3	2		*Nymania capensis*	(*132*)

Dregeana 2	3		*Trichilia dregeana*	(136)
Prieurianin	4		*T. prieuriana* *G. guidona* *N. capensis*	(96, 136) (131) (132)
Rohituka 4	5	12-(3′-methylbutyrate) corresponding to prieurianin (entry 4)	*Aphanamixis polystacha*	(35)
—	6	14β,15β-epoxide corresponding to prieurianin (entry 4)	*G. guidona*	(131)

Table 10 *(continued)*

Name		Structure	Origin	References
Rohituka 8	7		*A. polystacha*	(35, 109)
Rohituka 6	8	1,2-dihydro-1α,11β-oxide analog of entry 8	*A. polystacha*	(35, 109)
Rohitukin	9		*A. polystacha*	(78)
"A"	10	12-(2′-OH-3-methylvaleryloxy) corresponding to rohitukin (entry 9)	*G. cedrata*	(17, 200)
"B"	11	7,29-lactone corresponding to entry 6	*G. thompsonii* *G. kunthiana*	(63) (200)

Name	Entry	Structure / description	Source	Ref.
"D₄"	12		T. prieuriana	(63)
Rohituka 7	13	14β-hydroxy-15β-acetoxy analog of entry 12	A. polystacha / T. hispida / T. dregeana	(35, 109) / (107) / (136)
—	14	12-(3'-methylbutyryloxy) analog of entry 13	A. polystacha	(200)
Hispidin B	15	15-tigloyloxy analog of entry 13	T. hispida	(107)
Rohituka 5	16		A. polystacha	(35, 109)
Rohituka 3	17	15-oxo compound corresponding to entry 16	A. polystacha	(35)
"C"	18	12-formyloxy-14β,15β-oxide corresponding to entry 16	G. thompsonii	(63)

Table 10 (continued)

Name		Structure	Origin	References
"E"	19	14β,15β-oxide corresponding to entry 16	G. thompsonii	(200)
Polystachin	20	1α,14β-oxide corresponding to rohitukin (entry 9)	A. polystacha	(135)
Dregeana I	21	12-(2'-hydroxy-3'-methylbutyryloxy analog of polystachin (entry 20)	T. dregeana	(136)
Hispidin A	22		T. hispida	(107)
Nymania 1	23	15-oxo derivative corresponding to hispidin A (entry 22)	N. capensis	(132)
"F"	24	14β,15β-oxide corresponding to hispidin A (entry 22)	G. thompsonii	(200)

Table 10 (*continued*)

Name		Structure	Origin	References
Dregeanin	25		*T. dregeana* *T. heudelottii*	(78, 186, 109) (154, 156)
12-(2'-deacetyl)- dregeanin	26		*G. thompsonii* *T. prieuriana* *G. kunthiana*	(77, 78) (63) (200)
"D₅"	27		*T. prieuriana*	(63, 133)
Rohituka 2	28	14β-OH-15β-OAc analog of entry 27	*A. polystacha*	(35, 133)
Rohituka 1	29	12-(3'-methylbutyryloxy) analog of entry 28	*A. polystacha*	(35, 133)

Table 11. *Unknown or Doubtful Structures*

Name	Source	Comment	References
30-Acetoxy dihydrokhayasin	*Khaya madagascariensis* *K. senegalensis*	Structure in error. Actually impure 2-hydroxykhayasin	(190, 200)
KIW "B"	*K. ivorensis*	Structure unknown	(15)
KIRB 1780	*K. ivorensis*	Structure doubtful	(195)
12-Acetoxyswietenolide acetate	*K. senegalensis*	Structure in error. 11β-Acetoxy isomer	(3)
Nyasin	*K. nyasica*	Structure in error. 11β-hydroxykhivorin	(80, 187)
Mahoganin	*Swietenia mahogani*	Structure in error. Mixture of methyl angolensate and methyl-6-hydroxyangolensate	(189)
Substance I	*Xylocarpus moluccensis*	Identified as methyl angolensate	(200)
$C_{27}H_{34}O_8$	*X. granatum*	Structure unknown	(158)
$C_{27}H_{32}O_6$	*K. senegalensis*	Structure unknown	(13)
Tabularin	*Chukrasia tabularis*	Structure doubtful. Possibly 1-ketone	(34)

References

1. ADEOYE, S. A., and D. A. BEKOE: The Molecular Structure of *Cedrela odorata* Substance B. J. Chem. Soc. Chem. Commun. **1965**, 301.
2. ADESIDA, G. A., E. K. ADESOGAN, D. A. OKORIE, B. T. STYLES, and D. A. H. TAYLOR: The Limonoid Chemistry of the Genus *Khaya* (Meliaceae). Phytochemistry **10**, 1845 (1971).
3. ADESIDA, G. A., E. K. ADESOGAN, and D. A. H. TAYLOR: Extractives from *Khaya senegalensis* A. Juss. J. Chem. Soc. Chem. Commun. **1967**, 790.
4. ADESIDA, G. A., and D. A. OKORIE: Heudebolin, a New Limonoid from *Trichilia heudelotii*. Phytochemistry **12**, 3007 (1973).
5. ADESIDA, G. A., and D. A. H. TAYLOR: The Chemistry of the Genus *Entandrophragma*. Phytochemistry **6**, 1429 (1967).
6. — — Extractives from *Soymida febrifuga*. Phytochemistry **11**, 1520 (1972).
7. — — Isolation of Obacunol from *Lovoa trichiliodes*. Phytochemistry **11**, 2644 (1972).
8. ADESOGAN, E. K., C. W. L. BEVAN, J. W. POWELL, and D. A. H. TAYLOR: West African timbers. Part XVIII. Some Reactions of *Cedrela odorata* Substance B and Khayasin. J. Chem. Soc. (C) **1966**, 2127.
9. ADESOGAN, E. K., D. A. OKORIE, and D. A. H. TAYLOR: Limonoids from *Khaya anthotheca* (Welw) C. DC. J. Chem. Soc. (C) **1970**, 205.
10. ADESOGAN, E. K., J. W. POWELL, and D. A. H. TAYLOR: Extractives from the Seed of *Khaya senegalensis*. J. Chem. Soc. (C) **1967**, 554.
11. ADESOGAN, E. K., and D. A. H. TAYLOR: Grandifoliolin, a New Limonoid from *Khaya grandifoliola* C. DC. J. Chem. Soc. Chem. Commun. **1967**, 225.
12. — — Extractives from the Seed of *Khaya anthotheca* (Welw). C. DC. J. Chem. Soc. Chem. Commun. **1967**. 379.
13. — — Extractives from *Khaya senegalensis* (Desr.) A. Juss. J. Chem. Soc. (C) **1968**, 1974.
14. — — Methyl Ivorensate, an A-seco-Limonoid from *Khaya ivorensis*. J. Chem. Soc. Chem. Commun. **1969**, 889.
15. — — Limonoid Extractives from *Khaya ivorensis*. J. Chem. Soc. (C) **1970**, 1710.
16. AHMED, F. R., A. S. NG, and A. G. FALLIS: 7α-Acetoxydihydronomilin: Isolation, Spectra. and Crystal Structure. Can. J. Chem. **56**, 1020 (1978).
17. AKINNIYI, J. A., J. D. CONNOLLY, D. S. RYCROFT, B. L. SONDENGAM, and N. P. IFEADIKE: Tetranortriterpenoids and Related Compounds. Part 25. Two 3,4-Secotirucallane Derivatives and 2′-Hydroxyrohitukin from the Bark of *Guarea cedrata* (Meliaceae). Can. J. Chem. **58**, 1865 (1980).
18. AKINNIYI, J. A., J. D. CONNOLLY, D. S. RYCROFT, and D. A. H. TAYLOR: Unpubl. work.
19. AKISANYA, A., C. W. L. BEVAN, J. HIRST, T. G. HALSALL, and D. A. H. TAYLOR: West African Timbers. Part III. Petroleum Extractives from the Genus *Entandrophragma*. J. Chem. Soc. **1960**, 3827.
20. AKISANYA, A., C. W. L. BEVAN, T. G. HALSALL, J. W. POWELL, and D. A. H. TAYLOR: West African Timbers. Part IV. Some Reactions of Gedunin. J. Chem. Soc. **1961**, 3705.
21. ARENE, E. O., C. W. L. BEVAN, J. W. POWELL, and D. A. H. TAYLOR: West African Timbers. Part XI. The Structure of Carapin, an Extractive from *Carapa procera*. J. Chem. Soc. Chem. Commun. **1965**, 302.
22. ARNDT, R. R., and W. H. BAARSCHERS: The Structure of Phragmalin, a Meliacin with a Norbornane Part Skeleton. Tetrahedron **28**, 2333 (1972).
23. ARIGONI, D., D. H. R. BARTON, L. CAGLIOTI, E. J. COREY, S. DEV, P. G. FERRINI, E. R. GLAZIER, O. JEGER, A. MELERA, S. K. PRADHAM, F. SCHAFFNER, S. STERNHELL, J. F. TEMPLETON, and S. TOBINAGA: The Constitution of Limonin. Experientia **16**, 41 (1960).
24. BARTON, D. H. R., S. K. PRADHAN, S. STERNHELL, and J. F. TEMPLETON: Triterpenoids. Part XXV. The Constitutions of Limonin and Related Bitter Principles. J. Chem. Soc. **1961**, 255.

25. BANERJI, R., and S. K. NIGAM: Studies on the Heartwood of *Melia birmanica*. Fitoterapia **52**, 3 (1981).
26. BEVAN, C. W. L., D. E. U. EKONG, T. G. HALSALL, and P. TOFT: West African Timbers. Part XIV. The Structure of Turreanthin, a Triterpene Monoacetate from *Turreanthus africanus*. J. Chem. Soc. Chem. Commun. **1965**, 636.
27. BEVAN, C. W. L., D. E. U. EKONG, T. G. HALSALL, and P. TOFT: West African Timbers. Part XX. The Structure of Turreanthin, an Oxygenated Tetracyclic Triterpene Monoacetate. J. Chem. Soc. (C). **1967**, 820.
28. BEVAN, C. W. L., D. E. U. EKONG, and D. A. H. TAYLOR: Extractives from West African Members of the Family Meliaceae. Nature **206**, 1323 (1965).
29. BEVAN, C. W. L., T. G. HALSALL, M. N. NWAJI, and D. A. H. TAYLOR: West African Timbers. Part V. The Structure of Khivorin, a Constituent of *Khaya ivorensis*. J. Chem. Soc. **1962**, 768.
30. BEVAN, C. W. L., J. W. POWELL, and D. A. H. TAYLOR: West African Timbers Part VI. Petroleum Extracts from Species of the Genera *Khaya, Guarea, Carapa* and *Cedrela*. J. Chem. Soc. **1963**, 980.
31. BIRCH, A. J., D. J. COLLINS, S. MUHAMMED, and J. P. TURNBULL: The Structure of Flindissol. Some Remarks on the Elemi Acids. J. Chem. Soc. **1963**, 2762.
32. BREDELL, L. D.: An Investigation into Extractives from the Meliaceae. M. Sc. Thesis. Durban, 1977.
33. BREEN, G. J. W., E. RITCHIE, W. T. L. SIDWELL, and W. C. TAYLOR: The Chemical Constituents of Australian *Flindersia* species XIX. Triterpenoids from the Leaves of *F. bourjotiana* F. Muell. Australian J. Chem. **19**, 455 (1966).
34. BROWN, D. A., and D. A. H. TAYLOR: ^{13}C Nuclear Magnetic Resonance Spectra of Some Limonoids. Part IV. Extractives from *Chukrasia tabularis*. A. Juss. J. Chem. Res. (S) 20, (M) 0301 (1978).
35. — — Limonoid Extractives from *Aphanamixis polystacha*. Phytochemistry **17**, 1995 (1978).
36. BUCHANAN, J. G. St. C., and T. G. HALSALL: The Synthesis of Possible Intermediates in the Biogenesis of Tetranortriterpenes by the Conversion of the Sidechain of Turreanthin into a β-Substituted Furan. J. Chem. Soc. Chem. Commun. **1969**, 48.
37. — — The Synthesis of the Simplest Meliacins (Limonoids) from Tetranortirucallane Triterpenoids Containing a β-Substituted Furyl Sidechain. J. Chem. Soc. Chem. Commun. **1969**, 242.
38. — — Conversion of a Simple Meliacin (7α-Acetoxy meliaca-14,20,22-trien-3-one) into Azadirone and of Khayanthone into Khivorin. J. Chem. Soc. Chem. Commun. **1969**, 1493.
39. — — The Conversion of Turreanthin and Turreanthin A into Simple Meliacins by a Route Involving an Oxidative Rearrangement of Probable Biogenetic Importance. J. Chem. Soc. (C). **1970**, 2280.
40. BURKE, B. A., W. R. CHAN, K. E. MAGNUS, and D. R. TAYLOR: Extractives of *Cedrela odorata* II. The Structure of Photogedunin. Tetrahedron **25**, 5007 (1969).
41. CAMERON, A. F., J. D. CONNOLLY, A. MALTZ, and D. A. H. TAYLOR: Tetranortriterpenoids and Related Compounds. Part 21. The Crystal and Molecular Structure of a Rearranged Tetranortriterpenoid Spiro-lactone from the Bark of *Carapa procera* (Meliaceae). Tetrahedron Letters **1979**, 967.
42. CASON, S. C., and K. S. BROWN: Biogenetically Significant Triterpenes in a Species of Meliaceae, *Cabralea polytricha*. Tetrahedron **28**, 315 (1972).
43. CHAKRABORTY, D. P., K. C. DAS, and C. F. HAMMER: Chemical Taxonomy X. Mahoganin. Tetrahedron Letters **1968**, 5015.
44. CHAN, W. R., J.·A. GIBBS, and D. R. TAYLOR: The Limonoids of *Trichilia havanensis* Jacq.: an Epoxide Rearrangement. J. Chem. Soc. Chem. Commun. **1967**, 720.

45. CHAN, W. R., K. E. MAGNUS, and B. S. MOOTOO: Extractives from *Cedrela odorata* L. The Structure of Methyl Angolensate. J. Chem. Soc. (C) **1967**, 171.
46. CHAN, W. R., N. L. HOLDER, D. R. TAYLOR, G. SNATZKE, and H. W. FEHLHABER: Extractives of *Cedrela odorata* L. Part II. The Structures of the *Cedrela* Tetracyclic Triterpenes, Odoratol, Iso-odoratol and Odoratone. J. Chem. Soc. (C) **1968**, 2485.
47. CHAN, W. R., and D. R. TAYLOR: Hirtin and Deacetylhirtin: New "Limonoids" from *Trichilia hirta*. J. Chem. Soc. Chem. Commun. **1966**, 206.
48. CHAN, W. R., D. R. TAYLOR, and T. YEE: Triterpenoids from *Entandrophragma cylindricum* Sprague. Part I. Structures of Sapelins A and B. J. Chem. Soc. (C) **1970**, 311.
49. — — — Triterpenoids from *Entandrophragma cylindricum* Sprague. Part II. The Structures of Sapelins C, D, E and F. J. Chem. Soc. (C) **1971**, 2662.
50. CHAKRABORTTY, T., J. D. CONNOLLY, R. MCCRINDLE, K. H. OVERTON, and J. C. D. SCHWARZ: Tetranortriterpenoids — VI. [Bicyclononanolides IV] Swietenolide — the Functional Groups. Tetrahedron **24**, 1503 (1968).
51. CHANG, F. C., and C. CHIANG: Kulinone, a Euphane-type Triterpenoid from *Melia azedarach*. J. Chem. Soc. Chem. Commun. **1968**, 1156.
52. — — Kulinone, a Euphane-type Triterpenoid from *Melia azedarach*. J. Chem. Soc. Chem. Commun. **1968**, 1156.
53. — — Tetracyclic triterpenes from *Melia azedarach* II. Trans-2-oxabicyclo[3.3.0]octanones. Tetrahedron Letters **1969**, 891.
54. CHATTERJEE, A., T. CHAKRABORTTY, and S. CHANDRASEKHARAN: Chemical Investigation of *Cedrela toona*. Phytochemistry **10**, 2533 (1971).
55. CHATTERJEE, A., and A. B. KUNDU: Isolation, Structure and Stereochemistry of Aphanamixin, a New Triterpene from *Aphanamixis polystacha*. Tetrahedron Letters **1967**, 1471.
56. CHIANG, C., and F. C. CHANG: Tetracyclic Triterpenoids from *Melia azedarach* (III). Tetrahedron **29**, 1911 (1973).
57. CONNOLLY, J. D.: Personal communication.
58. CONNOLLY, J. D., K. L. HANDA, and R. MCCRINDLE: Further Constituents of Nim Oil: the Constitution of Meldenin. Tetrahedron Letters **1968**, 437.
59. CONNOLLY, J. D., K. L. HANDA, R. MCCRINDLE, and K. H. OVERTON: Grandifolione: a Novel Tetranortriterpenoid. J. Chem. Soc. Chem. Commun. **1966**, 867.
60. — — — — Tetranortriterpenoids and related substances. Part XI. Odoratol and its Congeners from *Cedrela glaziovii*. J. Chem. Soc. (C) **1968**, 2230.
61. CONNOLLY, J. D., R. HENDERSON, R. MCCRINDLE, K. H. OVERTON, and N. S. BHACCA: Tetranortriterpenoids. Part I. [Bicyclononanolides. Part I.] The Constitution of Swietenine. J. Chem. Soc. **1965**, 6935.
62. CONNOLLY, J. D., C. LABBÉ, and D. S. RYCROFT: Tetranortriterpenoids and Related Substances. Part 20. New Tetranortriterpenoids from the Seed of *Chukrasia tabularis* (Meliaceae). Simple Esters of Phragmalin and 12α-Acetoxyphragmalin. J. Chem. Soc. Perkin Trans. I, **1978**, 285.
63. CONNOLLY, J. D., C. LABBÉ, D. S. RYCROFT, D. A. OKORIE, and D. A. H. TAYLOR: Tetranortriterpenoids and Related Compounds. Part 23. Complex Tetranortriterpenoids from *Trichilia prieuriana* and *Guarea thompsonii* (Meliaceae) and the Hydrolysis Products of Dregeanin, Prieurianin and Related Compounds. J. Chem. Res. (S) 256, (M) 2858 (1979).
64. CONNOLLY, J. D., C. LABBÉ, D. S. RYCROFT, and D. A. H. TAYLOR: Tetranortriterpenoids and Related Compounds. Part 22. New Apo-tirucallol Derivatives and Tetranortriterpenoids from the Wood and Seed of *Chisocheton paniculatus* (Meliaceae). J. Chem. Soc. Perkin Trans. I, **1979**, 2959.
65. CONNOLLY, J. D., and C. LABBÉ: Tetranortriterpenoids and Related Compounds. Part 24. The Interrelation of Swietenine and Swietenolide, the Major Tetranortriter-

penoids from the Seeds of *Swietenia macrophylla* (Meliaceae). J. Chem. Soc. Perkin
Trans. I, **1980**, 529.

66. CONNOLLY, J. D., and R. McCRINDLE: The Constitution of Grandifoliolenone, a Novel
Triterpenoid from *Khaya grandifolia*. J. Chem. Soc. Chem. Commun. **1967**, 1193.

67. — —: Tetranortriterpenoids and related substances. Part XIII. The Constitution of
Grandifoliolenone, an *apo*-Tirucallol Derivative from *Khaya grandifoliola* (Meliaceae).
J. Chem. Soc. (C) **1971**, 1715.

68. CONNOLLY, J. D., R. McCRINDLE, and K. H. OVERTON: Tetranortriterpenoids IV.
Bicyclononalides II. Constitution and Stereochemistry of Mexicanolide. Tetrahedron
24, 1489 (1968).

69. — — — Tetranortriterpenoids V [Bicyclononanolides III]. The P. M. R. spectrum of
Mexicanolide. Tetrahedron **24**, 1497 (1968).

70. CONNOLLY, J. D., R. McCRINDLE, K. H. OVERTON, and J. FEENEY: Tetranortriterpenoids. Part II. Heartwood Constituents of *Carapa guyanensis* Aubl. Tetrahedron **22**, 811 (1966).

71. CONNOLLY, J. D., R. McCRINDLE, K. H. OVERTON, and W. D. C. WARNOCK: Tetranortriterpenoids VII [Bicyclononanolides V]. The Constitution and Stereochemistry of
Swietenolide. Tetrahedron **24**, 1507 (1968).

72. CONNOLLY, J. D., M. MacLELLAN, D. A. OKORIE, and D. A. H. TAYLOR: Limonoids
from *Xylocarpus moluccensis* (Lam) M. Roem. J. Chem. Soc. Perkin Trans. I, **1976**, 1993.

73. CONNOLLY, J. D., K. H. OVERTON, and J. POLONSKY: The Chemistry and Biochemistry of
the Limonoids and Quassinoids. Progress in Phytochemistry **2**, 385 (1970).

74. CONNOLLY, J. D., I. M. S. THORNTON, and D. A. H. TAYLOR: Partial Synthesis of Methyl
Angolensate from 7-Oxo-7-acetoxykhivorin. J. Chem. Soc. Chem. Commun. **1970**, 1205.

75. — — — Partial Syntheses of Mexicanolide from 7-Oxo-7-deacetoxykhivorin. J. Chem.
Soc. Chem. Comm. **1971**, 17.

76. — — — Tetranortriterpenoids. Part XVI. Partial Syntheses of Andirobin, Methyl
Angolensate, Mexicanolide and 1-Deoxymexicanolide. J. Chem. Soc. Perkin Trans. I,
1973, 2407.

77. CONNOLLY, J. D., D. A. OKORIE, and D. A. H. TAYLOR: Limonoid Extractives from
Species of *Guarea*. An Unusual Shielding Effect on an Acetyl group. J. Chem. Soc. Perkin
Trans. I, **1972**, 1145.

78. CONNOLLY, J. D., D. A. OKORIE, L. D. DE WIT, and D. A. H. TAYLOR: Structure of
Dregeanin and Rohitukin, Limonoids from the Subfamily Melioideae of the Family
Meliaceae. An Unusually High Absorption Frequency for a Six-membered Lactone
Ring. J. Chem. Soc. Chem. Commun. **1976**, 909.

79. CONNOLLY, J. D., W. R. PHILLIPS, D. A. MULHOLLAND, and D. A. H. TAYLOR: Spicatin, a
Protolimonoid from *Entandrophragma spicatum*. Phytochemistry **20**, 2596 (1981).

80. CONNOLLY, J. D., and D. A. H. TAYLOR: Tetranortriterpenoids. Part XIV. The Structure
of Nyasin, a Limonoid from *Khaya nyasica*; a Correction. J. Chem. Soc. Perkin Trans. I,
1973, 686.

81. CONNOLLY, J. D., M. F. GRUNDON, and D. A. H. TAYLOR: Phytochemical Society
Symposium. Glasgow 1982.

82. COTTERRELL, G. P., T. G. HALSALL, and M. J. WRIGLESWORTH: A Chemical Model for a
Possible Oxidative Rearrangement in the Biosynthesis of Tetranortriterpenes. The
Preparation of Methyl 3α-Acetoxy-7-oxo-apo-tirucalla-14,24-dien-21-oate. J. Chem.
Soc. Chem. Commun. **1967**, 1121.

83. — — — A Chemical Model for a Possible Oxidative Rearrangement in the Biosynthesis
of Triterpenes. The Rearrangement of 7α,8α- and 8,9-Epoxytirucallanes. J. Chem. Soc.
(C) **1970**, 1503.

84. DREYER, D. L.: Limonoid Bitter Principles. Fortschritte der Chemie organischer
Naturstoffe **26**, 190 (1968).

85. EKONG, D. E. U.: Chemistry of the Meliacins (Limonoids). The Structure of Nimbolide, a New Meliacin from *Azadirachta indica*. J. Chem. Soc. Chem. Commun. **1967**, 808.

86. EKONG, D. E. U., C. O. FAKUNLE, A. K. FASINA, and J. I. OKOGUN: The Meliacins (limonoids). Nimbolin A and B, two new Meliacin Cinnamates from *Azadirachta indica* L. and *Melia azedarach* L. J. Chem. Soc. Chem. Commun. **1969**, 1166.

87. EKONG, D. E. U., S. A. IBIYEMI, and E. O. OLAGBEMI: The Meliacins (limonoids). Biosynthesis of Nimbolide in the Leaves of *Azadirachta indica*. J. Chem. Soc. Chem. Commun. **1971**, 1117.

88. EKONG, D. E. U., J. I. OKOGUN, and B. L. SONDENGAM: The Meliacins (Limonoids): Minor Constituents of *Khaya anthotheca*. Reduction of the Meliacins with Zinc-Copper Couple. J. Chem. Soc. Perkin Trans. I, **1975**, 2118.

89. EKONG, D. E. U., and E. O. OLAGBEMI: Novel Meliacins (Limonoids) from the Wood of *Pseudocedrela kotschyii*. Tetrahedron Letters **1967**, 3525.

90. EPE, B., and A. MONDON: Zur Kenntnis der Bitterstoffe aus Cneoraceen. XII. Tetrahedron Letters **1979**, 2015.

91. ESHIETT, I. T. U., A. AKISANYA, and D. A. H. TAYLOR: Diterpenes from *Annona senegalensis*. Phytochemistry **10**, 3294 (1971).

92. FERGUSON, G., P. A. GUNN, W. C. MARSH, R. McCRINDLE, R. RESTIVO, J. D. CONNOLLY, J. W. B. FULKE, and M. S. HENDERSON: Tetranortriterpenoids and Related Substances. Part XVII. A New Skeletal Class of Triterpenoids from *Guarea glabra* (Meliaceae). J. Chem. Soc. Perkin Trans. I, **1975**, 491.

93. GOPINATH, K. W., T. R. GOVINDACHARI, P. C. PARTHASARATHY, N. VISWANATHAN, D. ARIGONI, and W. C. WILDMAN: The Structure of Cedrelone. Proc. Chem. Soc. **1961**, 446.

94. GRANT, I. G., J. A. HAMILTON, T. A. HAMOR, R. HODGES, S. G. McGEACHIN, R. A. RAPHAEL, J. MONTEATH ROBERTSON, and G. A. SIM: The Structure of Cedrelone. Proc. Chem. Soc. **1961**, 444.

95. GRIJPMA, P.: Personal communication.

96. GULLO, V. P., I. MIURA, K. NAKANISHI, A. F. CAMERON, J. D. CONNOLLY, F. D. DUNCANSON, A. E. HARDING, R. McCRINDLE, and D. A. H. TAYLOR: Structure of Prieurianin, a Complex Tetranortriterpenoid; Nuclear Magnetic Resonance Analysis at Non-ambient Temperatures and X-Ray Structure Determination. J. Chem. Soc. Chem. Commun. **1975**, 345.

97. HARRIS, M., R. HENDERSON, R. McCRINDLE, K. H. OVERTON, and D. W. TURNER: Tetranortriterpenoids VIII. The Constitution and Stereochemistry of Nimbin. Tetrahedron **24**, 1517 (1968).

98. HALSALL, T. G., and J. A. TROKE: The Structure of Three New Meliacins Isolated from *Khaya anthotheca* heartwood. J. Chem. Soc. Perkin Trans. I, **1975**, 1758.

99. HALSALL, T. G., K. WRAGG, J. D. CONNOLLY, M. A. MacLELLAN, L. D. BREDELL, and D. A. H. TAYLOR: ^{13}C Nuclear Magnetic Resonance Spectra of Some Limonoids. Part III. The Spectra of Some Derivatives of Entandrophragmin and a Revised Structure for Candollein. J. Chem. Res. (S) 154, (M) 1727 (1977).

100. HANNI, R., and Ch. TAMM: Structure of the Tetranortriterpene Bussein, J. Chem. Soc. Chem. Commun. **1972**, 1253.

101. HANNI, R., Ch. TAMM, V. P. GULLO, and K. NAKANISHI: Modification of the Structure of Bussein. J. Chem. Soc. Chem. Commun. **1975**, 563.

102. HARRISON, H. R., O. J. R. HODDER, C. W. L. BEVAN, D. A. H. TAYLOR, and T. G. HALSALL: Crystallographic Structure Determination of Utilin, $C_{41}H_{52}O_{17}$, a Complex Meliacin with a 1,29 Cycloswietenan Skeleton. J. Chem. Soc. Chem. Commun. **1970**, 1388.

103. HENDERSON, R., R. McCRINDLE, A. MELERA, and K. H. OVERTON: Tetranortriterpenoids IX. The Constitution and Stereochemistry of Salannin. Tetrahedron **24**, 1525 (1968).

104. HOUSELY, J. R., F. E. KING, T. J. KING, and P. R. TAYLOR: The Chemistry of Hardwood Extractives. Part XXXIV. Constituents of *Guarea* species. J. Chem. Soc. **1962**, 5095.

105. JIBODU, K. O., N. S. OHOCHUKU, and D. A. H. TAYLOR: Chemical Shifts of the Tertiary Methyl Groups in the Nuclear Magnetic Resonance Spectra of Some Limonoids. Part II. J. Chem. Soc. (C) **1970**, 2396.

106. JOLAD, S. D., J. J. HOFFMANN, J. R. COLE, M. S. TEMPESTA, and R. B. BATES: Constituents of *Trichilia hispida* (Meliaceae) 2. A New Triterpene, Hispidone, and Bourjotinolone A. J. Org. Chem. **45**, 3132 (1980).

107. JOLAD, S. D., J. J. HOFFMANN, K. H. SCHRAM, J. R. COLE, M. S. TEMPESTA, and R. B. BATES: Constituents of *Trichilla hispida* (Meliaceae) 3. Structures of the Cytotoxic Limonoids, Hispidins A, B and C. J. Org. Chem. **46**, 641 (1981).

108. JOLAD, S. D., R. M. WIEDHOPF, and J. R. COLE: Cytotoxic Agents from *Bursera klugii* (Burseraceae) I. Isolation of Sapelins A and B. J. Pharm. Sci. **66**, 889 (1977). Chemical Abstracts **87**, 78324z (1977).

109. KING, T. J., and D. A. H. TAYLOR: Limonoid Extractives from *Aphanamixis polystacha* Part II. Phytochemistry **22**, 307 (1983).

110. KRAUS, W., and M. BOKEL: Neue Tetranortriterpenoide aus *Melia azedarach* Linn. (Meliaceae). Chem. Ber. **114**, 267 (1981).

111. KRAUS, W., and R. CRAMER: 17-Epi-Azadiradione und 17-β-Hydroxy-azadiradione; zwei neue Inhaltsstoffe aus *Azadirachta indica* A. Juss. Tetrahedron Letters **1978**, 2395.

112. — — Neue Tetranortriterpenoide mit insektenfraßhemmender Wirkung aus Neem-Öl. Liebigs Ann. Chem. **1981**, 187.

113. — — Pentanortriterpenoide aus *Azadirachta indica* A. Juss (Meliaceae). Chem. Ber. **114**, 2375 (1981).

114. KRAUS, W., R. CRAMER, and G. SAWITZKA: Tetranortriterpenoids from the Seeds of *Azadirachta indica*. Phytochemistry **20**, 117 (1981).

115. KRAUS, W., and K. KYPKE: Surenone and Surenin, two Novel Tetranortriterpenoids from *Toona sureni* [Blume] Merrill. Tetrahedron Letters **1979**, 2715.

116. KRAUS, W., K. KYPKE, M. BOKEL, W. GRIMMINGER, G. SAWITZKI, and G. SCHWINGER: Surenolactone, ein neues Tetranortriterpenoid-A/B-dilacton aus *Toona sureni* [Blume] Merrill (Meliaceae). Liebigs Ann. Chem. **1982**, 87.

117. KRAUS, W., and W. GRIMMINGER: 23-(R,S)-hydroxytoonacilid und 21-(R,S)-Hydroxy-toonacilid, zwei neue B-*seco*-Tetranortriterpenoide mit insektenfraßhemmender Wirkung aus *Toona ciliata* M. J. Roem var. Australis (Meliaceae). Nouveau Journal de Chimie **4**, 651 (1980).

118. — — Toonafolin, ein neues Tetranortriterpenoid-B-lacton aus *Toona ciliata* var *australis* (Meliaceae). Liebigs Ann. Chem. **1981**, 1838.

119. KRAUS, W., W. GRIMMINGER, and G. SAWITZKI: Toonacilin and 6-Acetoxytoonacilin, Two Novel B-Seco-tetranortriterpenoids with Antifeeding Activity. Angewandte Chemie International Edition **17**, 452 (1978).

120. KUBO, I., and J. A. KLOCKE: Les Mediateurs Chimiques. INRA Colloque **7**, 117 (1981).

121. LAVIE, D., and M. K. JAIN: Tetranortriterpenoids from *Melia azadirachta* L. J. Chem. Soc. Chem. Commun. **1967**, 278.

122. LAVIE, D., M. K. JAIN, and I. KIRSON: Terpenoids. Part VI. The Complete Structure of Melianone. J. Chem. Soc. (C) **1967**, 1347.

123. LAVIE, D., M. K. JAIN, and S. R. SHPAN-GABRIELITH: A Locust Phagorepellent from Two *Melia* Species. J. Chem. Soc. Chem. Commun. **1967**, 910.

124. LAVIE, D., and E. C. LEVY: Studies on Epoxides IV. Rearrangements in Triterpenoids. Tetrahedron Letters **1968**, 2097.

125. — — Compounds Linking Meliones with Melianes. Tetrahedron Letters **1969**, 3525.

126. — — Oxidative Reactions of Biogenetic Interest. Tetrahedron Letters **1970**, 1315.

127. — — Meliane-Meliacin Relation. Tetrahedron **27**, 3941 (1971).

128. LAVIE, D., E. C. LEVY, and M. K. JAIN: Limonoids of Biogenetic Interest from *Melia azadirachta*. Tetrahedron **27**, 3927 (1971).
129. LAVIE, D., E. C. LEVY, and R. ZELNIK: Constituents of *Carapa guianensis* and their Biogenetic Relationship. Bioorganic Chemistry **2**, 59 (1972). Chemical Abstracts **78**, 58635n (1973).
130. LAVIE, D., E. C. LEVY, C. ROSITO, and R. ZELNIK: Studies on Tetranortriterpenoids from *Cedrela angustifolia* Sessee and Moc. Tetrahedron **26**, 219 (1970).
131. LUKACOVA, V., J. POLONSKY, C. MORETTI, G. R. PETTIT, and J. M. SCHMIDT: Isolation and Structure of 14β,15β-Epoxyprieurianin from the South American Tree *Guarea guidona*. J. Natural Products **45**, 288 (1982). Chemical Abstracts **97**, 69260e (1982).
132. MacLACHLAN, L. K., and D. A. H. TAYLOR: Limonoids from *Nymania capensis*. Phytochemistry **21**, 1701 (1982).
133. — — A Revision of the Structures of Three Limonoids. Phytochemistry **21**, 2426 (1982).
134. MITRA, C. R., H. S. GARY, and G. H. PANDEY: Constituents of *Azadirachta indica* III. Phytochemistry **10**, 857 (1971).
135. MULHOLLAND, D. A., and D. A. H. TAYLOR: A New Limonoid from *Aphanamixis polystacha*. J. Chem. Res. (S) 291, (M) 3101 (1979).
136. — — Limonoids from the Seed of the Natal Mahogany, *Trichilia dregeana*. Phytochemistry **19**, 2421 (1980).
137. NAKANISHI, K.: Personal communication.
138. NAKATANI, M., J. C. JAMES, K. NAKANISHI: Isolation and Structures of Trichilins, Antifeedants against the Southern Army Worm. J. Amer. Chem. Soc. **103**, 1228 (1981).
139. NARAYANAN, C. R., R. V. PACHAPURKAR, S. K. PRADHAN, V. R., SHAH, and N. S. NARASIMHAN: Structure of Nimbin. Chemistry and Industry **1964**, 322. Chemical Abstracts **60**, 12063d (1964).
140. NARAYANAN, C. R., and K. N. IYER: Isolation and Characterisation of Deacetylnimbin. Indian J. Chem. **5**, 460 (1967). Chemical Abstracts **68**, 27509h (1968).
141. NARAYANAN, C. R., R. V. PACHAPURKAR, B. M. SAWANT, and M. S. WAIDA: Vepinin, a New Constituent of Neem Oil. Indian J. Chem. **7**, 187 (1969). Chemical Abstracts **71**, 50267d (1969).
142. NG, A. S., and A. G. FALLIS: Comment: 7α-Acetoxydihydronomilin and Mexicanolide: Limonoids from *Xylocarpus granatum* (Koenig). Canadian J. Chem. **57**, 3088 (1979).
143. OBASI, M. E., J. I. OKOGUN, and D. E. U. EKONG: The Meliacins (Limonoids). Partial Synthesis of Mexicanolide. J. Chem. Soc. Chem. Commun. **1971**, 727.
144. — — — The meliacins (Limonoids). Some Transformations and Interconversions of the Meliacins. J. Chem. Soc. Perkin Trans. I, **1972**, 1943.
145. OCHI, M., H. KOTSUKI, K. HIROTSU, and T. TOKOROYAMA: Sendanin, a New Limonoid from *Melia azedarach* L. var. *japonica* Makino. Tetrahedron Letters **1976**, 2877.
146. OCHI, M., H. KOTSUKI, M. IDO, H. NAKAI, M. SHIRO, and T. TOKOROYAMA: Limonoids from *Melia azedarach* L. var. *japonica* Makino V. The Structures of Ohchinolide A and B. Chemistry Letters **1979**, 1137. Chemical Abstracts **91**, 189780x (1979).
147. OCHI, M., H. KOTSUKI, H. ISHIDA, and T. TOKOROYAMA: Limonoids from *Melia azedarach*. L. var. *japonica* Makino. The Natural Hydroxyl Precursor of Sendanin. Chemistry Letters **1978**, 99. Chemical Abstracts **88**, 136807t (1978).
148. OCHI, M., H. KOTSUKI, T. KATAOKA, T. TADA, and T. TOKOROYAMA: Limonoids from *Melia azedarach* L. var. *japonica* Makino III. The Structures of Ohchinal and Ohchinin Acetate. Chemistry Letters 1978, 331. Chemical Abstracts **89**, 43845j (1978).
149. OCHI, M., H. KOTSUKI, and T. TOKOROYAMA: Limonoids from *Melia azedarach* L. var. *japonica* Makino. Sendanal, a New Limonoid. Chemistry Letters **1978**, 621. Chemical Abstracts **89**,129751n (1978).
150. OCHI, M., H. KOTSUKI, T. TOKOROYAMA, and T. KUBOTA: The Structure of

Sendanolactone; a New Triterpenoid from *Melia azedarach* L. var. *japonica* Makino. Bull. Chem. Soc. Japan **50,** 2499 (1977).

151. OHOCHUKU, N. S., and D. A. H. TAYLOR: Chemical Shift of the Tertiary Methyl Groups in the Nuclear Magnetic Resonance Spectra of Some Limonoids. J. Chem. Soc. (C) **1969,** 864.

152. — — 6β-Hydroxygedunin. J. Chem. Soc. (C) **1970,** 421.

153. OKOGUN, J. I., C. O. FAKUNLE, D. E. U. EKONG, and J. D. CONNOLLY: Chemistry of the Meliacins (Limonoids). The Structure of Melianin A, a New Protomeliacin from *Melia azedarach.* J. Chem. Soc. Perkin Trans I **1975,** 1352.

154. OKORIE, D. A., and D. A. H. TAYLOR: The Structure of Heudelottin, an Extractive from *Trichilia heudelottii.* J. Chem. Soc. Chem. Commun. **1967,** 83.

155. — — The Mechanism of the Rearrangement of Havanensin. J. Chem. Soc. Chem. Comm. **1968,** 737.

156. — — Limonoids from the Timber of *Trichilia heudelottii* Planch ex Oliv. J. Chem. Soc. (C), **1968,** 1828.

157. — — Extractives from the Seed of *Cedrela odorata* L. Phytochemistry **7,** 1683 (1968).

158. — — Limonoids from *Xylocarpus granatum.* J. Chem. Soc. (C) **1970,** 211.

159. — — Limonoids from *Swietenia humilis.* Phytochemistry **10,** 469 (1971).

160. — — Limonoids from *Trichilia heudelottii* Part II. J. Chem. Soc. Perkin Trans I, **1972,** 1488.

161. — — Triterpenes from the Seed of *Entandrophragma* Species. Phytochemistry **16,** 2029 (1977).

162. OLLIS, W. D., A. D. WARD, H. M. DE OLIVIERA, and R. ZELNIK: Andirobin. Tetrahedron **26,** 1037 (1970).

163. OLLIS, W. D., A. D. WARD, and R. ZELNIK: Andirobin. Tetrahedron Letters **1964,** 2607.

164. PACHAPURKAR, R. V., P. M. KORNULE, and C. R. NARAYANAN: New Hexacyclic Tetranortriterpenoid. Chemistry Letters **1974,** 357. Chemical Abstracts **80,** 146356 u (1974).

165. PEGEL, K. H.: Personal communication.

166. POLONSKY, J., Z. VARON, B. ARNOUX, C. PASCARD, G. R. PETTIT, J. H. SCHMIDT, and L. M. LANGE: Isolation and Structure of Aphanastatin. J. Amer. Chem. Soc. **100,** 2575 (1978).

167. POLONSKY, J., Z. VARON, B. ARNOUX, C. PASCARD, G. R. PETTIT, and J. H. SCHMIDT: Isolation and Structure of Amoorastatin. J. Amer. Chem. Soc. **100,** 7731 (1978).

168. POLONSKY, J., Z. VARON, C. MARAZANO, B. ARNOUX, G. R. PETTIT, J. H. SCHMIDT, M. OCHI, and H. KOTSUKI: The Structure of Amoorastatone and the Cytotoxic Limonoid 12-Hydroxyamoorastatin. Experientia **35,** 987 (1979).

169. PURUSHOTHAMAN, K. K., and S. CHANDRASEKHARAN: Occurrence of Methyl Angolensate and Deoxyandirobin in *Soymida febrifuga.* Indian J. Chem. **12,** 207 (1974). Chemical Abstracts **81,** 101847j (1974).

170. PURUSHOTHAMAN, K. K., S. CHANDRASEKHARAN, J. D. CONNOLLY, and D. S. RYCROFT: Tetranortriterpenoids and Related Substances Part 18. Two New Tetranortriterpenoids with a Modified Furan Ring from the Bark of *Soymida febrifuga* A. Juss (Meliaceae). J. Chem. Soc. Perkin Trans I. **1977,** 1873.

171. RAGETTLI, T., and Ch. TAMM: Die Chukrasine A, B, C, D und E, fünf neue Tetranortriterpene aus *Chukrasia tabularis* A. Juss. Helv. Chim. Acta **61,** 1814 (1978).

172. RAO, M. M., A. S. GUPTA, P. P. SINGH, and E. M. KRISHNA: Structure of Febrinin A, a New Tetranortriterpenoid from the Heartwood of *Soymida febrifuga.* Indian J. Chem. **17B,** 158 (1979). Chemical Abstracts, **92,** 22648 u (1980).

173. RAO, M. M., E. M. KRISHNA, P. S. GUPTA, and P. P. SINGH: A New Tetranortriterpenoid Isolated from the Heartwood of *Soymida febrifuga.* Indian J. Chem. **16B,** 823 (1978). Chemical Abstracts **90,** 168783q (1979).

174. RAO, M. M., H. MESHULAM, R. ZELNIK, and D. LAVIE: Cabralea eichleriana DC. (Meliaceae)-1. Structure and Stereochemistry of Wood Extractives. Tetrahedron **31**, 333 (1975).

175. — — — — Cabralea eichleriana extractives II. Structure and Stereochemistry of Limonoids of Cabralea eichleriana. Phytochemistry **14**, 1071 (1975).

176. SAIKIA, B., J. C. S. KATAKY, R. K. MATHUR, and J. N. BARUAH: New Meliacins from the Fruits of Chisocheton paniculatus Hiern. Indian J. Chem. **16B**, 1042 (1978). Chemical Abstracts **91**, 2507y (1979).

177. SHIENGTHONG, D., U. KOKPOL, P. KARNTIANG, and R. A. MASSEY-WESTROPP: Triterpenoid Constituents of Thai medicinal plants — II. Isomeric Aglaitriols and Aglaiondiol. Tetrahedron **30**, 2211 (1974).

178. SCHULTE, K. E., G. RUECKER, and H. U. MATERN: Some Constituents of the Fruit and Roots of Melia azedarach L. Planta Medica **35**, 76 (1979). Chemical Abstracts **90**, 164741b (1979).

179. SHU, G., and X. LIANG: Correction of the Structure of Chuanliansu. Hua Hsueh Hsueh Pao **38**, 196 (1980). Chemical Abstracts **93**, 239693k (1982).

180. SIDDIQUI, S., S. FUCHS, J. LUCKE, and W. VOELTER: Struktur eines neuen Naturstoffes aus Melia azadirachta L. 17-Hydroxyazadiradion. Tetrahedron Letters **1978**, 611.

181. SINGH, S., H. S. GARG, and N. M. KHANNA: Dysobinin, a New Tetranortriterpene from Dysoxylum binectariferum. Phytochemistry **15**, 2001 (1978).

182. SONDENGAM, B. L., C. S. KAMGA, and J. D. CONNOLLY: Evodulone, a New Tetranortriterpenoid from Carapa procera. Tetrahedron Letters **1979**, 1357.

183. — — — Personal communication.

184. SONDENGAM, B. L., C. S. KAMGA, S. F. KIMBU, and J. D. CONNOLLY: Proceranone, a New Tetranortriterpenoid from Carapa procera. Phytochemistry **20**, 173 (1981).

185. STRAKA, H., F. ALBERS, and A. MONDON: Die Stellung und Gliederung der Familie Cneoraceae (Rutales). Biologie der Pflanzen **52**, 267 (1976).

186. TAYLOR, D. A. H.: Extractives from East African Timbers. Part 1. J. Chem. Soc. **1965**, 2495.

187. — Nyasin, an Extractive from Khaya nyasica Stapf. J. Chem. Soc. Chem. Commun. **1967**, 500.

188. — 11β-Acetoxykhivorin, a New Limonoid. J. Chem. Soc. Chem. Commun. **1968**, 1172.

189. — Extractives from Swietenia mahogani (L) Jacq. J. Chem. Soc. Chem. Commun. **1969**, 58.

190. — Limonoids from Khaya nyasica Stapf. ex Bak. J. Chem. Soc. (C). **1969**, 2439.

191. — Limonoids from Khaya madagascariensis Jumelle et Perrier. J. Chem. Soc. (C) **1970**, 336.

192. — The Structure of Fissinolide. Tetrahedron Letters **1970**, 2797.

193. TAYLOR, D. A. H., and F. W. WEHRLI: The Structure of Fissinolide and Angustidienolide, Limonoids from Species of Meliaceae. J. Chem. Soc. Perkin Trans 1. **1973**, 1599.

194. TAYLOR, D. A. H.: ^{13}C Nuclear Magnetic Resonance Spectra of Some Limonoids. Part I. The Structure of Procerin, an Extractive from Carapa procera. J. Chem. Soc. Perkin Trans I **1974**, 437.

195. — The Structure of an Extractive from Khaya ivorensis. Phytochemistry **16**, 1847 (1977).

196. — A Limonoid, Pseudrelone B, from Pseudocedrela kotschyii. Phytochemistry **18**, 1574 (1979).

197. — Ekebergin, a Limonoid Extractive from Ekebergia capensis. Phytochemistry **20**, 2263 (1981).

198. — A New Structural Proposal for the Tetranortriterpenoid Dregeanin. J. Chem. Res. (S) **1982**, 55.

199. — Flora Neotropica. Monograph Number 28. Meliaceae. Chemotaxonomy 450 (1981).

200. Taylor, D. A. H.: Unpublished work.
201. Taylor, D. R.: New Limonoids from *Trichilia trifolia* (Meliaceae). Revista Latinoamericana de Quimica **2,** 87 (1971). Chemical Abstracts **75,** 129969a (1971).
202. Zanno, P. R., I. Miura, K. Nakanishi, and D. L. Elder: Structure of the Insect Phagorepellent Azadirachtin. Application of PRFT/CWD Carbon-13 Nuclear Magnetic Resonance. J. Amer. Chem. Soc. **97,** 1975 (1975).
203. Zelnik, R.: Tetranortriterpenoids from *Cabralea eichleriana*. Phytochemistry **11,** 1866 (1972).
204. Zelnik, R., and C. M. Rosito: Le Fissinolide. Tetrahedron Letters **1966,** 6441.

(Received January 3, 1983)

Recent Progress in
the Chemistry of Lichen Substances

By J. A. ELIX and A. A. WHITTON, Chemistry Department, The Faculties,
Australian National University, Canberra, Australia, and M. V. SARGENT,
Organic Chemistry Department, University of Western Australia,
Nedlands, Australia

Contents

1. Introduction

Since the last major review of the chemistry of lichen substances by
HUNECK in 1971 (*171*) there has been an accelerating rate of development in
this field. As with all areas of natural product chemistry this impetus has

been provided by more rapid and improved methods for detecting, isolating and purifying lichen metabolites and in the structural elucidation of these compounds. The techniques of thin layer chromatography and high performance liquid chromatography have provided rapid and efficient methods for the detection and purification of lichen substances and the development of ^{13}C-n.m.r. spectroscopy and sophisticated techniques in ^{1}H-n.m.r. spectroscopy have greatly aided structural studies.

The more classical chemical procedures of degradation and in particular, total synthesis, have also developed apace with the use of newer reagents and synthetic methods. For instance in 1971 no natural depsidones had been synthesised but in the intervening decade, 15 such compounds have been prepared. Similarly the condensing reagents trifluoroacetic anhydride and dicyclohexylcarbodiimide have made the preparation of lichen depsides a relatively straightforward procedure, so that total synthesis is now a common means of structural confirmation.

Several new classes of compounds have been discovered in lichens during this period. For example six diphenyl ethers have been isolated and it would appear that they may arise biogenetically either by catabolism of the more common depsidones or by direct phenolic oxidation resulting in carbon-oxygen coupling of the mononuclear aromatic units [e.g. in leprolomin (118)]. A novel group of nonacyclic diketopiperazines have been isolated (14) and it would appear that these compounds are derived biogenetically from the aromatic amino-acid, tyrosine.

Chemical investigations on lichens have assumed increased taxonomic significance although the interpretation of the chemical data is still subject to some controversy (113).

New biological activity has also been discovered for some lichen metabolites. Thus a variety of compounds have been found to be plant growth regulators (188), with the common dibenzofuranoid derivative, usnic acid, being particularly active. The scabrosin esters (nonacyclic diketopiperazines) (14) on the other hand, have been found to be highly toxic to rats!

The development in the chemistry of structurally related groups of lichen metabolites will now be discussed in detail, in approximate biogenetic order.

2. Carbohydrates

2.1 Mono- and Oligosaccharides

Simple sugars such as glycerol, erythritol, ribitol, arabinitol, fructose, glucose, sucrose and trehalose are common lichen metabolites. Their occurrence is listed in CULBERSON's compendia (65, 78). Recent surveys have been conducted on the occurrence of these compounds in lichens by use of the techniques of paper chromatography (*299*) and g.l.c. of their acetyl, trifluoroacetyl, and trimethylsilyl derivatives (*254*).

The mobile carbohydrates produced by photosynthesis in lichens have been studied and this work has been reviewed by RICHARDSON (*272*). HILL and AHMADJIAN (*161*) have shown that four genera of lichen algae isolated in pure culture were able to produce the polyol which is known to move from alga to fungus in lichens containing these algae. Similar results were obtained by KOMIYA and SHIBATA (*213*) by working with cultured phycobionts and mycobionts.

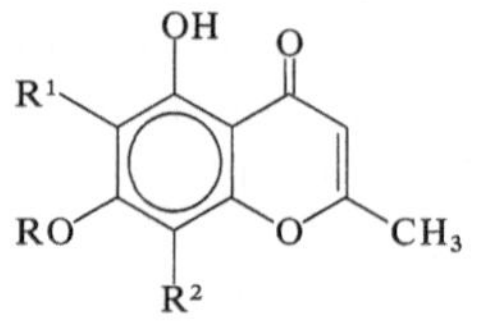

(**1**) R = β-D-diacetylglucosyl, R^1 = Me, R^2 = H
(**2**) R = β-D-diacetylglucosyl, R^1 = R^2 = Me
(**3**) R = β-D-triacetylglucosyl, R^1 = Me, R^2 = H
(**4**) R = β-D-triacetylglucosyl, R^1 = H, R^2 = Me

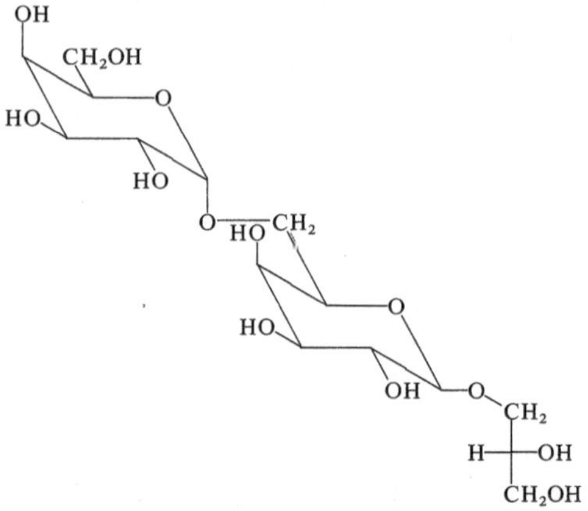

(**5**)

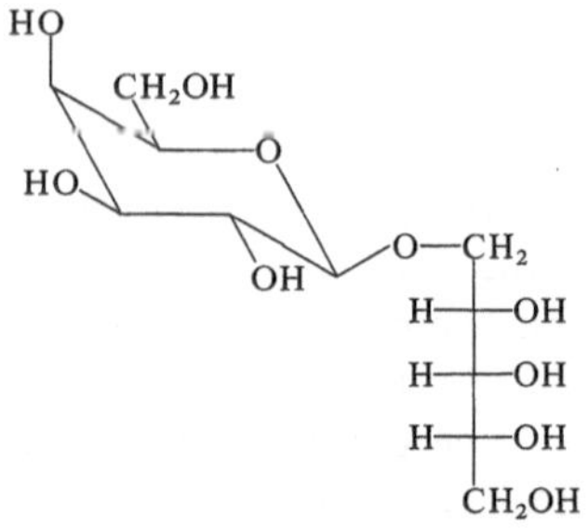

(**6**)

A number of partially acetylated chromone glucosides have been isolated by HUNECK. Extraction of *Roccellaria mollis, Schismatomna accedens,* and *Roccella galapagoensis* has yielded roccellin, galapagin, and mollin (*172*). Degradative and spectroscopic studies have established the partial structures (**1**) and (**2**) for roccellin and galapagin. Roccellin and mollin afforded the same product on acetylation indicating mollin to be a monoacetyl derivative of roccellin, i.e. (**3**). Lobodirin, from *Lobodirina* *cerebriformis* is a 7-O-β-D-triacetylglucosylisoeugenitol (**4**). Hydrolysis afforded glucose and isoeugenitol and the partial structure (**4**) was confirmed by the synthesis of acetyllobodirin from α-acetobromoglucose and subsequent acetylation (*175*).

O-α-D-Galactopyranosyl-(1→6)-O-β-D-galactopyranosyl-(1→1)-D-glyceritol (**5**), a trisaccharide which occurs in glycolipids, has been isolated from *Xanthoria parietina* (*143*). This lichen also produced the new disaccharide 1-O-(β-D-galactopyranosyl)-D-ribitol (**6**). Its structure followed from acid hydrolysis which produced D-galactose and ribitol; the rate of hydrolysis and the low optical rotation indicated a β-D-galactopyranoside and this was confirmed by the hydrolysis of the new disaccharide with β-D-galactosidase. Periodate oxidation showed that the compound was a 1-O-β-D-galactopyranosyl-D- or L-ribitol and the structure was finally proved by synthesis using the Kochetkov glycosylation method. This method has also been used in a new synthesis of peltigeroside (*214*). Thus the condensation (Scheme 1) of 1,2,5,6-tetra-O-benzoyl-D-mannitol (**7**) with 3,5,6-tri-O-acetyl-1,2-O-methylorthoacetyl-α-D-galactofuranoside (**8**) in nitromethane in the presence of mercuric bromide gave the protected peltigeroside (**9**) which on mild hydrolysis with triethylamine in chloroform and methanol gave 3-O-(β-D-galactofuranosyl)-D-mannitol (peltigeroside) (**10**) identical to the product isolated from *Peltigera horizontalis* (*230*).

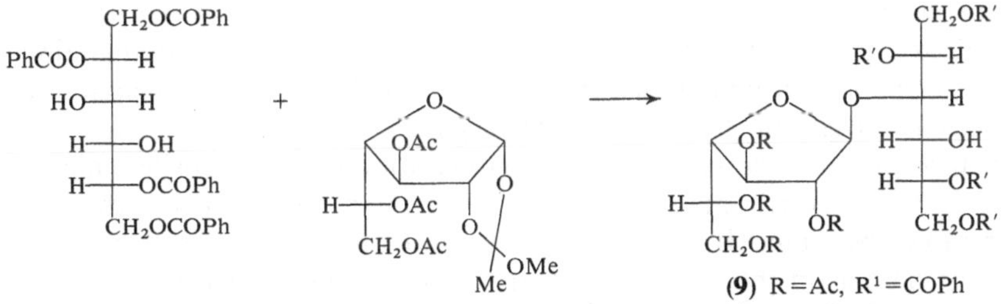

Scheme 1. Synthesis of peltigeroside

2.2 Polysaccharides

There has been considerable activity in this field since the last review (*171*) which has no doubt been stimulated by reports of the anti-tumour activity of many of these compounds (*255, 292, 294*), although the mechanism of action is not known (*330*). The anti-tumour properties of some chemically modified lichen polysaccharides have also been studied (*257*).

Takahashi, Takeda, and Shibata (*322*) have made a comparative study of the water-soluble polysaccharides isolated from laboratory cultures of lichen phycobionts and mycobionts. Those of the mycobionts were closely related to those of the parent lichen so that it is likely that the water-soluble lichen polysaccharides are produced by the mycobiont like the lichen metabolites of lower molecular weight.

Most of the polysaccharides of lichens are homo-D-glucans and the standard methods of polysaccharide chemistry are applied to their structural elucidation. Homogeneity is first ensured by rigorous purification which usually starts with precipitation from aqueous solution by ethanol and is followed by some form of chromatography often of the gel filtration type. Acid hydrolysis allows the component monosaccharides to be estimated and identified. The degree of polymerisation can be estimated from the molecular weight which is in turn obtained by ultracentrifugation, gel filtration against markers or by end group assay from the reducing end of the glucan. The types of linkage involved may be identified by methylation, often by the Hakomori method, followed by hydrolysis. These data are usually supplemented by partial hydrolysis studies using acidic or enzymic methods and by periodate oxidation, often in conjunction with the Smith degradation. Identification of degradation products is usually carried out by derivatisation and g.l.c. analysis. Partially acetylated polysaccharides are usually identified by their characteristic infrared spectra and the degree of acetylation can be estimated after hydrolysis. The site of attachment is obtained by Bouveng's method in which the free hydroxy groups are protected as their phenyl carbamates. Methylation, treatment with lithium aluminium hydride, and hydrolysis of the partially methylated glucan then yields a readily identifiable O-methyl monosaccharide. The determination of the configuration of the glycoside linkages often depends on the magnitude of the molecular rotation and a study of the infrared spectrum. High resolution ^{1}H-n.m.r. spectroscopy is also able to assist here as in the case of lichenan a linear β-D-glucan with a ratio of $(1{\rightarrow}3):(1{\rightarrow}4)$ linkages of about 3:7. The 250 MHz spectrum of D_2O exchanged lichenan in dimethyl sulfoxide-d_6 at 100° C exhibits the anomeric protons as three doublets at δ 4.34, 4.29, and 4.25 each with J 7.5 Hz characteristic of the β-configuration (*40*). It was therefore postulated that lichenan is a regular copolymer with the repeating unit (**11**) in agreement with degradation by the enzyme cellulase which affords O-β-D-glucopyranosyl-(1→3)-O-β-D-glucopyranosyl-(1→4)-D-glucopyranose as the major trisaccharide.

$\rightarrow$4)-β-D-Glcp-(1$\longrightarrow$3)-β-D-Glcp-(1$\longrightarrow$4)-β-D-Glcp-(1$\longrightarrow$4)-β-D-Glcp-(1$\longrightarrow$3)-β-D-Glcp-(1$\longrightarrow$

(**11**)

References, pp. 219—234

The ratios of the integrated peak intensities associated with the C-3, C-4, and C-1 signals in the ^{13}C-n.m.r. spectra of homo-D-glucans have been used as a measure of the ratio of the $(1→3):(1→4):(1→6)$ linkages (*336*).

A review by SHIBATA (*292*) summarises much of the early work on lichen polysaccharides including lichenan; isolichenan, a linear α-D-homoglucan with $(1→3)$ and $(1→4)$ linkages; GE-3, a $(1→6)$-β-D-glucan with about 10% of the C-3 hydroxy groups acetylated (*159, 256, 293*); PC-3, which is similar to isolichenan but differs in the sequence of linkages (*328*); EP-6, which is similar to isolichenan except for the degree of polymerisation (*327*); EP-7, which is a β-D-homoglucan with a ratio of $(1→3):(1→4)$ linkages of 3 : 1 (*327*); and acrosyphan, an α-$(1→3)$ $(1→4)$ $(1→6)$-homoglucan (*327*). Branched chain α-glucans have been isolated from *Stereocaulon paschale* (*154*), *Stereocaulon japonicum* (*335*), and *Evernia prunastri* (*168*). The major polysaccharide of *S. japonicum* is an α-$(1→3)$ $(1→4)$-glucan with some branching at either the 3,4- or the 2,3-positions; that of *S. paschale* is similar.

Heteropolysaccharides have been isolated from *Evernia prunastri* (*239*), *Cetraria islandica* (*169*), *Usnea rubescens* and some *Cladonia* species (*255*). The *E. prunastri* sugar is a highly branched galactomannan with a basic chain of $(1→2)$ and $(1→6)$ linked mannose residues. Galactopyranose and galactouronic acid residues are present as non-reducing terminal groups; the galactopyranose residues were assigned the β-D configuration and the mannose residues the α-D configuration. The *C. islandica* sugar contains D-glucose and D-glucuronic acid.

Glycopeptides have been isolated from *Parmelia michauxiana* (*167*) and lichens of the genus *Stictaceae* (*323*). The former is able to agglutinate several types of human and animal erythrocytes and the latter exhibit anti-tumour activity.

3. Amino-acid Derivatives

The numerous free amino-acids detected in and isolated from lichens are not listed in this review but the reader is referred to the publications by C. F. CULBERSON *et al.* (*65, 78*) for this material.

BERNARD *et al.* (*15*) have isolated sticticin (**12**) from the thallus of *Lobaria laetevirens;* the structure was determined by spectroscopic methods and comparison with a synthetic sample. The synthesis of (**12**) from L-DOPA and the degradation reactions of this compound are depicted in Scheme 2.

The nonacyclic diketopiperazine derivatives scabrosin 4,4′-diacetate (**13**), scabrosin 4-acetate-4′-butyrate (**14**), scabrosin 4,4′-dibutyrate (**15**) and scabrosin 4-acetate-4′-hexanoate (**16**) have been isolated from a chemical strain of *Parmelia scabrosa* by BEGG, ELIX, and JONES (*14*).

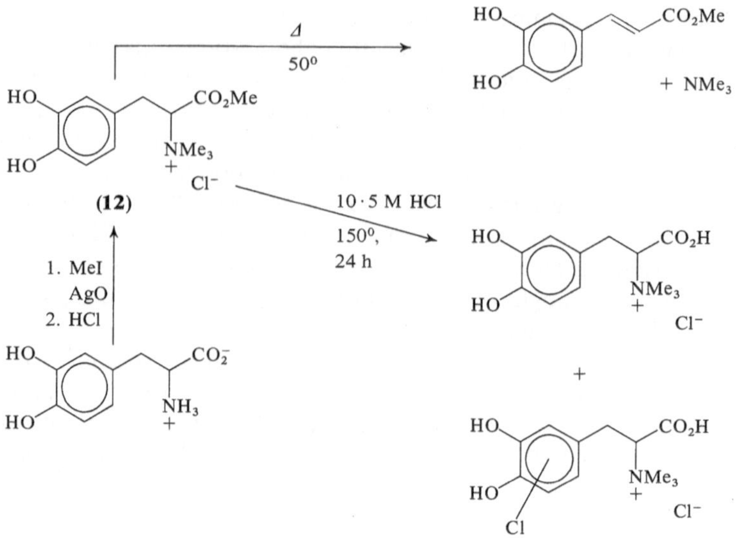

Scheme 2. Synthesis and degradation of sticticin

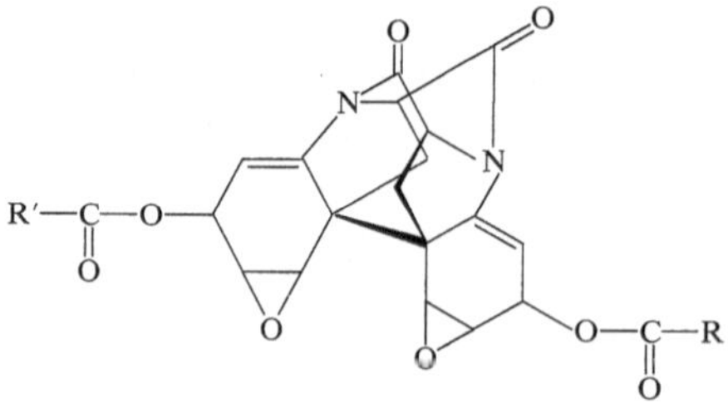

	R	R¹
(13)	Me	Me
(14)	Me	n-C₃H₇
(15)	n-C₃H₇	n-C₃H₇
(16)	Me	n-C₅H₁₁

The structure of these compounds followed from the spectroscopic properties, in particular the ^{1}H-n.m.r. and ^{13}C-n.m.r. spectra. It has been postulated (*14*) that the carbon skeleton of (**13 – 16**) arises biosynthetically from the intermolecular condensation of two tyrosine molecules to form a diketopiperazine with subsequent intramolecular oxidative coupling and cyclisation.

A cyclic tetrapeptide, roccanin (**17**), has been isolated by BOHMAN-LINDGREN from *Roccella canariensis* (*22*). Acid hydrolysis of (**17**) yielded two amino-acids, *L*-proline (**18**) and *R*-β-phenyl-β-alanine (**19**), and quantitative amino-acid analysis confirmed that (**18**) and (**19**) were formed in equimolar amounts. Mass spectrometry and the observed chemical reactivity confirmed that (**17**) was a cyclic peptide. Selective reduction of (**17**) at the more reactive acylproline linkage by treatment with lithium aluminium hydride gave the aminoaldehyde (20) as the sole product, thus confirming the symmetrical nature of this cyclic peptide (Scheme 3). The structure of roccanin (17) was ultimately confirmed by synthesis (23).

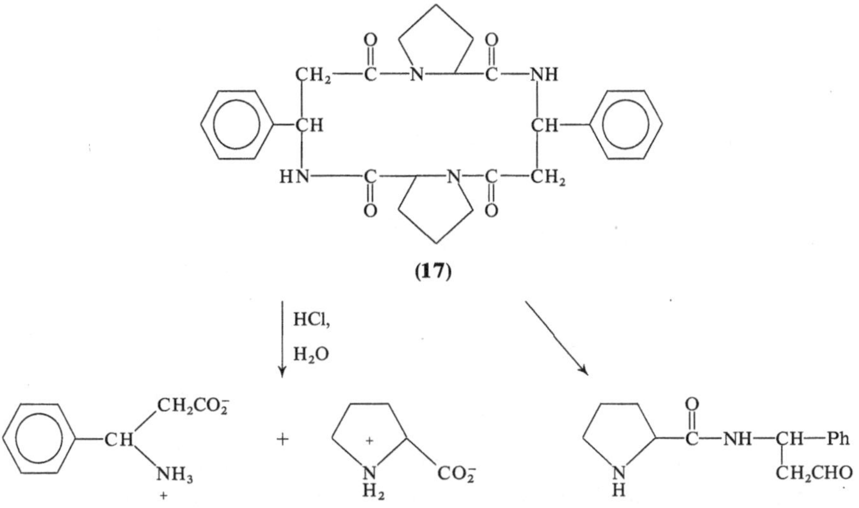

(17)

(19)   (18)   (20)

Scheme 3. Degradation of roccanin

Allantoin (21) and xantholamine (22) have been isolated by SOLBERG from the lichen *Xanthoria parietina* (300, 301). The former compound was identified by comparison of the observed spectrum with those of authentic samples while the structure of (22) was based on elemental analysis and spectroscopic data.

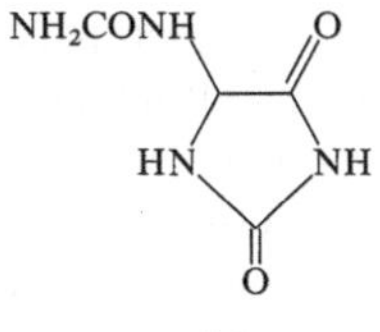

(21) 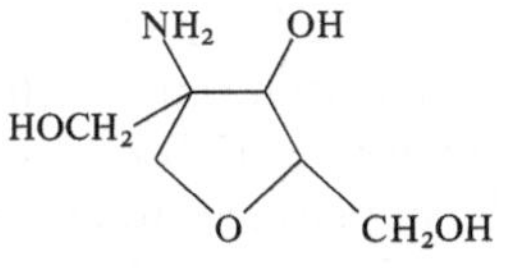   (22)

4. 4-Ylidenetetronic Acids

Synthetic and biosynthetic work on the lichen pulvinic acids, *e.g.* pulvinic acid (23), calycin (24), and pinastric acid (25), has been reviewed recently by PATTENDEN (265). MAASS (231) has studied the biosynthesis of

(23)   (24)   (25)

(26)   (27)   (28)

some of these phenylpropanoids and has shown that ^{14}C labelled pulvinic acid (23), vulpinic acid (26) and ethyl pulvinate (27) are efficient precursors for calycin (24) in *Pseudocyphellaria crocata*. Full details of KNIGHT and PATTENDEN's synthesis (*212*) (Scheme 4) of the O-methyl derivative of pinastric acid have appeared. The readily available maleic anhydride (29) on metal hydride reduction undergoes regioselective attack and furnished the O-methyl tetronic acid (30). This was metallated at $-78°$ C with lithium N-cyclohexyl-N-isopropylamide and treated with methyl benzoylformate thus yielding the tertiary alcohol (31) which on dehydration furnished methyl pinastrate (32). WEINSTOCK *et al.* (*331*) have reported a regiospecific synthesis of pulvinic acid derivatives which they used for the preparation of synthetic analogues for testing as anti-inflammatory agents. In a typical example (Scheme 5) dimethyl phenyloxalacetate (33), easily prepared from methyl phenylacetate and methyl oxalate in presence of sodium methoxide, gave the enol ester (34) on treatment with 4-chlorophenylacetyl chloride and triethylamine in acetone at room temperature. On addition of a further molar equivalent of triethylamine and heating at $60°$ C a low yield of the ylidenetetronic acid (35) resulted. The stereochemistry about the exocyclic double bond was established by infrared and n.m.r. spectroscopic evidence for the existence of chelate type H-bonding.

Reaction of singlet oxygen with pulvinic acid (23) by irradiation of a solution of pulvinic acid in methanol with visible light in presence of oxygen and rose bengal as a sensitizer yielded leprapinic acid (28). The reaction is considered (*43*) to involve electrophilic attack by singlet oxygen at the electron-rich *ortho* position to produce an intermediate hydroperoxide.

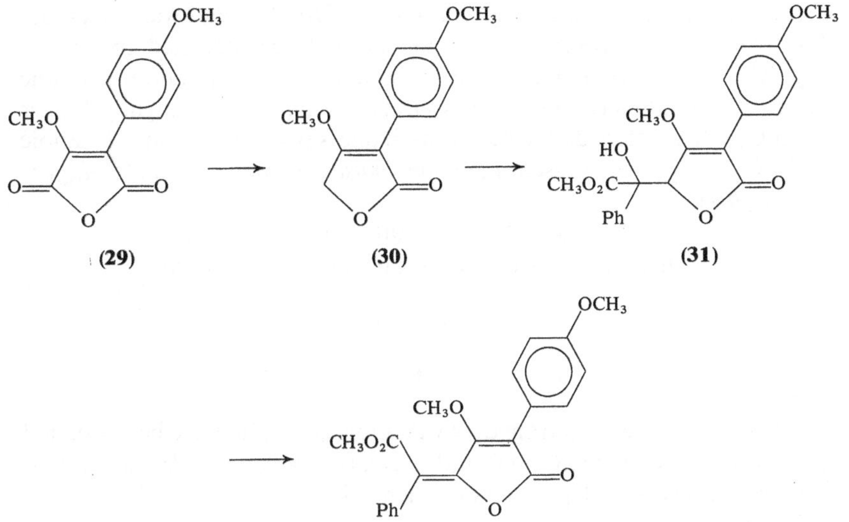

(29) (30) (31) (32)

Scheme 4. Synthesis of methyl pinastrate

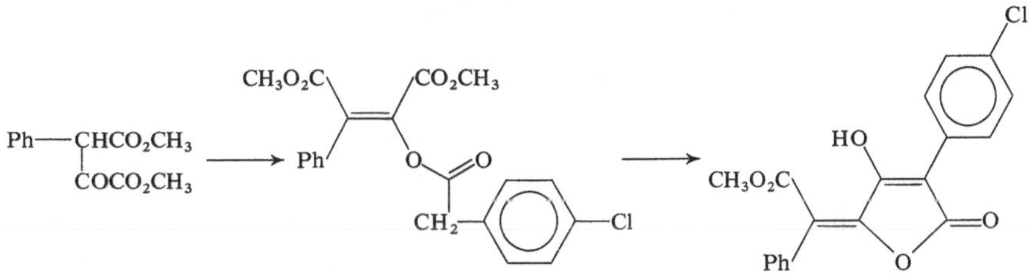

(33) (34) (35)

Scheme 5. Synthesis of a pulvinic acid analogue

5. Aliphatic Acids and Related Compounds

5.1 Straight Chain Aliphatic Acids and Related Esters, Alcohols and Lactones

Saturated n-alkanoic acids isolated recently from various lichen species include palmitic acid, stearic acid, nonadecanoic acid, eicosanoic acid (*33*) and the $C_{26}-C_{34}$ acids (*305*). The unsaturated acids oleic acid (*33*), linoleic (*304*), linolenic and arachidonic acids (*333*) have also been identified in lichen extracts. Free n-alkanols ($C_{16}-C_{28}$) have been isolated from

Cetraria cucullata (*33*) and triglycerides (*33, 216, 305*) and wax esters (*33, 304*) from several species. The available evidence indicated that while $C_{40} - C_{50}$ wax esters derived from n-alkanols ($C_{16} - C_{28}$) and n-alkanoic acids ($C_{16} - C_{26}$) were present in *Cetraria cucullata* (*33*), slightly longer chain $C_{54} - C_{58}$ esters derived from n-alkanols ($C_{24} - C_{28}$) and n-alkanoic acids ($C_{16} - C_{30}$) were present in *Omphalodiscus spodochrous* and *Peltigera canina* (*304*).

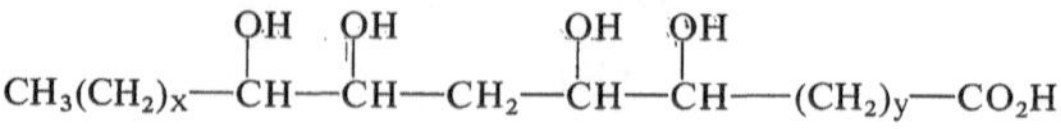

$$CH_3(CH_2)_x - \underset{\underset{OH}{|}}{CH} - \underset{\underset{OH}{|}}{CH} - CH_2 - \underset{\underset{OH}{|}}{CH} - \underset{\underset{OH}{|}}{CH} - (CH_2)_y - CO_2H$$

$$x = 6 - 10, \quad y = 7 - 10$$

(36)

Further mixtures of tetrahydroxyalkanoic acids (**36**) have been isolated from several species (*303, 304*) and *Alectoria fremontii* (*303*) appears to produce n-alkyl ($C_6 - C_{12}$) esters of such acids.

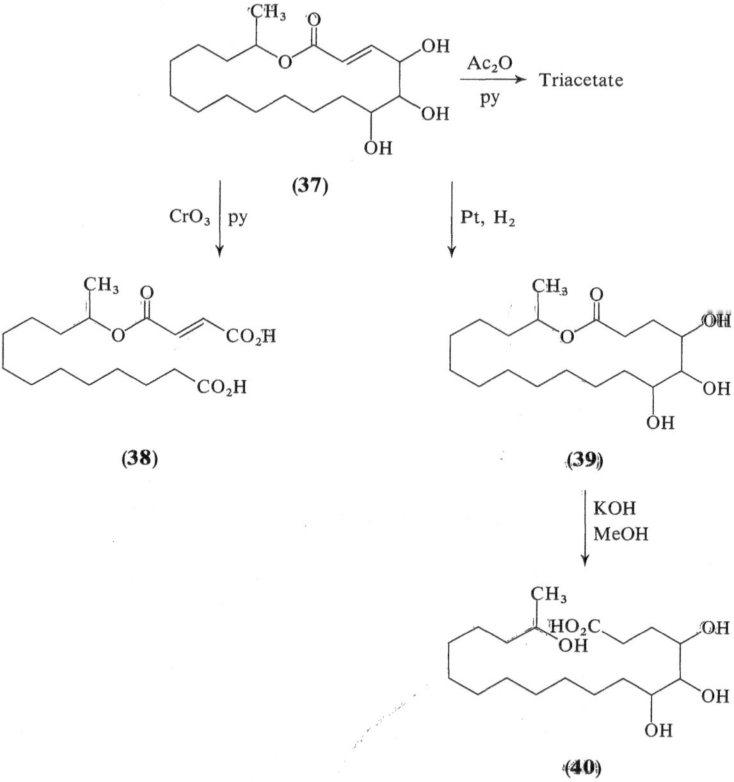

Scheme 6. Reactions of aspicilin

Aspicilin (**37**), the first macrocyclic lactone isolated from a lichen (*189*), would appear to be biogenetically related to the hydroxyacids (**36**). HUNECK and SCHREIBER (*189*) isolated aspicilin from *Aspicilia gibbosa* and established the structure (**37**) by analysis of the spectroscopic properties, chemical interconversions and properties of the derivatives so obtained (Scheme 6). Hence oxidation of aspicilin (**37**) gave the diacid (**38**) while catalytic reduction gave dihydroaspicilin (**39**). Subsequent saponification of (**39**) gave 4,5,6,17-tetrahydroxystearic acid (**40**).

5.2 Cn + C₄ Derived Aliphatic Acids and Lactones

5.2.1 Acylic Acids

A comparison of the mass spectral fragmentation patterns of the dimethyl ester of caperatic acid (**41**) and the corresponding di(trideuteromethyl) ester located the position of the ester function in this molecule and established the structure as methyl 3,4-dicarboxy-3-hydroxyoctadecanoate (**41**) (*25*). In a corroborative chemical degradation (**41**) was oxidized by treatment with sodium bismuthate, whereupon methyl 3-oxo-octadecanoate (**42**), the decarboxylation product of the initially formed β-keto-carboxylic acid, was detected (*25*).

(41)

(42)

(43)

(44)

A homologous aliphatic acid, methyl 3,4-dicarboxy-3-hydroxy-19-oxoeicosanoate (**43**), has been isolated from *Usnea meridensis* (*210*). The structure of this compound followed from the spectroscopic and chemical properties, by comparison with the properties of caperatic acid (**41**) and by sodium bismuthate oxidation to methyl 3,19-dioxoeicosanoate (**44**) (*210*). Caperatic acid (**41**) and methyl 3,4-dicarboxy-3-hydroxy-19-oxoeicosanoate (**43**) have similar optical rotations but the absolute configuration of these acids remains to be determined.

8*

(45)

(46) R=Ac
(47) R=H

The structurally and biogenetically closely related aliphatic acid, 2-methylene-3-carboxy-18-hydroxynonadecanoic acid (45) has been isolated from *Usnea aliphatica* (*211*), *Parmelia xanthina* (*195*) and *P. madagascariacea* (*195*). KEOGH and ZURITA (*211*) determined the overall structure of (45) by spectroscopic and chemical methods. The relative position of the carboxyl groups was confirmed by treatment with acetic anhydride and pyridine, whereupon the maleic anhydride (46) was obtained. Pyrolysis of (45) gave the corresponding anhydride (47). Norcaperatic acid (48) obtained by hydrolysis of caperatic acid (41) (*211*) underwent decarboxylation, dehydration and an analogous thermal rearrangement on pyrolysis, to give 2-tetradecyl-3-methylmaleic anhydride (49).

(48)

(49)

HUNECK and SNATZKE (*195*) have established that the absolute configuration of (45) is 3 (*R*), 18 (*R*) from CD data and by Horeau's method. The following interconversions were undertaken to determine the configuration at C 3 (Scheme 7). Catalytic hydrogenation of (45) gave 2-methyl-3-carboxy-18-hydroxynonadecanoic acid (50) and subsequent reactions converted this compound to (+)-2(*S*)-ethyloctadecanoic acid-*N*-thiocarbonylmorpholinamide (52). The CD curves of (51) and (52) were compared with the corresponding (+)-2(*S*)-ethyltetradecanoic acid derivatives and shown to have the 2(*S*)-configuration. Hence (45) must have the 3(*R*)-configuration. The configuration of (45) at C-18 was determined by Horeau's method using the methyl ester of (45) and (±)-2-phenylbutyric acid and established as 18 (*R*).

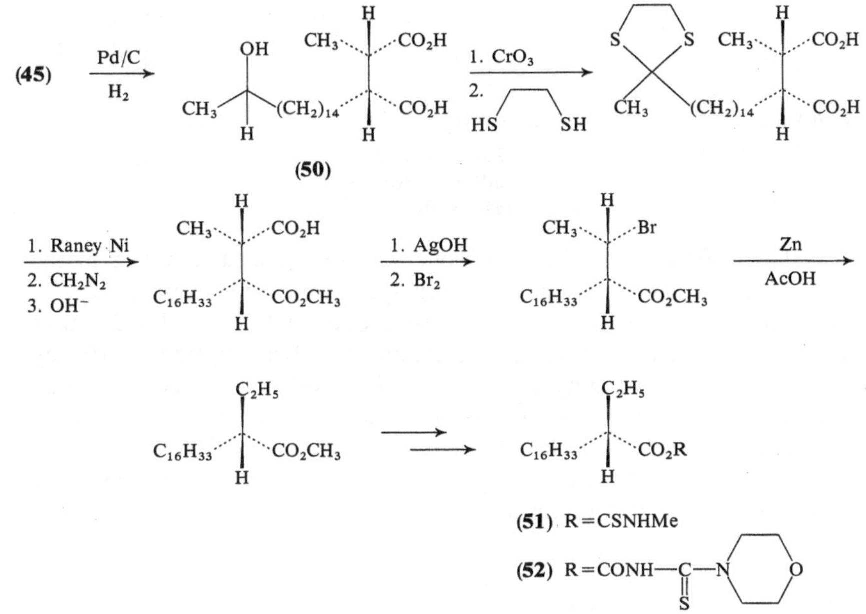

Scheme 7. Determination of the absolute configuration at C-3 of the acid (**45**)

5.2.2 γ-Lactonic Acids

A number of stereoisomeric lactonic acids, higher homologues of lichesterinic acid and protolichesterinic acid, have been isolated from several lichen genera. Thus (+)-murolic acid (**53**) and (+)-neodihydromurolic acid (**54**) have been isolated from *Lecanora muralis, L. melanophthalma* and *L. rubina* (*192*), isomuronic acid (**55**) and neuropogalic acid (**56**) from *Neuropogon trachycarpus* (*20*) and protoconstipatic acid (**58**), dehydroconstipatic acid (**59**) and constipatic acid (**60**) from various *Xanthoparmelia* species (*46*). The structure of compounds (**53**), (**54**), (**55**), (**56**), (**58**), (**59**) and (**60**) followed from spectroscopic and chemical data and in particular, by comparison with the properties of lichesterinic and protolichesterinic acids.

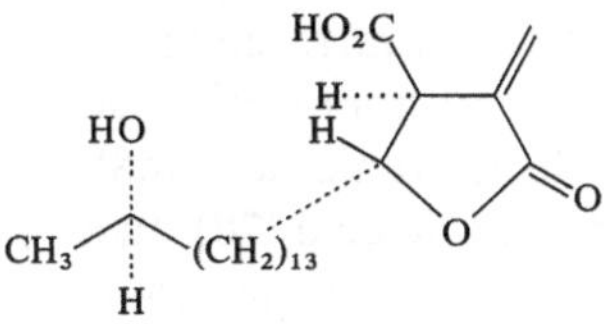

(**53**)

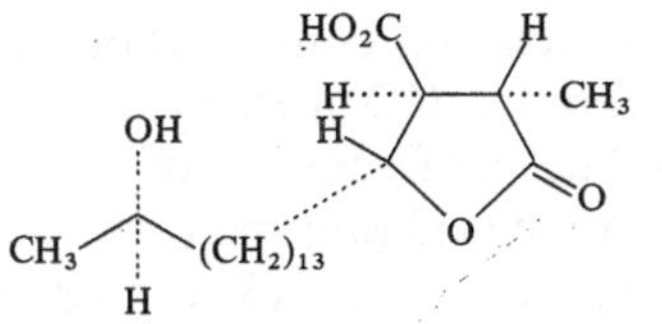

(**54**)

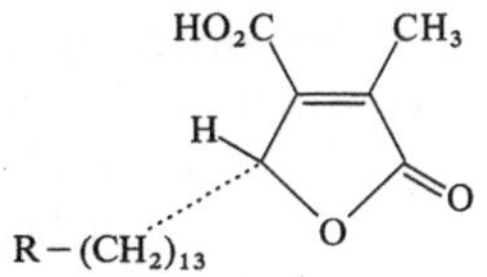

(**55**) R=COCH$_3$
(**56**) R=CHOHCH$_3$
(**57**) R=H

HO₂C····
H····
R—(CH₂)₁₃

(58) R=CHOHCH₃

HO₂C   CH₃
H···
R—(CH₂)₁₃

(59) R=COCH₃
(60) R=CHOHCH₃
(61) R=H

HO₂C
H·····
H·
C₁₃H₂₇

(62)

The configuration of (+)-murolic acid **(53)** was determined in the following manner (*192*). The ¹H-n.m.r. spectrum of **(53)** exhibited $J_{3,4}$ of 6 Hz comparable with that of (+)-protolichesterinic acid **(62)**, thus establishing a *trans*-relationship of these protons. Rearrangement of **(53)** by treatment with acetic anhydride and pyridine produced (+)-acetylisomurolic acid **(63)**. The CD curve of this compound was compared with that of the 4(*S*)-lactone **(64)** and shown to have the opposite configuration at C-4, i.e. 4(*R*). Hence from the ¹H-n.m.r. studies, (+)-murolic acid **(53)** must have the 3(*S*), 4(*R*) configuration shown.

AcO
CH₃
H
CH₃
(CH₂)₁₃
H
HO₂C   CH₃
H.

(63)

HO₂CCH₂
H···

(64)

The relative configuration of (+)-neodihydromurolic acid **(54)** similarly followed from the ¹H-n.m.r. spectrum and was shown to be *trans, trans* ($J_{3,2}=11$ Hz, $J_{3,4}=9$ Hz). Catalytic hydrogenation of (+)-murolic acid **(53)** gave a mixture of isomurolic acid **(65)** and dihydromurolic acid **(66)**. Subsequent esterification and equilibration with sodium methoxide gave methyl neodihydromurolate **(67)** thus confirming that **(53)** and **(54)** must have the same configuration at C-18, C-4 and C-3 (Scheme 8).

The configuration of C-18 in **(67)** was determined by Horeau's method and shown to be 18(*R*), and hence the absolute configuration of **(54)** was established as being 2(*S*), 3(*S*), 4(*R*), 18(*R*).

By deducing the relative configuration from ¹H-n.m.r. coupling constants and comparative CD measurements these authors (*192*) also established the absolute configuration of (+)-nephrosteranic acid **(68)**, (−)-alloprotolichesterinic acid **(69)** and (+)-nephrosterinic acid **(70)**. In a similar manner, using the known 2(*S*)-configuration of (−)-lichesterinic acid **(61)** as a reference, the absolute configuration of **(55)** and **(56)** was established as 4(*R*) (*20*), **(59)** and **(60)** as 4(*S*) and **(62)** as 3(*R*), 4(*S*) (*46*).

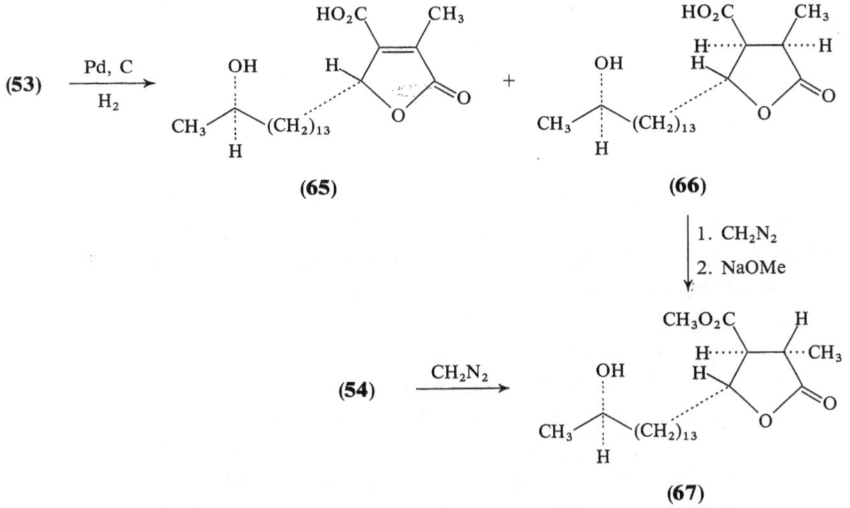

Scheme 8. Correlation of the structure of murolic acid and neodihydromurolic acid

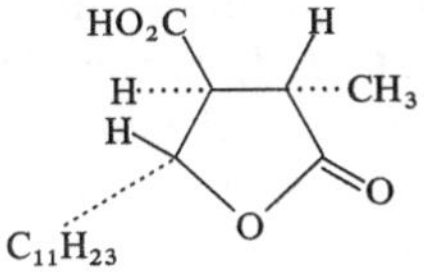

(68)
3(S), 4(R)

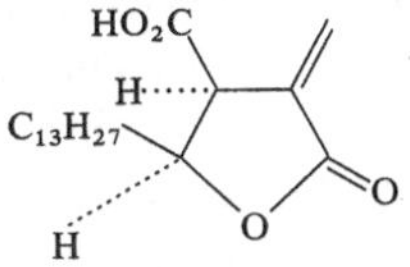

(69)
3(S), 4(S)

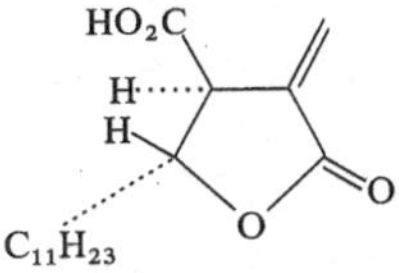

(70)
3(S), 4(R)

5.2.3 δ-Lactonic Acids

Acaranoic acid and acarenoic acid isolated from *Acarospora chloro-phana* have now been assigned the six-membered δ-lactone structures **(71)** and **(73)** respectively (*182*) rather than the previously accepted γ-lactone structures **(74)** and **(75)**.

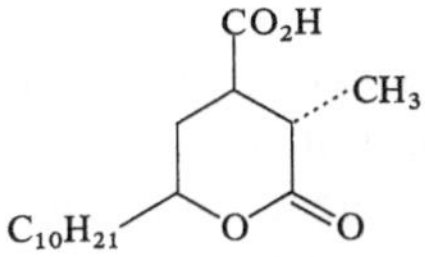

(71)

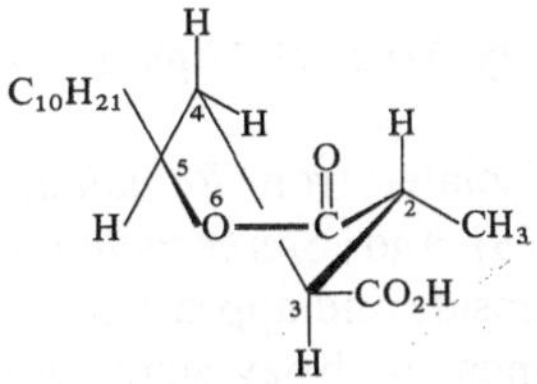

(72)

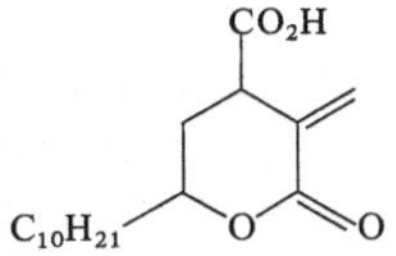

(73)

The ^{1}H-n.m.r. spectra of (71) and (73) established the correct carbon skeleton and the relative stereochemistry of C-2, C-3, C-4 and C-5 in (71). With $J = 11\,\text{Hz}$ and $12\,\text{Hz}$, the protons on these carbons must occupy antiperiplanar positions in a six-membered ring, a conclusion confirmed by ^{13}C-n.m.r. data.

The negative Cotton effect exhibited by ($-$)-acaranoic acid (71) fixed the chirality as shown in (72) and the absolute configuration as ($-$)-2(S)-methyl-3(R)-carboxypentadecyl-1→5(S)-olide. Catalytic hydrogenation of ($-$)-acarenoic acid (73) produced ($-$)-acaranoic acid (71) and hence established the configuration of the former compound as 3(R).

Acarospora chlorophana has been found to exist in two chemical races (*177*), one producing ($-$)-acaranoic acid (71) and ($-$)-acarenoic acid (73) in addition to rhizocarpic acid, and the other with ($+$)-roccellic acid (76) and rhizocarpic acid. *Acarospora oxytona* (*177*) however produces rhizocarpic acid and ($+$)-lichesterinic acid (77). The acids (71), (73), (76) and (77) have the same chirality at C-2 and C-3 and this circumstantial evidence led HUNECK (*177*) to propose the biogenetic pathway outlined in Scheme 9, elaborating that proposed earlier by MOSBACH (*244*).

In general terms it seems probable that all the aliphatic acids of the group are interrelated in an analogous manner to that depicted in Scheme 9. Thus the biogenesis is believed to involve the intermediacy of n-alkanoylcoenzyme-A, the α-methylene group of which undergoes an aldol type condensation with oxalacetic acid. Subsequent dehydration, decarboxylation and oxidative cyclisation can then lead to the corresponding acylic acids and the γ- and δ-lactonic acids. It is interesting to note that the alkanoyl group is normally fully reduced when the chain length is C_{16} or less, but that the C_{18} compounds invariably have a penultimate ketonic or alcoholic function in the chain.

5.3 Biogenetically Atypical Aliphatic Acids and Esters

Bourgeanic acid (78) isolated from *Ramalina bourgeana, R. evernioides* ($\equiv$ *Niebla evernioides*) (*18*) and *Stereocaulon tomentosum* (*32*) can be considered an aliphatic depside and appears to be a biogenetic oddity since by inspection the component hydroxy acids would appear to arise from three propionate and one acetate units. However labelling experiments

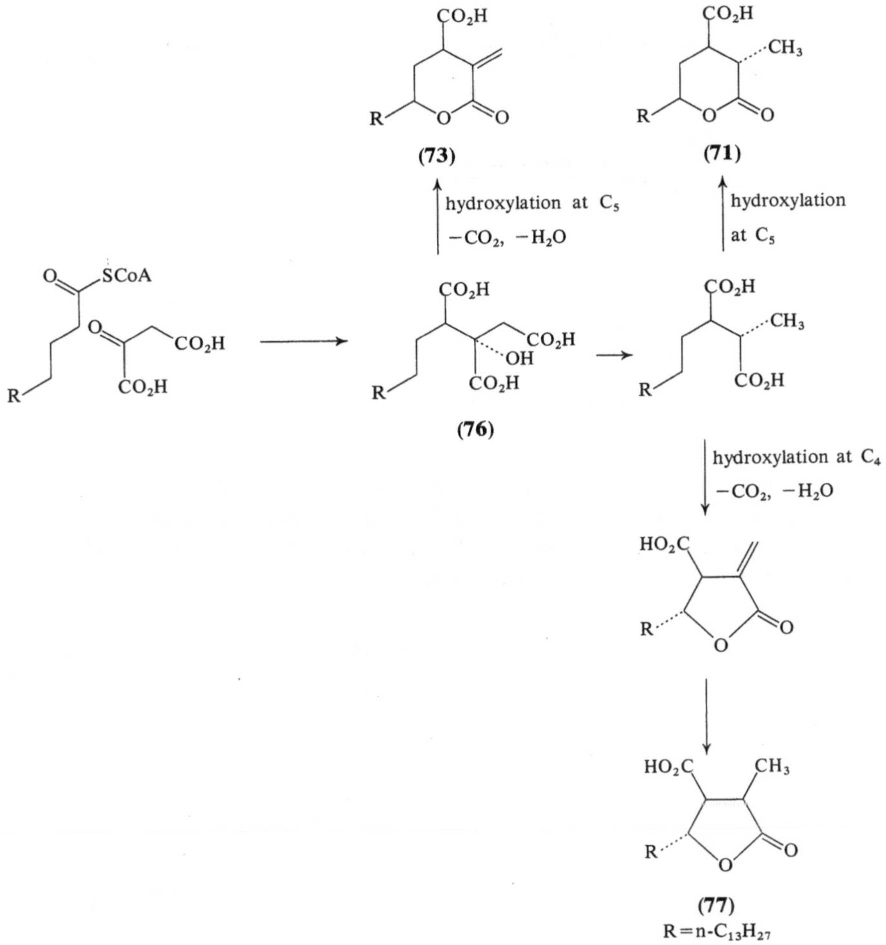

(71) (73) (76): R=n-$C_{10}H_{21}$

Scheme 9. Suggested routes for the biogenesis of aliphatic acids

(*17, 21*) confirmed that bourgeanic acid (**78**) is biosynthesised via C-methylation of a polyketide formed from four acetates.

Hydrolysis of (**78**) produced 3-hydroxy-2,4,6-trimethyloctanoic acid (**79**) while pyrolysis of the potassium salt of (**78**) gave 2,4-dimethylhexanal (**80**) and a hydrocarbon ($C_{20}H_{20}$). Treatment of (**79**) with acetic anhydride and pyridine gave a mixture of 2,4,6-trimethyl-2-octenoic acid (**81**) and the cyclic dilactone (**82**) (Scheme 11). These degradation experiments together with the observed spectroscopic properties established the structure of (**78**),

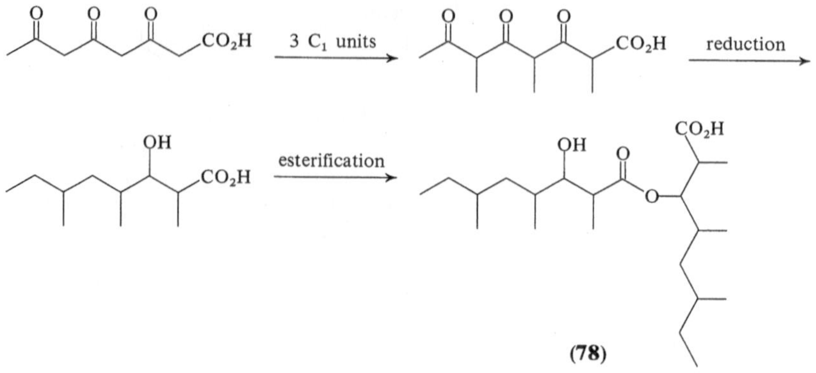

Scheme 10. Schematic biosynthetic route to bourgeanic acid

which was ultimately confirmed by total synthesis (*17*) (Scheme 12). Further the absolute configuration of (**78**) has been established and bourgeanic acid was shown to be 2(*S*), 4(*R*), 6(*R*)-trimethyl-3(*S*)-(3′(*S*)-hydroxy-2′(*S*), 4′(*R*), 6′(*R*)-trimethyl)octanoyloxy-octanoic acid (*17*).

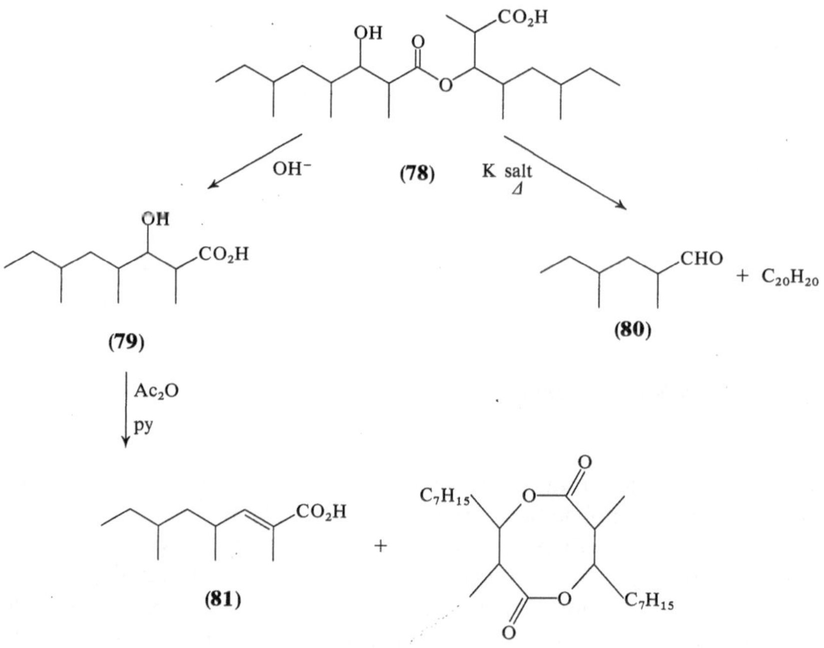

Scheme 11. Reactions of bourgeanic acid

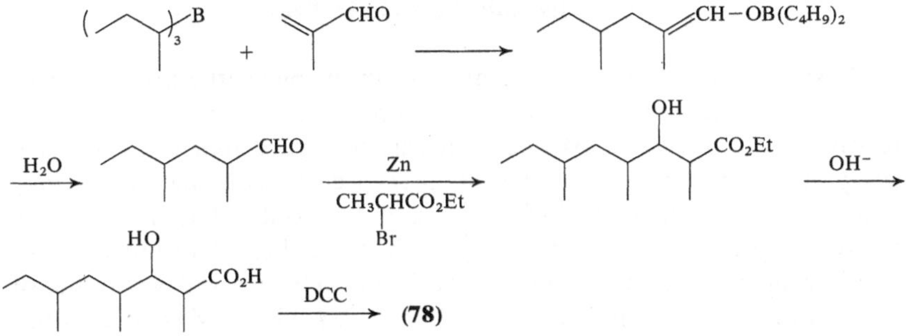

Scheme 12. Synthesis of bourgeanic acid

A structurally rather bizarre lipid is the monoacetylpentol (83) isolated from *Alectoria ochroleuca* (*305*). Besides spectroscopic evidence, periodate oxidation of (83) apparently produced a mixture of 8-acetoxyoctanal (84) and 2-ethyl-2-pentylpropan-1,3-dial (85) which were isolated as the corresponding 2,4-dinitrophenylhydrazones and characterised by mass spectroscopy (*305*). The polyester, methyl hexa(α-hydroxyvalerate) (86) has been isolated from *Cetraria nivalis* and *Cladonia gonecha* (*33*).

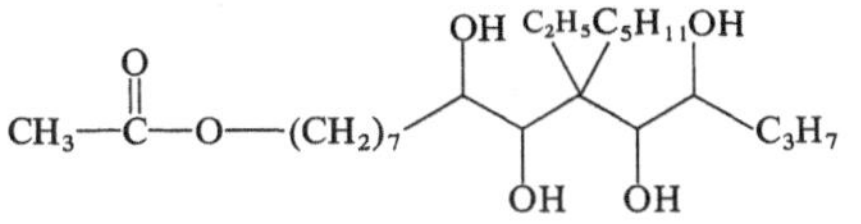

(83)

AcO—(CH$_2$)$_7$CHO

(84)

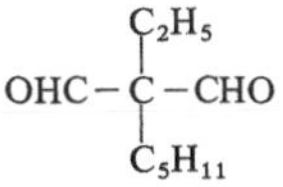

$$OHC-\underset{\underset{C_5H_{11}}{|}}{\overset{\overset{C_2H_5}{|}}{C}}-CHO$$

(85)

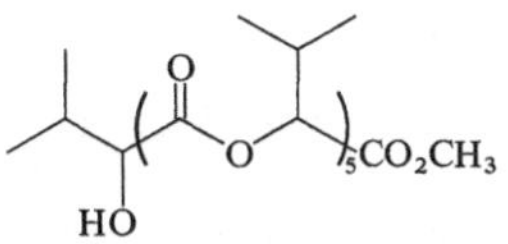

(86)

6. Mononuclear Phenolic Compounds from Lichens

The mononuclear phenol compounds isolated from lichens can be subdivided into three major categories, namely the orsellinic acid, the phloroacetophenone and the phthalide derivatives. These are discussed in turn.

6.1 Orsellinic Acid Derivatives

In spite of their apparent intermediacy in the biosynthesis of depsides, simple orsellinic acid derivatives have only been found in lichens on relatively rare occasions. Moreover it can be argued that traces of such compounds could arise as artefacts of the co-occurring depsides, which may undergo partial hydrolysis during the extraction and isolation procedure. However special care has been taken in a number of instances (with monitoring by two-dimensional t.l.c.) to confirm the natural occurrence of simple orsellinic acid derivatives (232, 233, 234). Thus Maass (232) has shown that the lichen *Pseudocyphellaria crocata* produces methyl orsellinate (**87**), orsellinic acid (**88**) and 4-O-methylorsellinic acid (**89**) together with the cogeneric tridepsides tenuiorin (**91**), methyl gyrophorate (**92**), 4-O-methylgyrophoric acid (**93**) and gyrophoric acid (**94**) and depsides methyl evernate (**95**), methyl lecanorate (**96**), evernic acid (**97**) and lecanoric acid (**98**)! The structure of the orsellinic acid derivatives (**87** – **89**) followed comparison with synthetic materials. Methyl orsellinate (**87**) has also been detected in *Peltigera aphthosa* (234), orsellinic acid (**88**) in *Omphalodiscus spodochrous* (304) while *Lobaria linita* (233) has been shown to contain both (**87**) and (**88**).

	R^1	R^2
(**87**)	H	Me
(**88**)	H	H
(**89**)	Me	H
(**90**)	Me	Me

	R^1	R^2
(**91**)	Me	Me
(**92**)	H	Me
(**93**)	Me	H
(**94**)	H	H

	R^1	R^2
(**95**)	Me	Me
(**96**)	H	Me
(**97**)	Me	H
(**98**)	H	H

In all instances these mononuclear compounds were accompanied by the structurally related depsides and tridepsides.

Evernia prunastri has proved to be a particularly rich source of lichen compounds and extraction and g.c.—m.s. studies have revealed the presence of orcinol (**99**) (*268, 269*), 3-methoxy-5-methylphenol (**100**) (*144*), 3,5-dimethoxytoluene (**101**) (*144*), 2-chloro-3,5-dimethoxytoluene (**102**) (*144*), phenol (*144*), thymol (*144*), 2-chloro-3-methoxy-5-methylphenol (**103**) (*144*), methyl salicylate (*144*), methyl 4-O-methylorsellinate (**90**) (*268, 269*), and methyl β-orcinolcarboxylate (**104**) (*253*). Rhizonic acid (**105**) has been isolated from *Parmelia subnuda* (*215*).

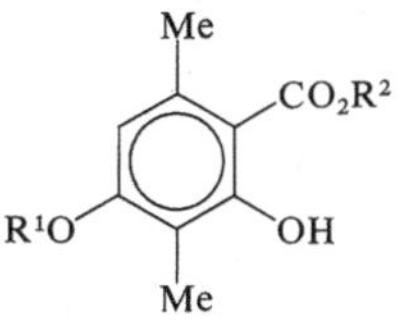

	R¹	R²
(**104**)	H	Me
(**105**)	Me	H

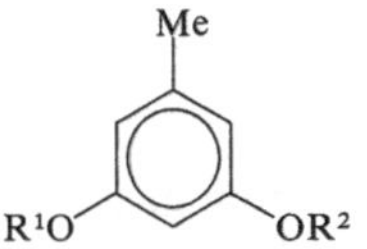

	R¹	R²
(**99**)	H	H
(**100**)	Me	H
(**101**)	Me	Me

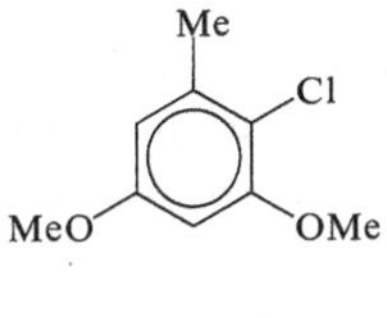

(**102**)

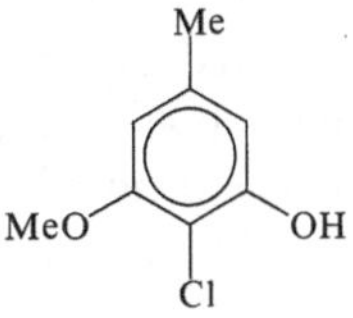

(**103**)

A mononuclear phenolic compound tentatively assigned the structure of ethyl 2-hydroxy-4-methoxy-6-pentylbenzoate (**106**) has been isolated by SOLBERG from *Icmadophila ericetorum* (*305*) but the possibility that this compound may be an artefact of the ethanolic work-up procedure is not discussed.

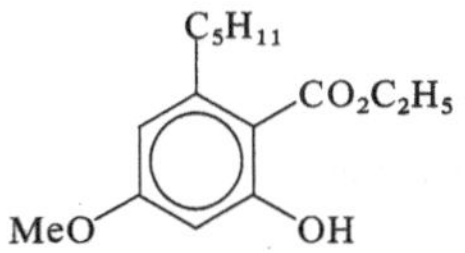

(**106**)

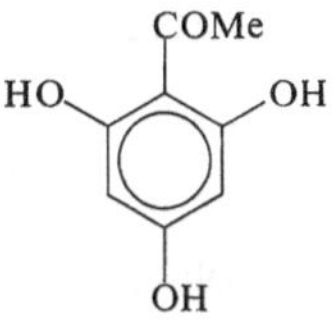

(**107**)

6.2 Phloroglucinol Derivatives

Orsellinic acid and its homologues form the most common mononuclear structural units of polyketide-derived lichen metabolites, but compounds derived from phloroacetophenone (**107**) (through an alternative cyclisation of a linear C_8 polyketide) are very limited in number. Two such mono-nuclear compounds have been reported, 2-O-methyl-3-methylphloroaceto-phenone (108) and 2-O-ethyl-3-formylphloroacetophenone (109). The former compound (108) has been isolated by BOLOGNESE, CHIOCCARA and SCHERILLO from *Stereocaulon vesuvianum* (24) and by AHMAD and HUSSAIN from *Pseudevernia furfuracea* (2) and the structure indicated from spec-troscopic and physical data, and by comparison with literature values. The ketone (109) was obtained from *Pseudevernia furfuracea* (2) and the structure derived in a similar manner.

COMe   COMe   Me   Me

HO   OMe   HO   OC$_2$H$_5$ CO—O   OH

Me   CHO   HO   OH   CO$_2$Me

OH   OH   CHO   Me

(108)   (109)   (110)

In both lichens the ketones co-occur with atranorin (110) and it has been pointed out by CULBERSON, CULBERSON and JOHNSON (78) that from the data available, (108) could not be distinguished from methyl β-orsellinate (111), a possible artefact of atranorin.

Similarly ethanolysis of atranorin (110) during the isolation procedure could lead to the formation of ethyl haematommate (112), a possible alternative formulation for (109)!

Me   Me

HO   OH   CO$_2$C$_2$H$_5$

CO$_2$Me   HO   OH

Me   CHO

(111)   (112)

6.3 Phthalide Derivatives

SOLBERG isolated 5,7-dihydroxy-6-methylphthalide (113) from extracts of *Alectoria nigricans* (303) where it co-occurs with the benzyl ester, alectorialic acid (115).

T.l.c. comparisons indicated that both the phthalide (113) and 5,7-dihydroxy-6-formylphthalide (114) occur in extracts of *Alectoria capillaris* together with the cogeneric benzyl esters, alectorialic acid (115) and barbatolic acid (116). It seems highly probable that these phthalides are artefacts of the isolation procedure rather than metabolites of the lichens (303).

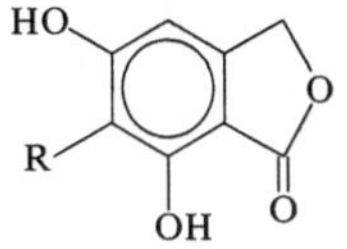

(113) R=Me
(114) R=CHO

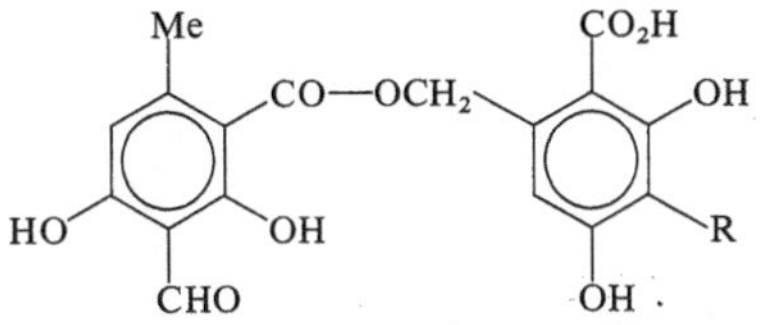

(115) R=Me
(116) R=CHO

7. Chromones

Sordidone, a chromone occurring in the lichen *Lecanora rupicola* (5, 99), has been synthesized by HUNECK and SANTESSON (*185*) by the chlorination of eugenitol. More recently a total synthesis of sordidone has been reported (*4*), which incorporates the above-mentioned chlorination (Scheme 13). Thus treatment of methylphloroacetophenone with acetic anhydride-sodium acetate yielded a 1 : 1 mixture of the isomeric chromones (117) and (118). Alkaline hydrolysis and separation of the isomers by fractional crystalization yielded eugenitol (119), and this compound was subsequently chlorinated to give sordidone (120).

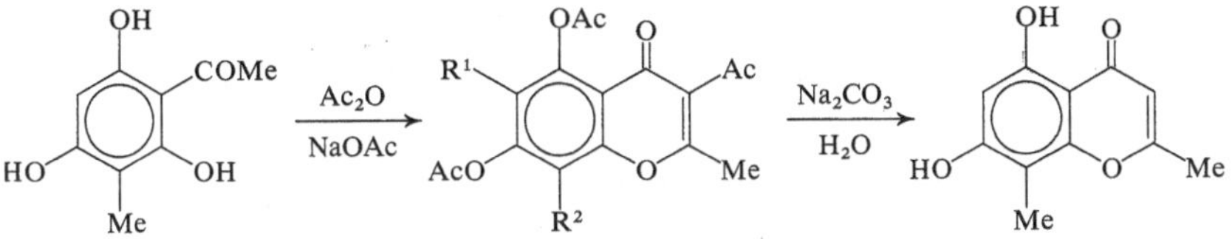

(117) R^1=H, R^2=Me
(118) R^1=Me, R^2=H

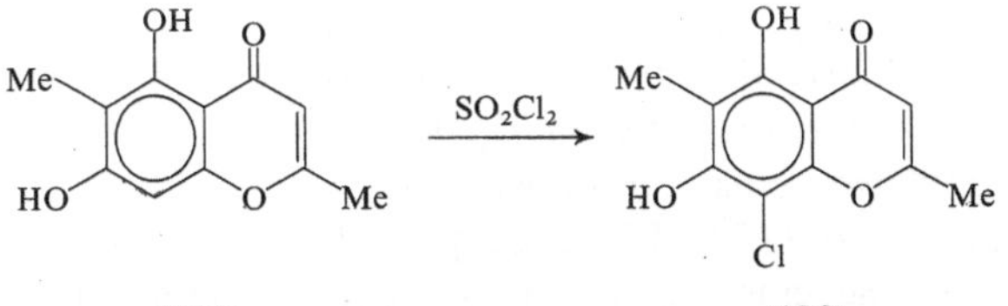

(119)   (120)

Scheme 13. Total synthesis of sordidone

In addition to the known chromone siphulin (121), SHIMADA et al. (*296*) described the isolation of two new derivatives, oxysiphulin (122) and protosiphulin (123), from *Siphula ceralites*. The structures of these metabolites were determined by comparison of the ^{1}H- and ^{13}C-n.m.r. spectra with those of siphulin. Protosiphulin (123) is regarded as the biosynthetic precursor of siphulin (121).

(121) R¹=R²=H
(122) R¹=H, R²=OH or R¹=OH, R²=H

(123)

HUNECK (*174*) isolated the new chromone 6-hydroxymethyleugenetin (**124**) from the lichen *Roccella fuciformis*. In addition the novel chromone glucosides galapagin (**2**), mollin (**3**) and roccellin (**1**) were isolated from *Roccellaria mollis*, *Schismatomma accedens* and *Roccella galapagoensis* while lobodirin (**4**) was isolated from *Lobodirina cerebriformis* (*172, 175*). The structure of these chromone glucosides have been discussed in Chapter 2.

(124)

8. Xanthones

Seven new lichen xanthones have been isolated and identified in the past decade, and these are listed in Table 1. Six of these new compounds (**125 – 130**) vary only in the degree of O-methylation and ring chlorination like all the previously identified lichen xanthones. However erythrommone (**131**), isolated from *Haematomma erythromma*, provides the first example of an O-acetylated lichen xanthone.

HAY and HARRIS (*153, 157*) have reported a biogenetic-type synthesis of lichexanthone (Scheme 14). To circumvent problems arising from numerous cyclisation pathways of a long polyketide chain these workers chose a starting compound with a pre-formed aromatic ring (**132**). Claisen condensation of the polyketide side chain of (**132**) in aqueous potassium hydroxide yielded a mixture of the benzophenone (**133**) and 3-O-methylnorlichexanthone (**134**). The benzophenone (**133**) could be converted into (**134**) by treatment with methanolic potassium hydroxide. Subsequent O-methylation of (**134**) with diazomethane yielded lichexanthone (**135**).

Table 1. *New Lichen Xanthones Isolated*

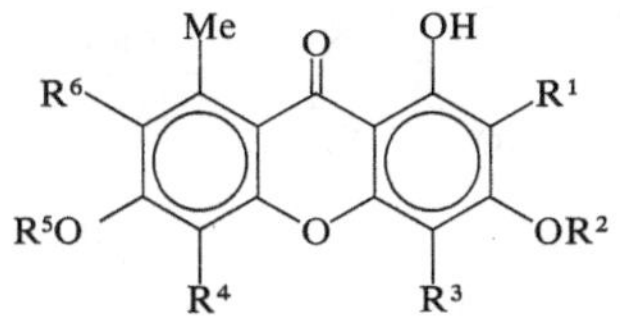

	R^1	R^2	R^3	R^4	R^5	R^6	Species	Reference	Identification Method
(125)	Cl	H	H	H	Me	H	*Pertusaria sulfurata*	*(128)*	m.s., ^{1}H-n.m.r.
(126)	H	H	Cl	H	Me	H	*Pertusaria sulfurata*	*(128)*	m.s., ^{1}H-n.m.r.
(127)	Cl	Me	H	H	Me	H	*Pertusaria sulfurata*	*(128)*	m.s., ^{1}H-n.m.r.
(128)	Cl	Me	H	Cl	Me	H	*Pertusaria* sp.	*(181)*	m.s., ^{1}H- and ^{13}C-n.m.r.
(129)	Cl	Me	Cl	H	Me	H	*Pertusaria* sp.	*(181)*	m.s., ^{1}H- and ^{13}C-n.m.r.
(130)	Cl	Me	Cl	Cl	Me	H	*Pertusaria* sp.	*(181)*	m.s., ^{1}H- and ^{13}C-n.m.r.
(131)	Cl	Ac	Cl	Cl	Ac	H	*Haematomma erythromma*	*(179, 181)*	m.s., ^{1}H-n.m.r.

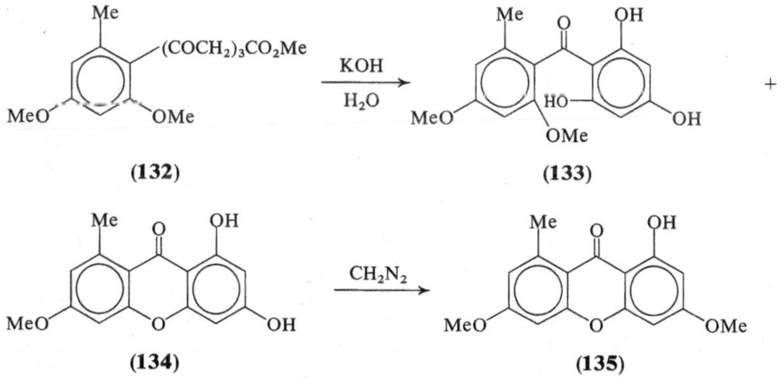

Scheme 14. Biomimetic synthesis of lichexanthone

Many lichen xanthones have recently been synthesised and these are listed in Table 2.

Synthetic routes A and B are exemplified here by the alternative routes to norlichexanthone reported by SANTESSON and SUNDHOLM *(285)* (Scheme 15).

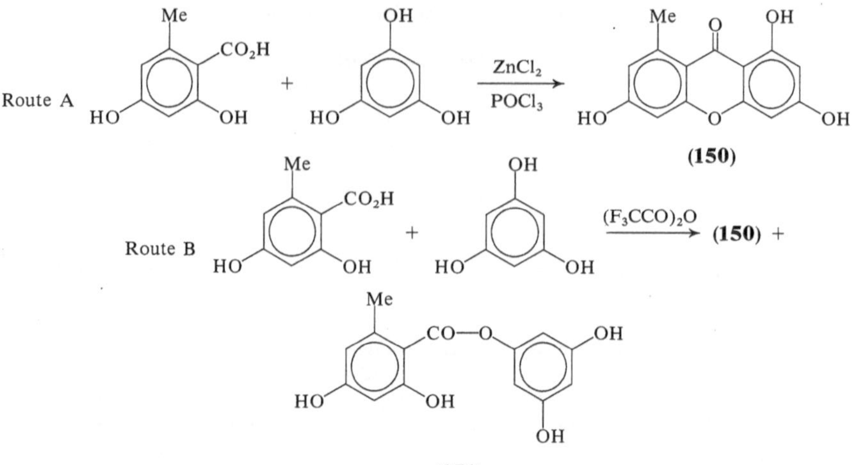

Route A

Route B

(150)

(151)

Scheme 15. Syntheses of norlichexanthone

Table 2. *Lichen Xanthones Recently Synthesised*

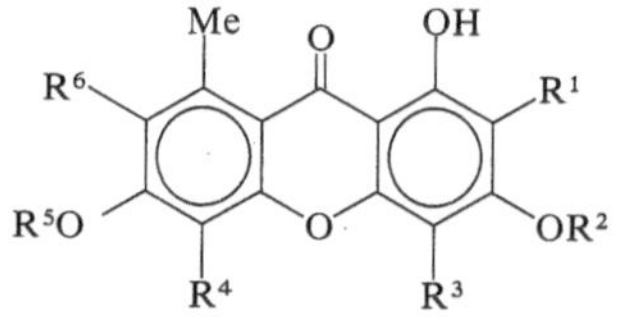

	R^1	R^2	R^3	R^4	R^5	R^6	Synthetic Route	References
(125)	Cl	H	H	H	Me	H	C	(128)
(126)	H	H	Cl	H	Me	H	C	(128)
(127)	Cl	Me	H	H	Me	H	C	(128)
(128)	Cl	Me	H	Cl	Me	H	B	(138, 310)
(130)	Cl	Me	Cl	Cl	Me	H	B	(138)
(136)	Cl	H	Cl	H	Me	H	C	(128)
(137)	Cl	H	H	H	H	H	B, C	(128, 308)
(138)	H	H	Cl	H	H	H	B, C	(128, 308)
(139)	H	H	H	Cl	H	H	B	(138, 310)
(140)	H	Me	H	Cl	H	H	B	(138, 310)
(141)	H	H	Cl	Cl	H	H	B	(138, 310)
(142)	H	Me	Cl	Cl	Me	H	B	(138, 310)
(143)	Cl	Me	Cl	Cl	H	H	B	(138)
(144)	Cl	H	Cl	Cl	H	H	B	(138)
(145)	Cl	Me	H	Cl	H	Cl	B	(310)
(146)	Cl	H	H	Cl	H	Cl	B	(310)
(147)	H	H	H	H	H	Cl	A, B	(180, 308)
(148)	Cl	Me	H	H	Me	Cl	B	(308)
(149)	Cl	H	H	H	H	Cl	B	(308)

Route A involves the condensation of appropriately substituted phloroglucinol derivatives with substituted orsellinic acid derivatives in the presence of anhydrous zinc chloride and phosphorus oxychloride. The analogous condensation in the presence of trifluoroacetic anhydride yielded the xanthone (150) (C-acylated product) as a minor byproduct together with the major aryl ester (151) (O-acylated product). However trifluoroacetic anhydride has become the reagent of choice (route B) since the use of O-methyl and O-benzyl protecting groups has enabled good yields of the xanthones to be obtained.

ELIX, MUSIDLAK, SALA and SARGENT (128) have reported a novel synthetic route to xanthones which used an Ullmann reaction to produce a diaryl ether and its subsequent condensation to yield the xanthones (route C). This route is exemplified by the synthesis of thiophaninic acid (136) (Scheme 16).

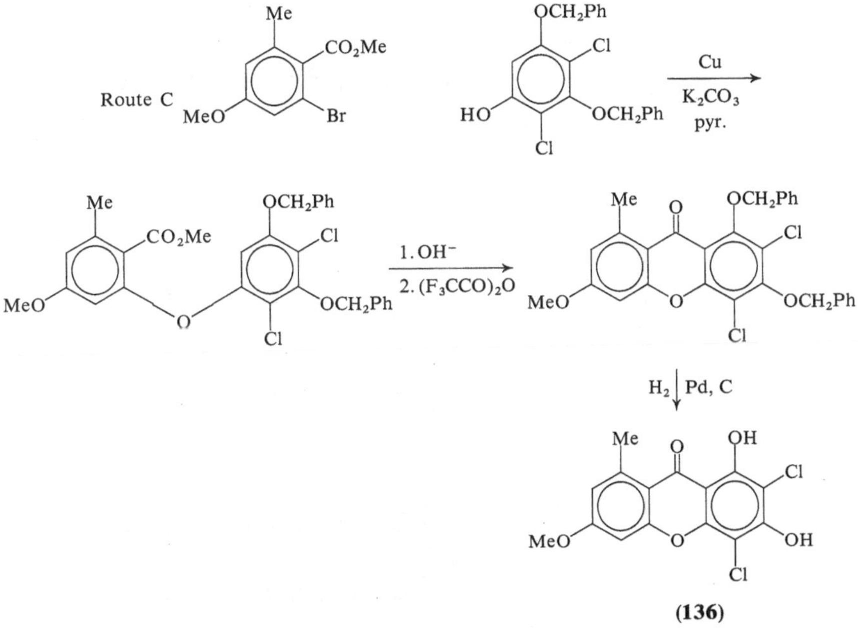

Scheme 16. Synthesis of thiophaninic acid

In addition to the synthesis of chlorinated lichen xanthones from suitably substituted mononuclear aromatic precursors, the elaboration of preformed xanthones through chlorination has also been reported (128, 138, 250).

9*

Table 3. *Structure of Lichen Xanthones*

	Original Structural Proposals								Presently Accepted Structures						
	R¹	R²	R³	R⁴	R⁵	R⁶	References		R¹	R²	R³	R⁴	R⁵	R⁶	References
(125)	Cl	Me	Cl	H	H	Cl	(186)	(143)	Cl	Me	Cl	Cl	H	H	(181, 308)
	Cl	H	Cl	H	H	Cl	(280)	(144)	Cl	H	Cl	Cl	H	H	(181, 308)
(137)	Cl	H	H	H	Me	H	(266)	(140)	H	Me	H	Cl	H	H	(181, 308)
	Cl	Ac	Cl	H	Ac	Cl	(179)	(131)	Cl	Ac	Cl	Cl	Ac	H	(181)
	Cl	H	H	H	H	H	(282)	(138) and (139)	H	H	H	H	H	H	(128, 308)
(128)	Cl	Me	H	Cl	Me	H	(284)	(148)	Cl	Me	H	H	Me	Cl	(308)
	Cl	Me	H	Cl	H	H	(284)		Cl	Me	H	H	Me	Cl	(308)
(149)	Cl	H	H	H	H	Cl	(281)	(141)	Cl	H	Cl	Cl	H	Cl	(308)
(148)	Cl	Me	H	H	Me	Cl	(283)	(142)	H	Me	Cl	Cl	Me	H	(308)

The structures of many lichen xanthones have been revised and the presently accepted structures have been summarized in Table 3. These revisions have stemmed from a reappraisal of the spectral data of the xanthones, in particular ^{1}H- and ^{13}C-n.m.r. data, and in many cases have been confirmed by synthesis (see Table 2). ^{13}C-n.m.r. has become an important tool in the identification and structural elucidation of such compounds (*181, 309, 310*).

9. Anthraquinones and Biogenetically Related Compounds

9.1 Anthraquinones

In 1972 NAKANO, KOMYA and SHIBATA (*247*) isolated two new anthraquinones from *Xanthoria* and *Caloplaca* species which were tentatively assigned the structures erythroglaucinic acid (**152**) and 2-chlorofallacinol (**153**). The structures were based on mass, ultraviolet and infrared spectra but could not be confirmed due to shortage of the natural materials. The position of the chlorine in (**153**) was assigned on biogenetic grounds.

(152)   (153)

Averantin (**154**) and 6-O-methylaverantin (**155**) have been isolated from the lichen *Solorina crocea* (*306*). Although averantin was originally isolated from fungi (*16*), 6-O-methylaverantin is a novel metabolite.

(**154**) R = H
(**155**) R = Me

MISHCHENKO and coworkers (*240*) have isolated the known fungal metabolites islandicin (**156**) and cynodontin (**157**) from a lichen species, *Asahinea chrysantha*, for the first time. In addition two new metabolites, 4-hydroxyemodin (**158**) and 4,5-dihydroxyemodin (**159**) were isolated from this lichen. The assigned structure (of 4-hydroxyemodin) of the new anthraquinone was untenable since the natural compound was found to differ from synthetic 4-hydroxyemodin.

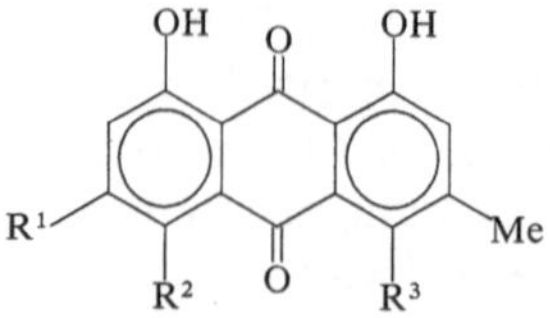

	R¹	R²	R³
(156)	H	H	OH
(157)	H	OH	OH
(158)	OH	OH	H
(159)	OH	OH	OH

A new synthesis of emodin has been reported (*165*) (Scheme 17). Thus the Friedel-Crafts acylation of methyl 3,5-dimethoxybenzoate with 2-methoxy-4-methylbenzoyl chloride followed by hydrolysis of the ester yielded the benzophenone (**160**). Cyclisation and demethylation was accomplished by treating (**160**) with an aluminium chloride-sodium chloride melt yielding emodin (**161**).

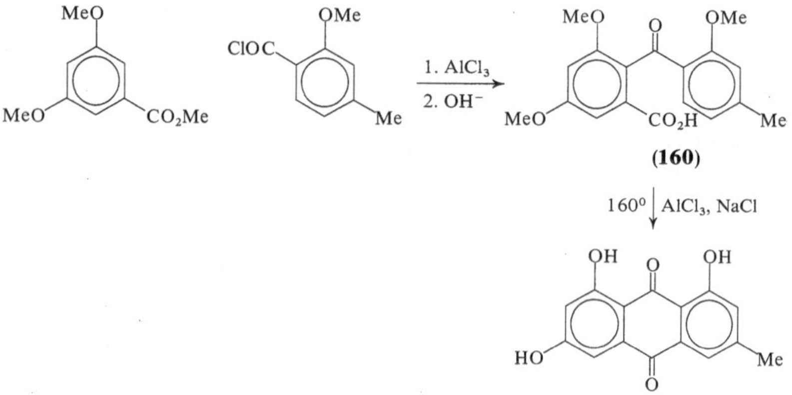

Scheme 17. Emodin synthesis

Lam *et al.* (*226*) have reported syntheses of valsarin and 2,4-dichloroemodin and a new synthesis of 2-chloroemodin (Scheme 18). Thus Friedel-Crafts acylation of a substituted resorcinol (**162**) with 3-methoxy-5-methylphthalic anhydride gave a crude benzophenone which was treated with boric oxide-sulfuric acid to yield the anthraquinones, 2-chloroemodin (**163**) and 2,4-dichloroemodin (**164**). 2,4-Dichloroemodin on treatment with boric acid and sulfuric acid gave valsarin (**165**). Tetra-O-methyl-valsarin was also synthesized via chlorination of xanthorin.

Several synthetic routes to islandicin have recently been described. Kende and coworkers (*206, 207*) have reported the novel, though lengthy, route depicted below (Scheme 19). The first step involved a photo-Fries rearrangement of 4-methoxyphenyl 2-cyano-3-methoxybenzoate (**166**), producing the benzophenone (**167**). Allylation of the phenol followed by Claisen rearrangement yielded (**168**). O-Methylation, nitrile hydrolysis and

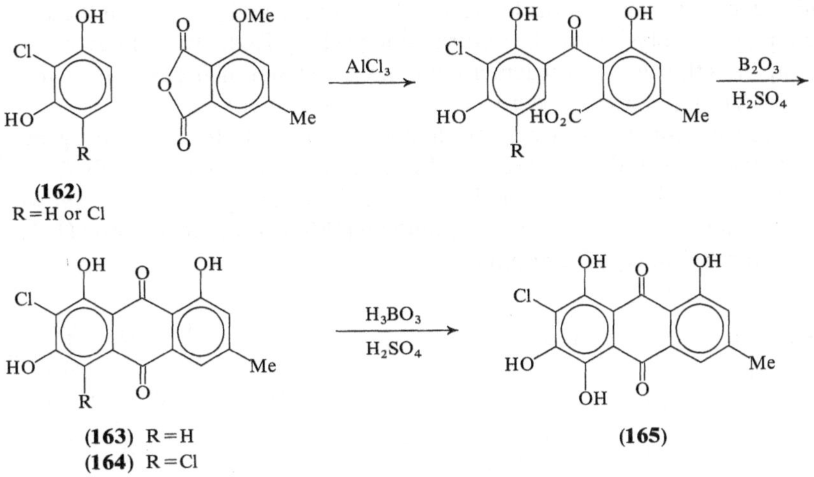

Scheme 18. Syntheses of 2,4-dichloroemodin and valsarin

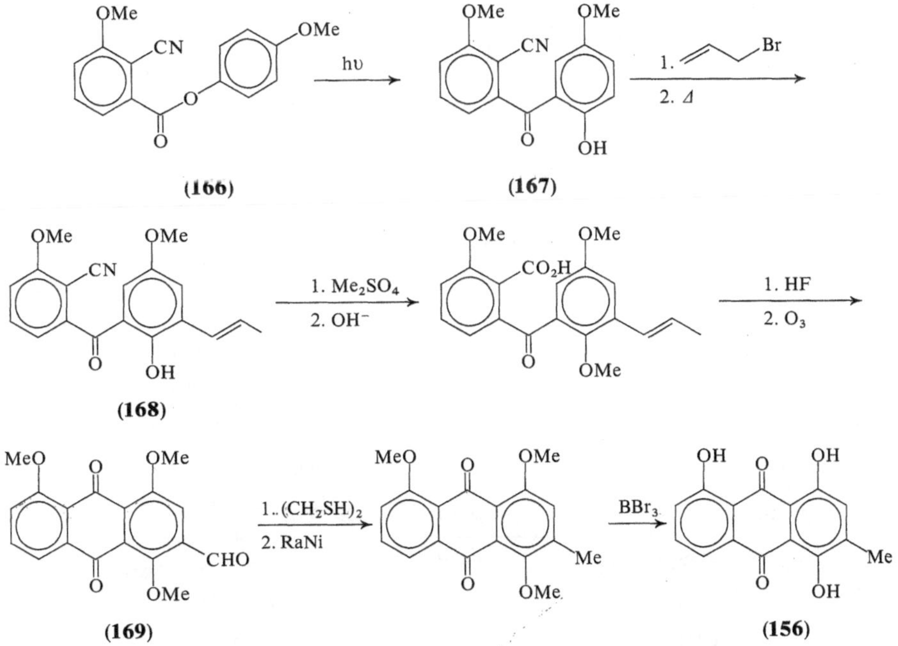

Scheme 19. Synthesis of islandicin after KENDE et al. (206, 207)

subsequent hydrogen fluoride catalysed cyclisation and cleavage of the allyl group by ozonolysis gave the anthraquinone (169). Reduction of the formyl group via a thio acetal and demethylation yielded islandicin (156). JUNG and LOWE (201) reported a synthesis of islandicin involving a Diels-Alder addition of 2-methyl-1,4-benzoquinone with the photochemically generated bisketene (170) (Scheme 20). However products from the Diels-Alder reaction were obtained in a very low yield and no regiospecificity was shown in the reaction, the two isomers, islandicin (156) and digitopurpone (171), being formed in equal amounts.

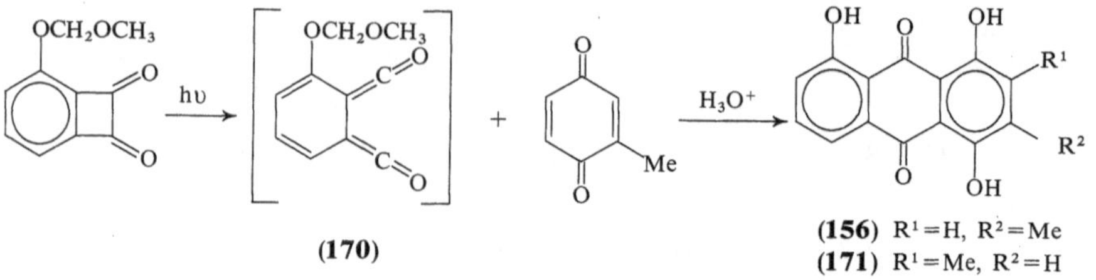

(156) $R^1 = H$, $R^2 = Me$
(171) $R^1 = Me$, $R^2 = H$

Scheme 20. Diels-Alder synthesis of islandicin

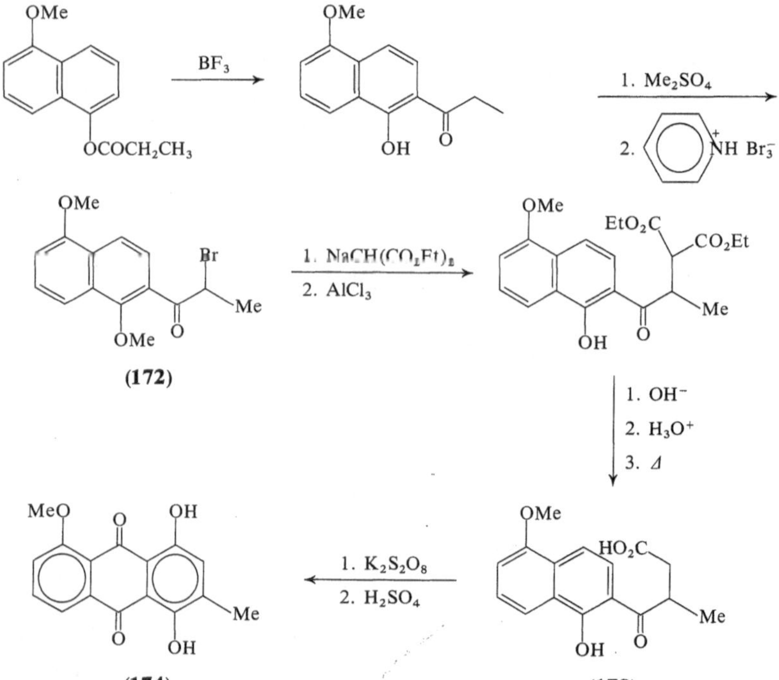

Scheme 21. Synthesis of 1-O-methylislandicin after GLEIN et al. (145)

A novel regiospecific synthesis of islandicin was described by GLEIN *et al.* (*145*) (Scheme 21). Thus Fries rearrangement of 1-propionyloxy-5-methoxynaphthalene followed by methylation and bromination yielded the ketone (**172**). Alkylation with sodium diethyl malonate followed by demethylation, ester hydrolysis and decarboxylation gave the acid (**173**). Elbs oxidation and sulfuric acid cyclisation of the acid yielded 1-O-methyl-islandicin (**174**).

A superior synthesis of islandicin reported by BRAUN (*26*) is outlined below (Scheme 22). The high regioselectivity of the Grignard reaction produced good yields of the product (**175**) derived from attack at the least hindered carbonyl group. Subsequent cyclisation and demethylation then yielded islandicin (**156**), completing this simple and highly efficient route.

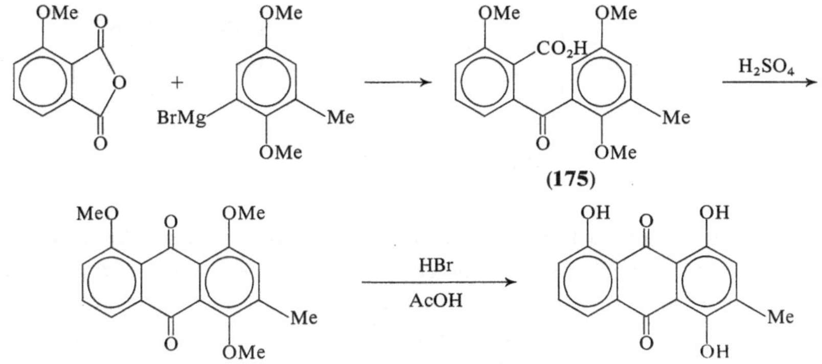

Scheme 22. Islandicin synthesis after BRAUN (*26*)

9.2 Bisanthraquinones and Related Bisxanthones

The novel bisanthraquinone disolorinic acid (**176**) was isolated from the lichen *Solorina crocea* (*306*). This metabolite co-occurred with its mono-meric precursor, solorinic acid, which facilitated structural elucidation.

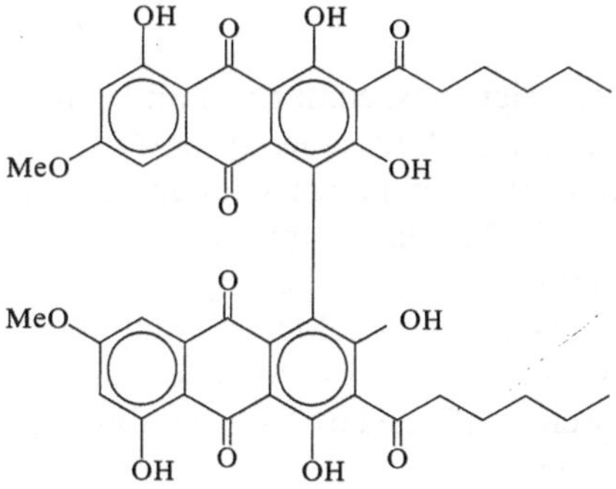

Three related bisanthraquinones (**177, 178, 179**) have been isolated from *Cladonia graciliformis* (*108*). The infrared, ultraviolet, optical rotatory dispersion and mass spectral data obtained from graciliformin (**177**) and (+)-rugulosin were almost identical, but differences in the ^{1}H-n.m.r. spectra suggested that graciliformin was an epimer of rugulosin. Hence graciliformin, monoacetylgraciliformin and diacetylgraciliformin were assigned structures (**177**), (**178**) and (**179**) respectively. Bellidiflorin, another metabolite isolated from this lichen, was thought to be an iron complex of diacetyl-graciliformin.

(**177**) $R^1=R^2=H$
(**178**) $R^1=Ac, R^2=H$
(**179**) $R^1=R^2=Ac$

The structures of the pigments secalonic acid A and secalonic acid C, isolated from the lichens *Parmelia entotheiochroa* (*340*) and *Cetraria ornata* (*346*), have been revised (*166*). On the basis of further spectral evidence secalonic acids A and C were formulated as the 4,4'-dimers (**180**) and (**181**), rather than the 2,2'-derivatives originally proposed.

(**180**) $R^1=\alpha COOMe, R^2=\beta Me$
(**178**) $R^1=\beta COOMe, R^2=\alpha Me$

Nuno (*260*) isolated two pigments, eumitrin A and eumitrin B_1 from a benzene extract of *Usnea baileyi*. Subsequently (*334*) it was found that eumitrin A was comprised of two components designated eumitrin A_1 and eumitrin A_2, which, although chromatographically inseparable, could be separated by fractional crystallisation. The physical and spectral data suggested that the eumitrins were homologous compounds structurally related to secalonic acid A. The structure proposed for eumitrins A_1, A_2 and B_1 (**182**), (**183**) and (**184**), were confirmed by X-ray analysis of a brominated derivative of eumitrin B_1 (*334*).

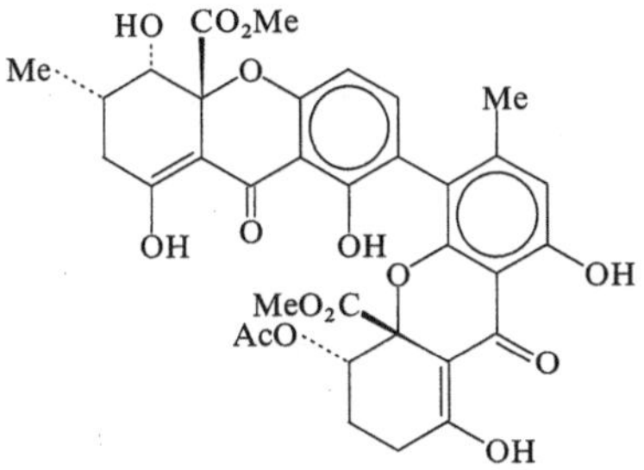

(182)

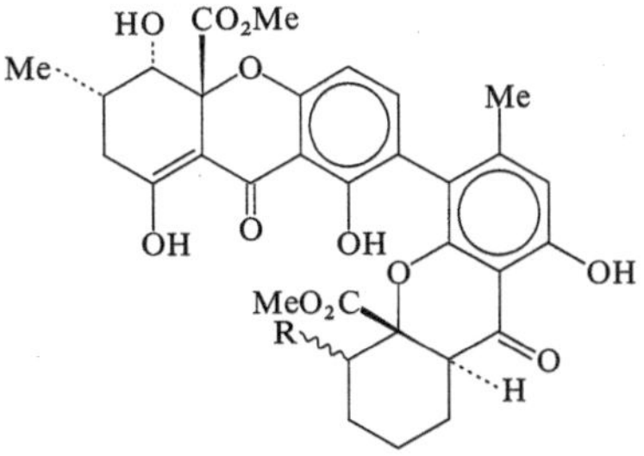

(183) R = βOAc

(184) R = αOAc

Some interesting results have been obtained in studies of the biosynthesis of bisanthraquinones.

SANKAWA, EBIZUKA and SHIBATA (279) studied the incorporation of ^{14}C into anthraquinones of the fungus *Penicillium brunneum*. Thus it was found that ^{14}C from ^{14}C-labelled emodin (161) and emodinanthrone (185) was incorporated into (+)-skyrin (189) and (+)-rugulosin (190) (stereochemistry not illustrated). However no ^{14}C was incorporated into (+)-rugulosin when the fungus was fed ^{14}C-(+)-skyrin confirming that skyrin was not a biological precursor of rugulosin.

SEO *et al.* (289, 290) noted that the fungal metabolite (−)-flavoskyrin (187) yielded (−)-rugulosin (190) on treatment with pyridine (although not noted, it appears that this transformation must involve oxidation of the flavoskyrin by atmospheric oxygen). These workers also found (290, 291) that dihydrochrysophanol (192), obtained from chrysophanol (191), dimerized upon standing to yield the flavoskyrin analogue (193) (Scheme 24).

This experimental information is correlated in the biosynthetic scheme (Scheme 23). The formation of rugulosin via an intramolecular Michael addition of the intermediate (188) has been proposed (329). This scheme is tentative as the intermediacy of flavoskyrin in the biosynthesis of skyrin or rugulosin has not been demonstrated. Hence it is possible that an intermediate such as (188) could be obtained directly from emodin (161) or emodinanthrone (185) and that skyrin and rugulosin may not share a common intermediate in the final step.

The biosyntheses of the secalonic acids and the eumitrins are discussed in a review by FRANCK (142). The biosynthetic scheme proposed, illustrated here for the biosynthesis of secalonic acid A (Scheme 25), was supported by studies involving labelled precursors and biomimetic studies. The biomimetic studies include a total synthesis of an analogue of the tetrahydroxanthone (194) (Scheme 26). Oxidation of the benzophenone (195) to the quinone (196) followed by an intramolecular Michael addition gave the

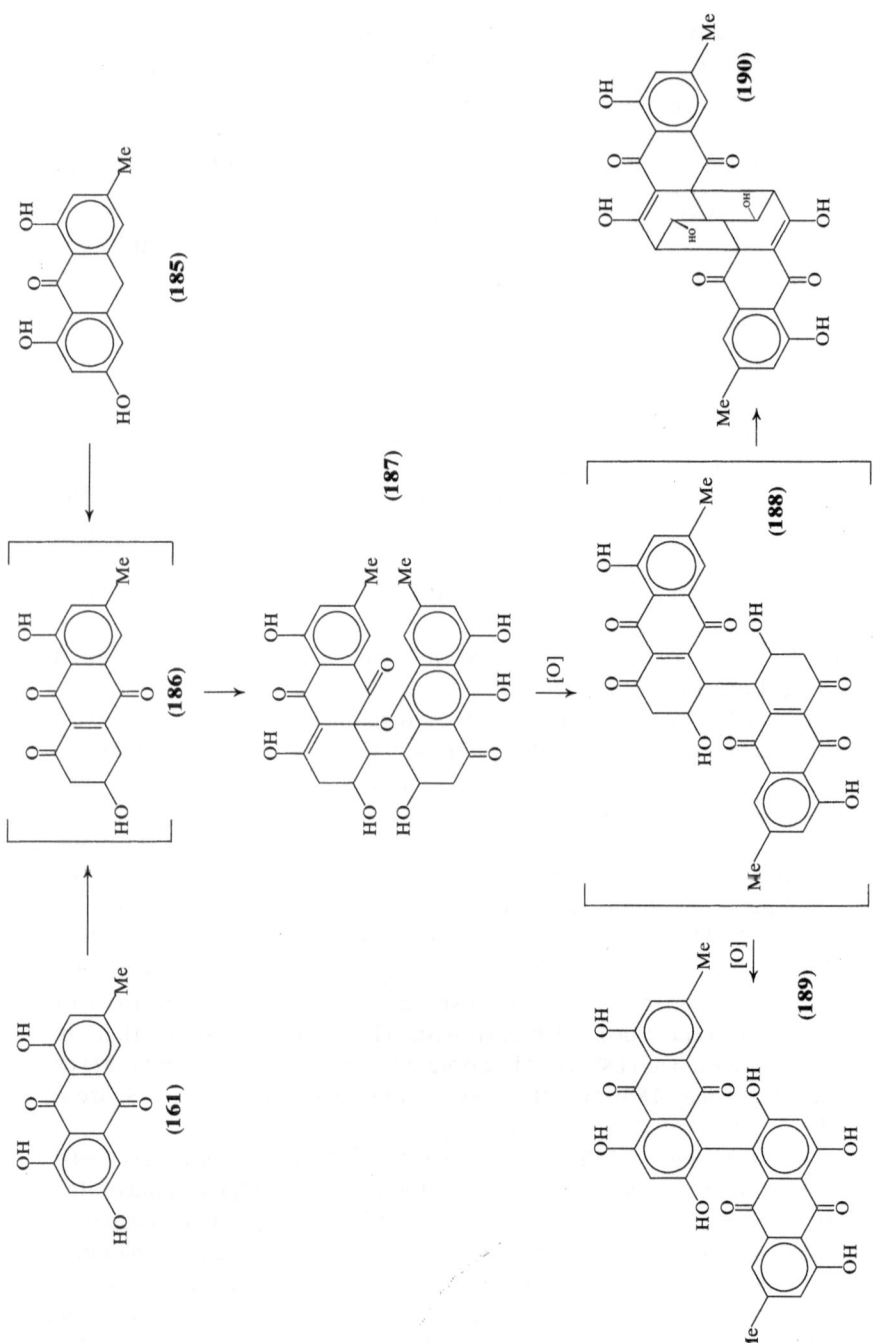

Scheme 23. Possible biosynthetic route to skyrin and rugulosin

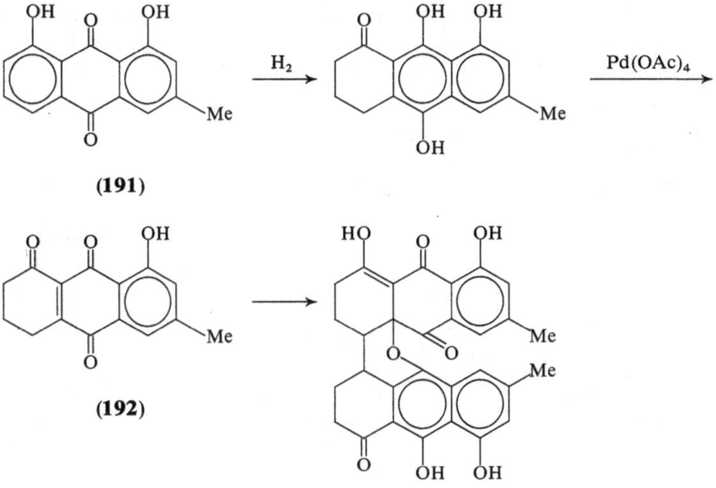

(191) **(192)** **(193)**

Scheme 24. Synthesis and dimerisation of dihydrochrysophanol

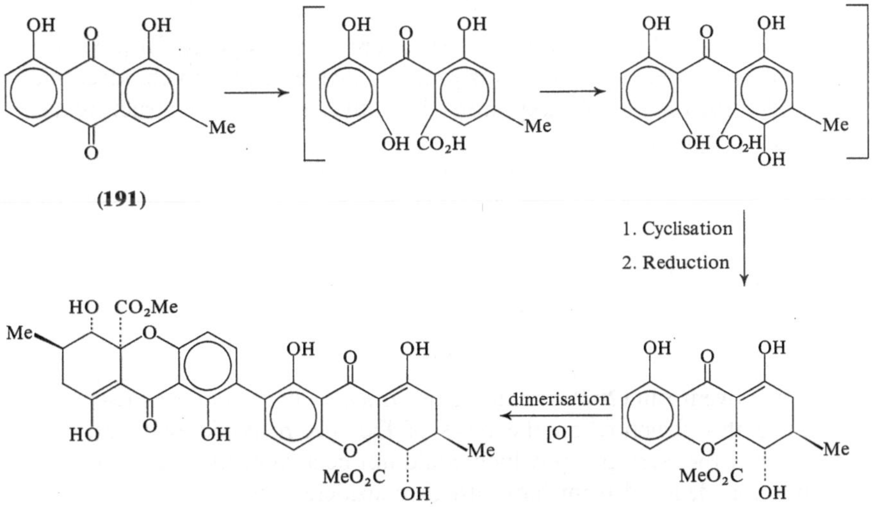

(191)

1. Cyclisation
2. Reduction

(180) **(194)**

dimerisation
[O]

Scheme 25. Biosynthesis of secalonic acid A

xanthone **(197)**. Sodium borohydride reduction showed good stereoselec-
tivity giving the *cis*-isomer **(198)** as the major product. Catalytic hydrogenation of **(198)** yielded the tetrahydroxanthone **(200)**.

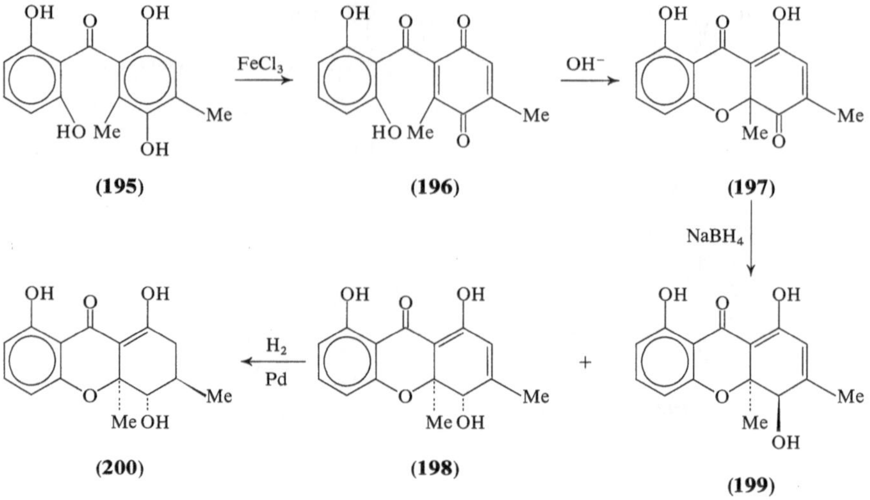

Scheme 26. Biomimetic synthesis of a tetrahydroxanthone

9.3 Naphthoquinones

HUNECK (*176*) reported that the structure (**201**) proposed for pyxiferin was incorrect. The metabolite was found to be identical with chiodectonic acid, identified as a naphthoquinone derivative, although no structure has yet been proposed.

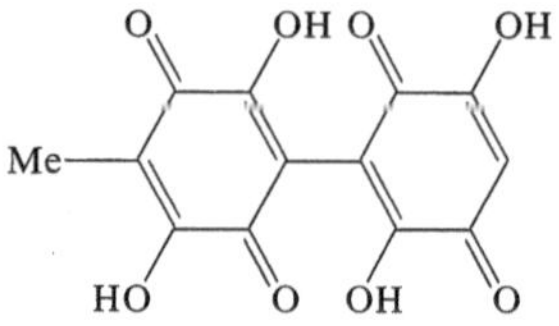

(**201**)

Haemoventosin (**202**), isolated from *Haematomma ventosum* (*34*), was assigned this structure on the basis of the spectroscopic evidence. The position of the ester group, which could not be assigned by spectrometric means, was deduced from biosynthetic considerations.

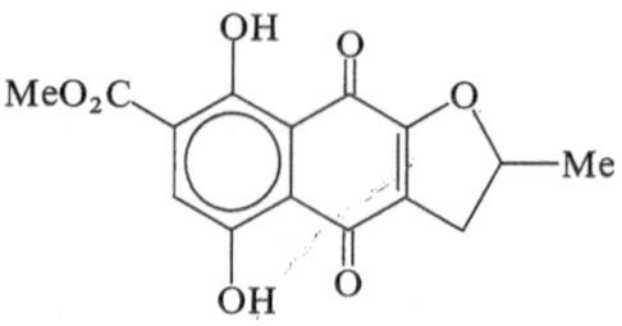

(**202**)

The structure of canarione (**203**), isolated from *Usnea canariensis,* was elucidated by u.v., ^{1}H-n.m.r., ^{13}C-n.m.r., mass spectrometry and chemical degradation (*196*).

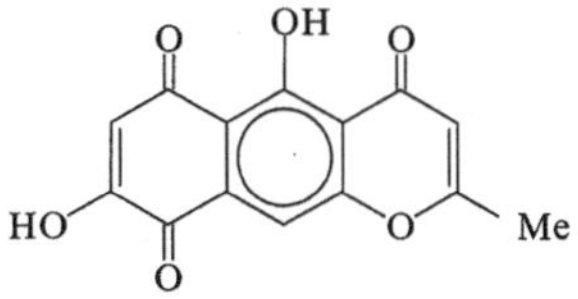

(203)

10. Depsides

Over the past decade 52 new depsides have been reported. Table 4 at the end of this section (p. 160) lists the structure and the lichen source and summarizes the method of separation and structural elucidation of these compounds.

In this review the detailed chemistry of each individual depside is not discussed but typical examples of each structural group with particular emphasis on innovative techniques, reactions, structure determination and syntheses are presented.

10.1 Structural Variation

10.1.1 Orcinol para-Depsides

The major structural variations observed in the orcinol *para*-depsides (**204**) are the length of the polyketide-derived side chains (R^1, R^2), the degree of oxidation of these side chains (CH_2COR or CH_2CH_2R) and the degree of methylation of the phenolic and carboxyl groups (R^3, R^4, R^5, $R^6 = Me$ or H).

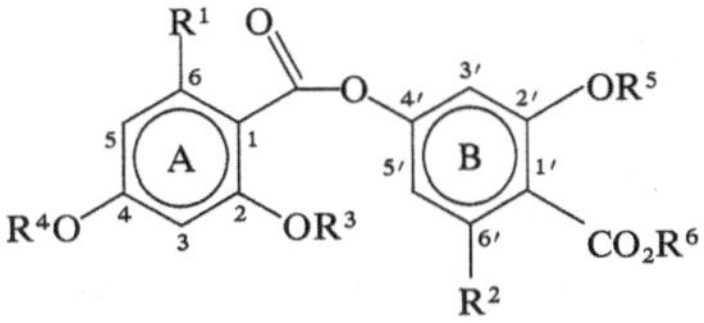

(204)

A depside with a reduced A-ring side chain and an oxidized B-ring side chain has been observed for the first time (*77*), although the alternative substitution pattern is quite common. The new depside oxostenosporic acid (**205**) was isolated from extracts of *Parmelia verruculifera* by preparative thin-layer chromatography (t.l.c.) and the structure established by micro-hydrolytic degradation (*77*).

(**205**)   (**206**)

Miriquidic acid (**206**) (*193*) is structurally unique since the side chain ketonic group is not present in a position predicted from the normal acetate-polymalonate biogenetic pathway to these compounds. The depside (**206**) has been isolated from several *Lecidea* species (*193*).

Chlorinated orcinol *para*-depsides occur very rarely in lichens and until recently timidulin (**207**) was the sole known representative. Two further chlorodepsides have now been isolated, 3,5-dichloro-2′-O-methylanziaic acid (**208**) from *Lecanora sulphurella* (*183*) and 3-chlorodivaricatic acid (**209**) from *Thelomma mammosum* (*197*).

(**207**)   (**208**)

(**209**)

10.1.2 Orcinol β-Orcinol para-Depsides

Until recently only two mixed orcinol β-orcinol depsides were known, namely obtusatic acid (**210**) and norobtusatic acid (**211**), but now a further four such compounds have been reported. These include 2-O-methylobtusatic acid (**212**) present in *Xanthoparmelia tucsonensis* (*45*) and evernine (**213**) (*253*), 3′-methylevernic acid (**214**) (*252*) and methyl 3′-methyllecanorate (**215**) (*252*) all obtained from *Evernia prunastri*.

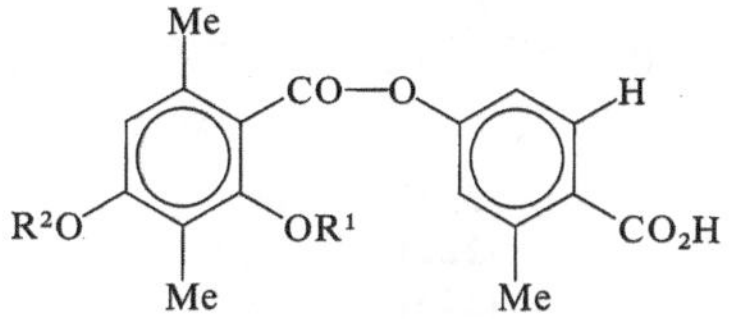

(**205**)

	R¹	R²
(**210**)	H	Me
(**211**)	H	H
(**212**)	Me	Me

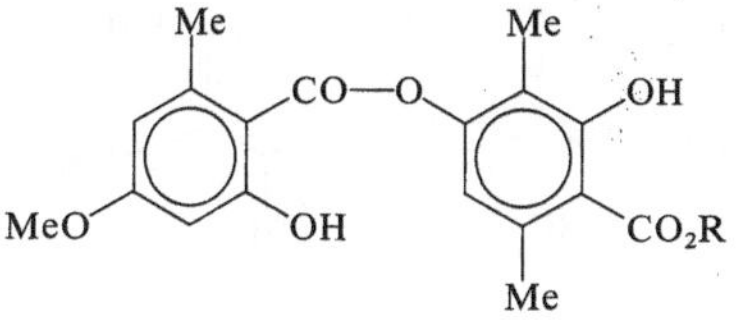

(**213**) R=Me
(**214**) R=H

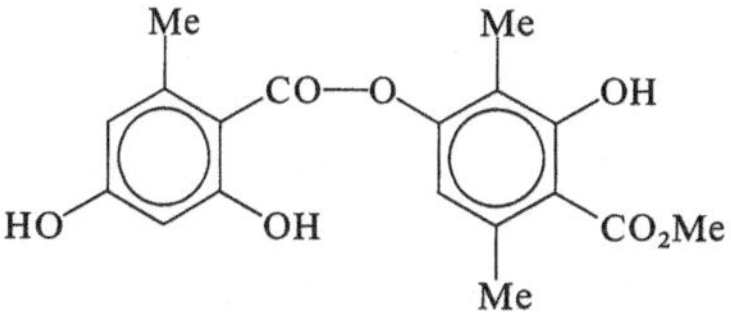

(**215**)

10.1.3 β-Orcinol para-Depsides and meta-Depsides

Structural variations in depsides based on β-orcinol moieties (**216, 217**) show variations in the degree of oxidation of the groups R^1 (CO_2H, CHO, CH_2OH, Me), R^2 (CHO, Me) as well as variation in the degree of methylation of the phenolic and carboxyl groups in (**216**) (R^3, R^4, R^5, R^6 =H, Me). Rarely the B-ring lacks a carboxy function as in decarboxythamnolic acid (**218**) and nephroarctin (**219**). Nephroarctin (**219**) (*262,31*) and phenarctin (**220**) (*31*) were isolated from the lichen *Nephroma arcticum* and are highly unusual because of the additional C_1-substituents observed in the 5 and 5′-positions. Phenarctin (**220**) is the only known fully substituted depside.

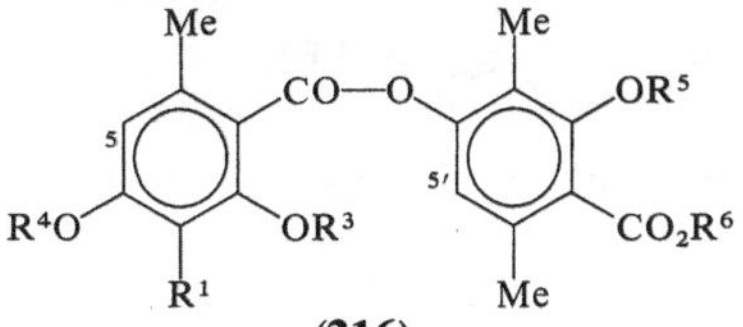

(**216**)

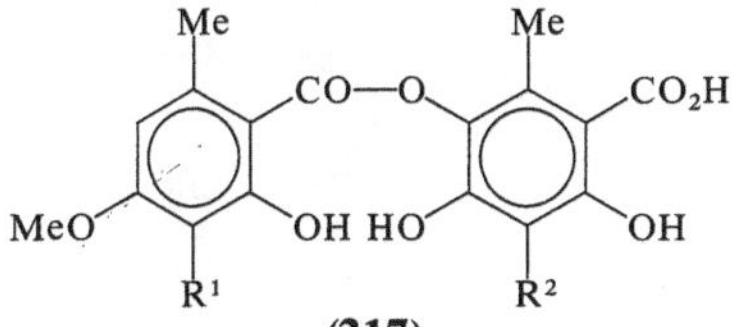

(**217**)

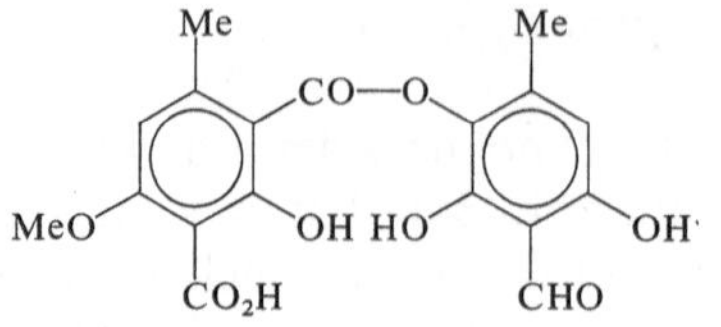

(218)

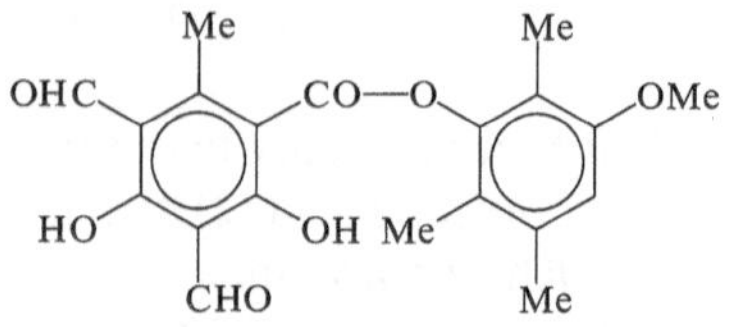

(219)

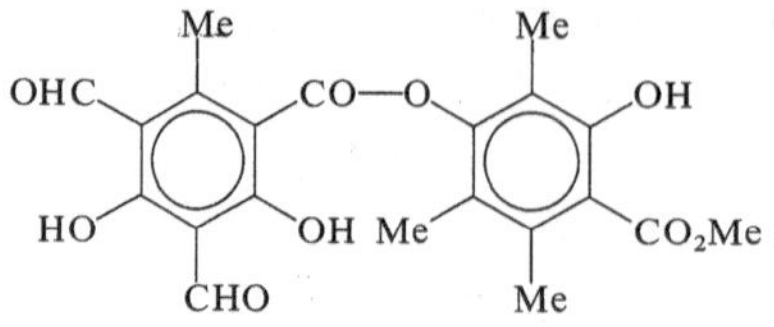

(220)

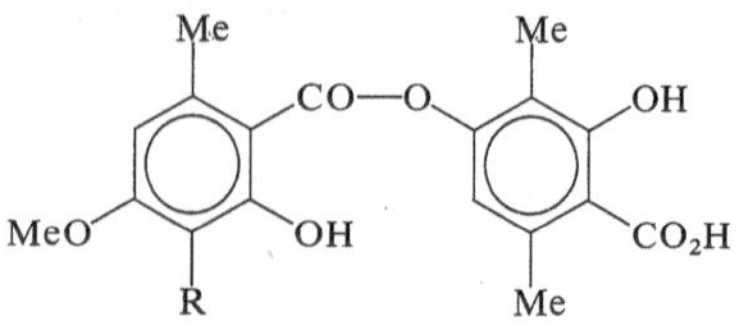

(221) R=Me
(222) R=CH$_2$OH
(223) R=CHO
(224) R=CO$_2$H

The joint occurrence of β-orcinol depsides differing only in the degree of oxidation of the 3-substituent was noticed as early as 1954 in lichens producing baeomycesic acid (223) and squamatic acid (224) (7). More recently NUNO (261) has reported the co-occurrence of barbatic acid (221) together with (223) and (224) in several *Cladonia* species. The intermediate depside, 3-α-hydroxybarbatic acid (222) has now been detected in several *Xanthoparmelia* and *Cladonia* species (87) and together with methyl 3-α-hydroxy-4-O-demethylbarbatic acid (225) isolated from *Oropogon loxensis* (73), these two compounds are the first depsides known to be substituted by a hydroxymethyl group.

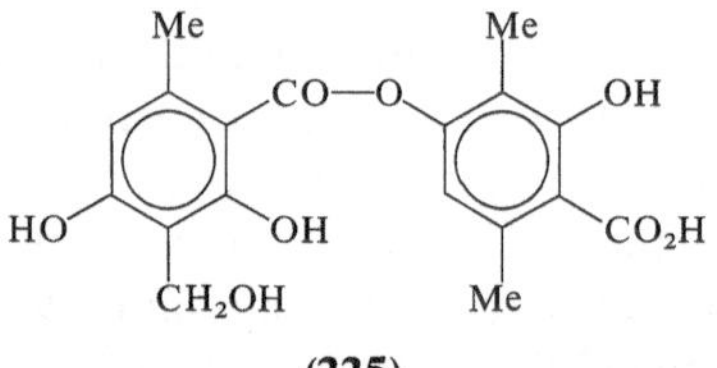

(225)

10.1.4 Orcinol Tridepsides

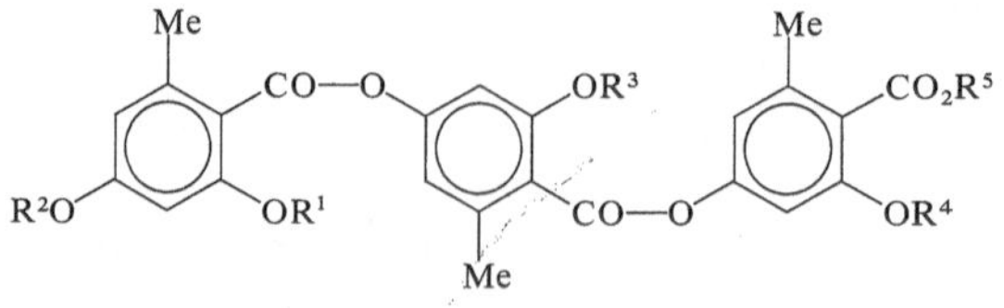

(226)

Besides the variation in the degree of methylation of the phenolic and carboxyl groups (R^1, R^2, R^3, R^4, R^5 = H or Me), the orcinol tridepsides commonly exhibit an additional hydroxy (or methoxy) group in the A-ring which is not inherent in the acetate-polymalonate biogenetic pathways to these compounds. Such substitution is apparently the result of nuclear hydroxylation which may or may not be followed by O-methylation. Thus 5-O-methylhiascic acid (**227**) (*122*), 4,5-di-O-methylhiascic acid (**228**) (*117*), 2,4,5-tri-O-methylhiascic acid (**229**) (*124*) and 3-methoxy-2,4-di-O-methylgyrophoric acid (**230**) (*123*) have been shown to co-occur with gyrophoric acid (**93**) in various *Parmelia* subg. *Hypotrachyna* species. The co-occurrence of the tridepsides (**227**), (**228**), (**230**) and gyrophoric acid (**93**) in one lichen (*Parmelia subfatiscens*) provided circumstantial evidence that the C-hydroxylation steps occur relatively late in the biosynthetic sequence and that free or bound gyrophoric acid would appear to be the most likely substrate (*123*). Orcinyl lecanorate (decarboxygyrophoric acid) (**231**) has also been isolated (*49*).

	R^1	R^2	R^3
(**227**)	H	H	Me
(**228**)	H	Me	Me
(**229**)	Me	Me	Me

(**230**)

(**231**)

10.1.5 Orcinol meta-Depsides

Like the orcinol *para*-depsides this group of compounds exhibit variations in the length of the side-chain (**232**, R^1, R^2 = C_3H_7, C_5H_{11}) and degree of methylation of the phenolic and carboxyl groups (**232**, R^3, R^4, R^5,

$R^6 = H$, Me). However the structural variations are more limited in the *meta*-depside group as no depsides with oxidised side-chains, or chain length other than C_3 or C_5 have yet been discovered.

(232)

10.2 Separation of Lichen Depsides

Well established techniques (*170, 171*) for the isolation and separation of depsides from mixtures of lichen acids include fractional crystallization, preparative layer chromatography and column chromatography. Vacuum liquid chromatography (*124*) provided a convenient and efficient modification of the latter technique and was used in separating a complex mixture of tridepsides from the crude extracts of *Parmelia damaziana*.

A major innovation in isolating depsides involves the benzylation of a crude mixture of lichen acids with phenyl diazomethane or benzyl bromide and potassium hydrogen carbonate followed by chromatographic separation of the benzyl esters formed. Far more efficient resolution is possible in chromatography of the less polar, highly soluble benzyl esters rather than the corresponding polar, sparingly soluble free acids. The natural depsides (acids) are subsequently regenerated by hydrogenolysis of the benzyl esters over palladium on carbon. In this way the new depside nordivaricatic acid (**233**) was separated from divaricatic acid (**234**) present in extracts from the lichen *Heterodea beaugleholei* (*132*). A number of other depsides including 2'-O-methylmicrophyllinic acid (*49*), 5-O-methylhiascic acid (**227**) (*122*), 4,5-di-O-methylhiascic acid (**228**) (*117*) and 3-methoxy-2,4-di-O-methylgyrophoric acid (**230**) (*123*) have been purified by this technique.

(233) R = H
(234) R = Me

10.3 Identification of Lichen Depsides

10.3.1 The Structure of Isolated Depsides

The spectral properties of depsides, in particular ^{1}H-n.m.r. and mass spectra, provide a powerful probe in structural elucidation. ^{13}C-n.m.r. spectroscopy has recently been used in the structural elucidation of ovoic acid (235) (194), 2″,4-di-O-methylgyrophoric acid (236) (252), evernine (213) (253), 3′-methylevernic acid (214) (252) and methyl 3′-methyl-ecanorate (215) (252) but following a generalised study this technique (311, 312) will undoubtedly prove to be particularly important in the future.

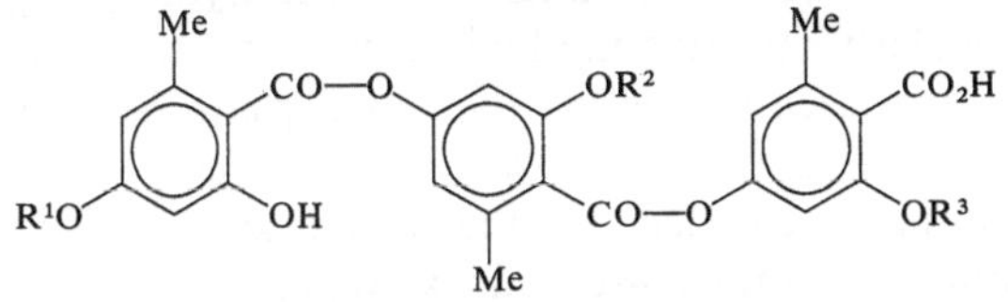

(235) $R_1 = R_3 = H$, $R_2 = Me$
(236) $R_1 = R_3 = Me$, $R_2 = H$

Chemically, the procedure of hydrolytic degradation of depsides and identification of the hydrolysis products remains a general and definitive procedure. For example the new depside 2-O-methylconfluentic acid (237) has been isolated from the lichen Lecidea fuscoatra and the structure deduced from spectroscopic data (190). Hydrolysis of the corresponding methyl ester, methyl 2-O-methylconfluentate (238) with methanolic potassium hydroxide gave 2,4-dimethoxy-6-(2-oxoheptyl)benzoic acid (239) and methyl 4-hydroxy-2-methoxy-6-pentylbenzoate (240) thus establishing the structure of this depside (Scheme 27).

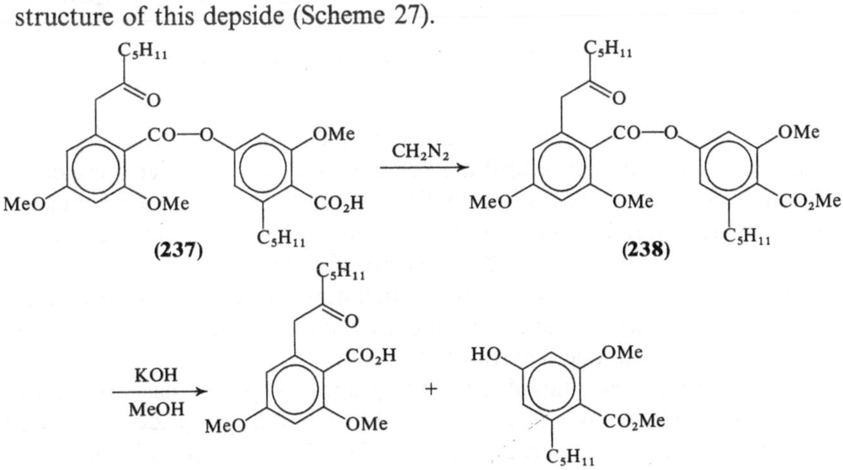

Scheme 27. Reactions of 2-O-methylconfluentic acid

10.3.2 Identification and Detection of Depsides Without Isolation

Further refinements of analytical t.l.c. procedures for detecting and comparing depsides in lichen extracts have been reported by C. F. Culberson and coworkers (64), (66), (86). Moreover two-dimensional t.l.c. is useful in microchemical studies on mixtures difficult to resolve by the standardized one-dimensional method (85).

In a number of cases where only fragments of lichen material were available (and hence only traces of the new depsides present often as mixtures), microhydrolysis of the depside with concentrated sulfuric acid followed by comparative t.l.c. analysis of the hydrolysis products has proved a particularly effective method of primary structural elucidation. This technique was used to identify the new depside 2-O-methylperlatolic acid (241) present in *Pertusaria tuberculifera* Nyl. (80). Microhydrolysis of (241) with sulfuric acid led to the formation of 2,4-dimethoxy-6-pentylbenzoic acid (242) and 2,4-dihydroxy-6-pentylbenzoic acid (243), and these acids were identified by comparison of the R_F values on t.l.c. with those of the authentic acids in three independent solvent systems (80) (Scheme 28).

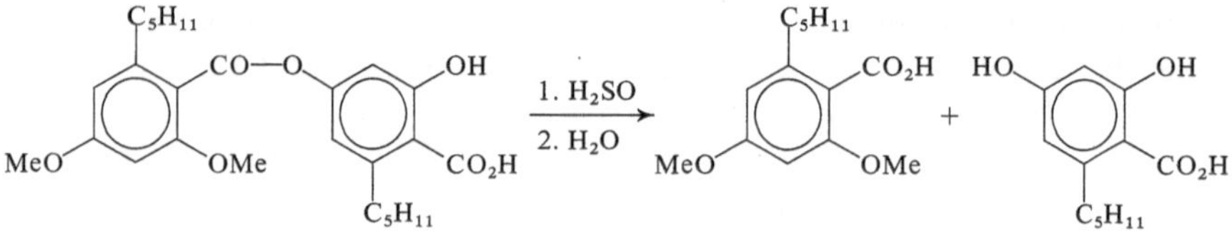

Scheme 28. Hydrolysis of 2-O-methylperlatolic acid

Although further transformations of the hydrolysis products (e.g. decarboxylation, lactonisation) are sometimes observed under these conditions, in general they cause no confusion. Although such experiments do not provide absolute proof of structure without isolation and characterisation of the new depsides and the hydrolysis products they certainly do provide compelling evidence for them. In a number of cases the structures assigned by this method have subsequently been confirmed by comparisons of natural and synthetic material (111, 120).

High performance liquid chromatography (h.p.l.c.) has also been effectively employed as an analytical tool in examining depsides and their hydrolysis products in microextracts (67, 73, 87). The major advantage of this technique is that it yields quantitative information about the components present in crude extracts.

The development of two dimensional t.l.c. (*85, 232, 233, 234*) has considerably improved the R_F differentiation of structurally similar compounds and has enabled the identification of minor constituents of established structure in complex mixtures of depsides. Furthermore new depsides have been tentatively identified from the known structures of their congeners and R_F correlations. For instance nordivaricatic acid (**233**) and 4-O-demethylstenosporic acid (**244**) were recognised in the following manner (*77*). Not only do (**233**) and (**244**) co-occur with the corresponding 4-O-methylated derivatives, divaricatic acid (**234**) and stenosporic acid (**245**) in various brown *Parmelia* species, but on co-chromatography with anziaic acid (**246**) the unknown depsides [i.e. (**233**) and (**244**)] behave as homologues of (**246**). Thus these three compounds gave spots with identical ultraviolet characteristics and colours after spraying, that lie nearly equidistant on a sloping straight line when chromatographed by the two-dimensional method.

(**233**) $R_1 = R_2 = C_3H_7$
(**244**) $R_1 = C_3H_7$, $R_2 = C_5H_{11}$
(**246**) $R_1 = R_2 = C_5H_{11}$
(**247**) $R_1 = C_5H_{11}$, $R_2 = C_3H_7$

(**234**) $R_1 = R_2 = C_3H_7$
(**245**) $R_1 = C_3H_7$, $R_2 = C_5H_{11}$

Such R_F correlations were also used to tentatively identify 4-O-demethylimbricaric acid (**247**) (*85*), 4-O-demethylmicrophyllinic acid (**248**) (*85*), 4-O-demethylglomellic acid (**249**) (*72*), 4-O-demethylglomelliferic acid (**250**) (*72*), and 4-O-demethylloxodellic acid (**251**) (*72*). The trace quantities present precluded the isolation and further investigation of these depsides.

(**248**) $R_1 = R_2 = CH_2COC_5H_{11}$
(**249**) $R_1 = R_2 = CH_2COC_3H_7$
(**250**) $R_1 = CH_2COC_3H_7$, $R_2 = C_5H_{11}$
(**251**) $R_1 = CH_2COC_3H_7$, $R_2 = C_3H_7$

10.4 Partial Synthesis of Depside Derivatives

The partial synthesis of a depside derivative occasionally provides a convenient means of structure correlation but this procedure is more commonly used in depsidone chemistry than in structural elucidation of depsides.

The structure of 3,5-dichloro-2′-O-methylanziaic acid (208) was assigned from the spectroscopic properties and the nature of the hydrolysis products (183). Further evidence for this structure was forthcoming from the partial synthesis of methyl 3,5-dichloro-2,4,2′-tri-O-methylanzeate (252) both from anziaic acid (253) and the new depside (208) (Scheme 29) (183).

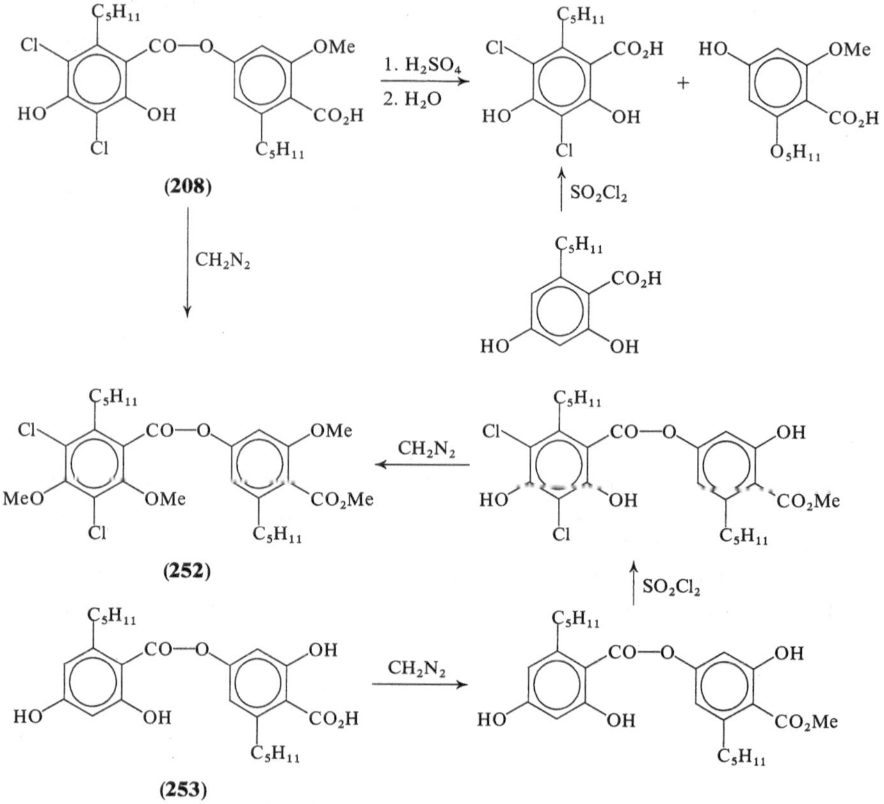

Scheme 29. Reactions of 3,5-dichloro-2′-O-methylanziaic acid

Similarly, correlation of the structure of the new depside 4′-O-methylnorsekikaic acid (254) isolated from *Ramalina* sp. with the common depside sekikaic acid (255) was achieved by permethylation of both depsides with diazomethane to form methyl 2,2'-di-O-methylsekikaiate (**256**) (*242*).

(**254**) R₁ = H, R₂ = Me
(**255**) R₁ = R₂ = Me

(**256**)

10.5 Total Synthesis of Depsides

Total synthesis has provided a definitive method for structural elucidation of all groups of depsides and provides a useful alternative to the classical hydrolytic degradation procedure. Further, in cases where complex mixtures of homologous depsides or simply lack of lichen material make isolation of the new depside impractical, t.l.c. comparisons of synthetic material with the natural mixtures has enabled the identity of a number of new depsides to be established (*44, 45*).

The use of the newer condensing reagents trifluoroacetic anhydride or dicyclohexylcarbodiimide has streamlined the synthesis of the depside linkage from the appropriately substituted benzoic acid and phenol. Where necessary the phenolic and carboxyl groups of the precursors are protected by O-benzylation until after depside-ester bond formation has been achieved. Catalytic hydrogenolysis of the so-formed O-benzyldepside esters subsequently gives the natural depsides.

10.5.1 Para-Orcinol Depsides

A number of well-known depsides were synthesised in the manner described above (*111*), as well as the recently reported compounds 2-O-methylperlatolic acid (**241**), 2'-O-methylperlatolic and 2'-O-methylanziaic acid (*111*). An earlier report of the isolation of the related depside, 4-O-demethylplaniaic acid (**257**) from *Stereocaulon ramulosum* (*41*) appears to be an error (*84*). However this depside has now been shown to co-occur with planiaic acid (**258**) in the lichens *Lecidea lithophila* and *L. plana* (*84*). The structure of (**257**) was indicated by microhydrolysis experiments and confirmed by comparison with synthetic material. The synthetic route to (**257**) is outlined in Scheme 30, and is typical of the approach to all the *para*-orcinol depsides.

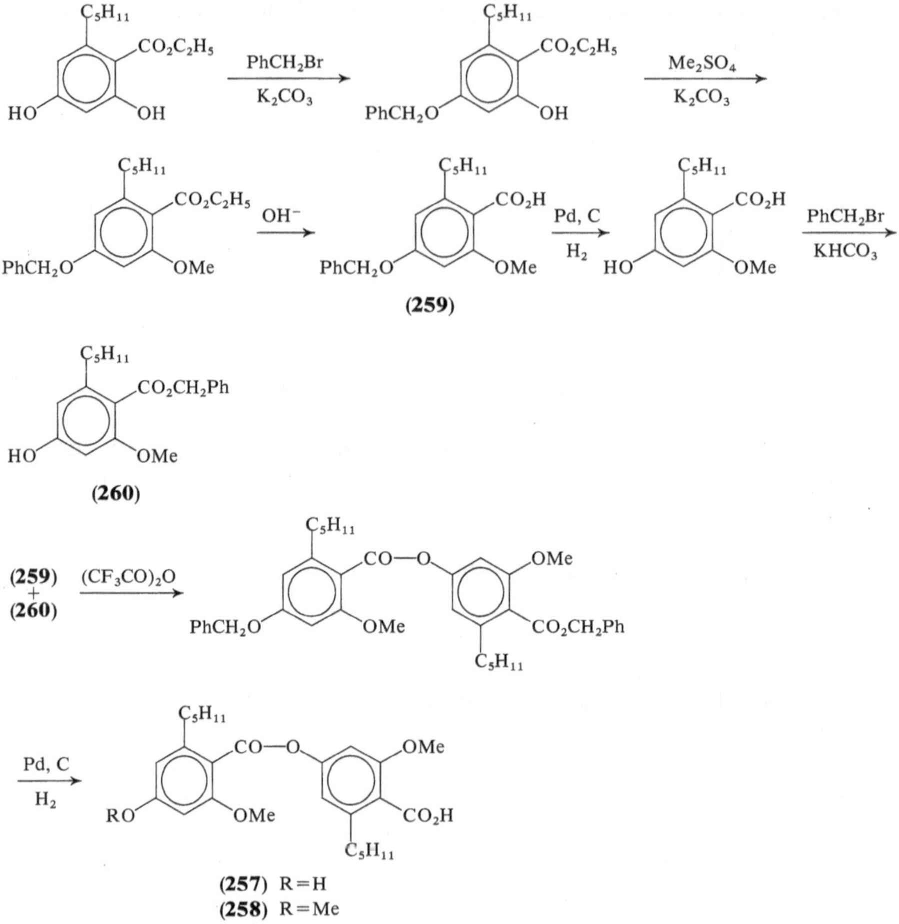

C_5H_{11} .CO$_2$C$_2$H$_5$ HO OH $\xrightarrow[K_2CO_3]{PhCH_2Br}$ C_5H_{11} .CO$_2$C$_2$H$_5$ PhCH$_2$O OH $\xrightarrow[K_2CO_3]{Me_2SO_4}$

C_5H_{11} .CO$_2$C$_2$H$_5$ PhCH$_2$O OMe $\xrightarrow{OH^-}$ C_5H_{11} .CO$_2$H PhCH$_2$O OMe **(259)** $\xrightarrow[H_2]{Pd, C}$ C_5H_{11} .CO$_2$H HO OMe $\xrightarrow[KHCO_3]{PhCH_2Br}$

C_5H_{11} .CO$_2$CH$_2$Ph HO OMe

(260)

(259) + **(260)** $\xrightarrow{(CF_3CO)_2O}$ C_5H_{11} .CO—O OMe PhCH$_2$O OMe CO$_2$CH$_2$Ph C_5H_{11}

$\xrightarrow[H_2]{Pd, C}$ C_5H_{11} .CO—O OMe RO OMe CO$_2$H C_5H_{11}

(257) R=H
(258) R=Me

Scheme 30. Synthesis of 4-O-demethylplaniaic acid

10.5.2 Olivetoric Acid and Related Depsides

The structural elucidation of the new depside 2-O-methylconfluentic acid **(237)** (*190*) via the nature of the hydrolysis products has been mentioned previously (Scheme 27). The total synthesis of this compound has also been reported (*119*) and was the first instance where a depside containing an oxidized side chain had been synthesized. Here additional difficulties arise because of the close proximity of the reactive benzyl ketone and depside ester functional groups. For example the reaction of olivetoric acid (261) with formic acid, potassium hydroxide, or even prolonged treatment with diazomethane resulted in the formation of enol lactones (Scheme 31). This necessitated the protection of the keto-group in precursors to (237).

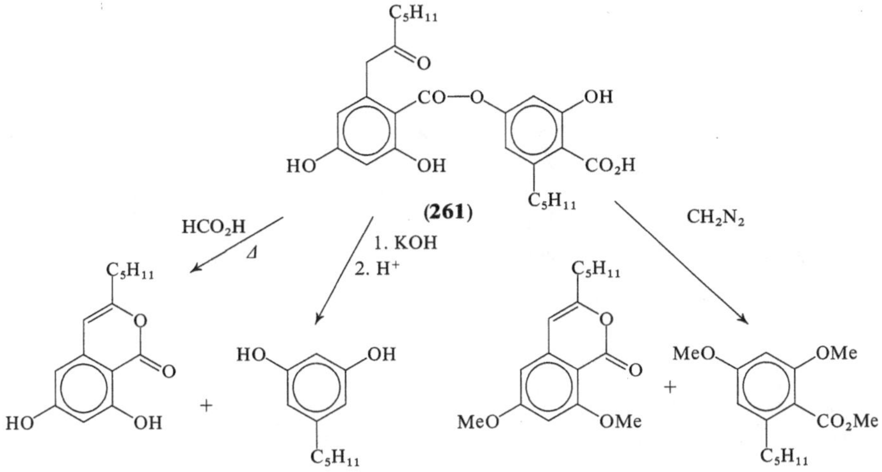

Scheme 31. Degradation of olivetoric acid

The preferred synthetic route to the depside (237) is outlined in Scheme 32 (*119*). Alkylation of 2-bromo-3,5-dimethoxybenzyl bromide (262) with (2-pentyl-1,3-dithian-2-yl)lithium gave 2-(2′-bromo-3′,5′-dimethoxybenzyl)-2-pentyl-1,3-dithiane (263). Lithiation of (263) with butyllithium followed by carbonation at low temperature then gave the required precursor benzoic acid (264). This acid condensed readily with benzyl 4-hydroxy-2-methoxy-6-pentylbenzoate (265) in the presence of dicyclohexylcarbodiimide to give the depside ester (266). The protecting dithioketal group was then removed under mild conditions by treating (266) with copper(II) chloride and copper(II) oxide in aqueous acetone at room temperature. The benzyl 2-O-methylconfluentate (267) so obtained was finally subjected to hydrogenolysis to yield the natural depside (237). Analogous procedures with the appropriately substituted precursors have been utilised in the total synthesis of olivetoric acid (261), confluentic acid and 4-O-methylolivetoric acid (*120*).

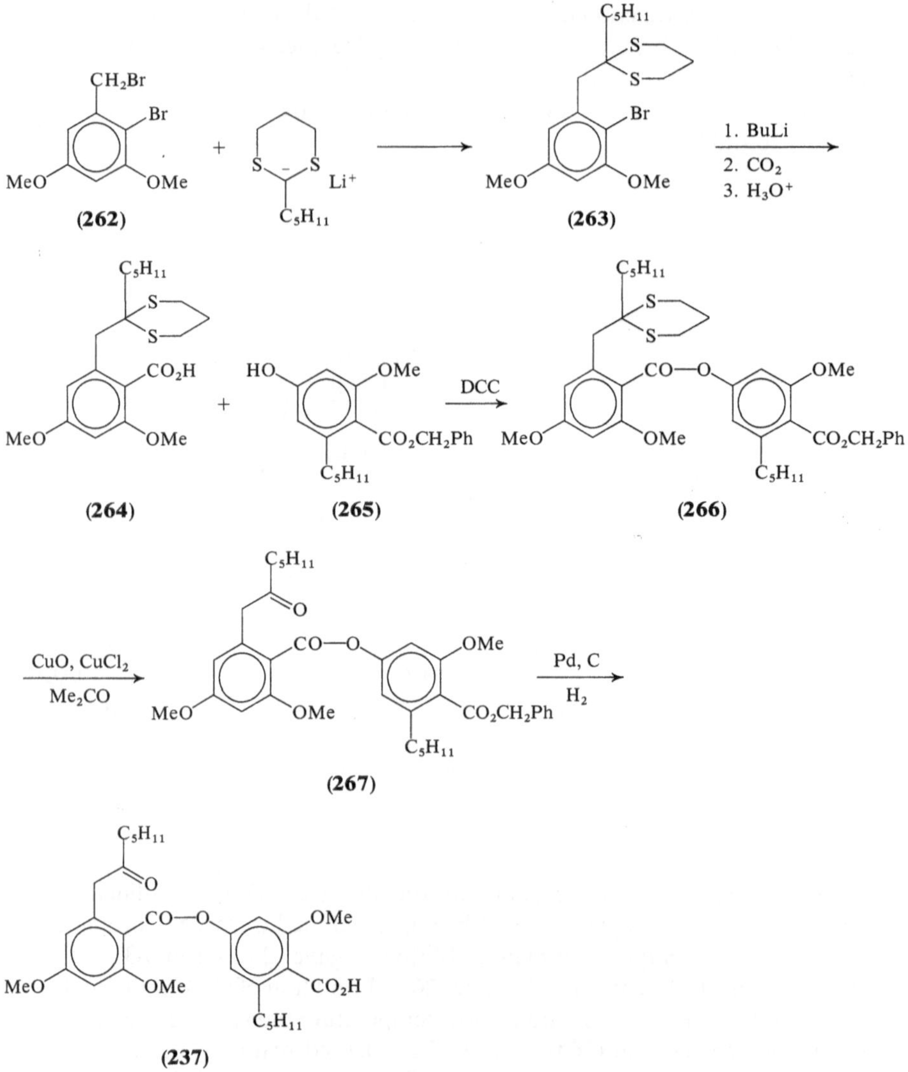

Scheme 32. Synthesis of 2-O-methylconfluentic acid

10.5.3 Synthesis of β-Orcinol para-Depsides, β-Orcinol meta-Depsides, meta-Orcinol Depsides and Orcinol Tridepsides

Starting with appropriately substituted phenols, procedures identical to those used for the synthesis of *para*-orcinol depsides have been employed in the synthesis of natural β-orcinol *para*-depsides (*130, 152, 251, 252*) β-orcinol *meta*-depsides (*131*), orcinol *meta*-depsides (*44, 129*) and orcinol tridepsides (*37, 49, 123, 124, 252*). Typically, HAMILTON and SARGENT (*152*) achieved a synthesis of nephroarctin (**219**) as depicted in Scheme 33. Gattermann formylation of methyl haematommate (**268**) produced the dialdehyde (**269**) and subsequent cleavage of the ester function by treatment with BBr$_3$ gave the desired A-ring precursor (**270**). Condensation of the acid (**270**) and the phenol (**271**) in the presence of trifluoroacetic anhydride then gave nephroarctin (**219**).

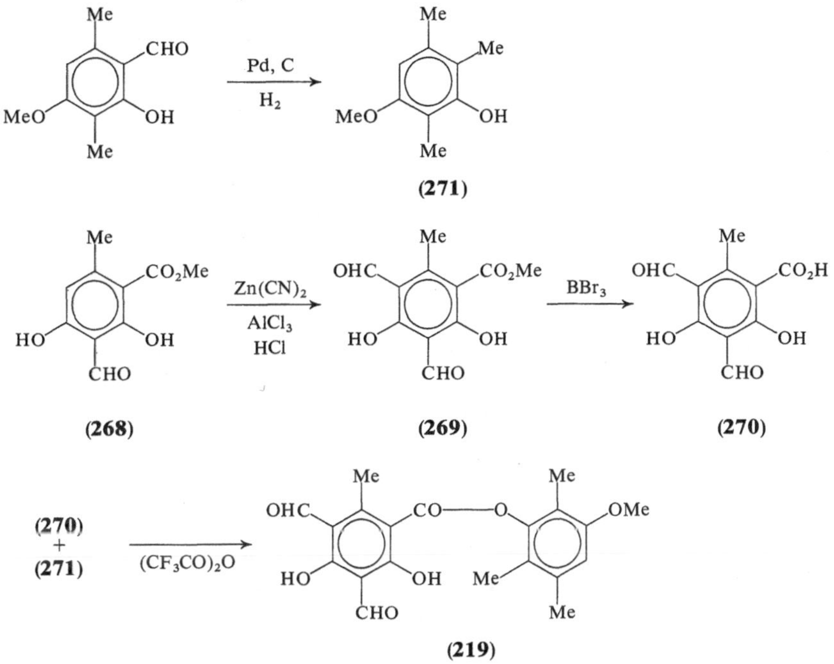

Scheme 33. Synthesis of nephroarctin

Although numerous orcinol *meta*-depsides have been synthesised (*44, 129*), the preparation of paludosic acid (**272**) is typical and is outlined in Scheme 34.

As a typical example of tridepside synthesis, the preparation of 5-O-methylhiascic acid (**227**) is depicted in Scheme 35 (*122*). Persulfate oxidation of ethyl 4-benzyloxy-2-hydroxy-6-methylbenzoate (**273**) in potassium hydroxide solution gave the corresponding 5-hydroxy derivative (**274**) from which 2,4-dibenzyloxy-5-methoxy-6-methylbenzoic acid (**275**) was prepared. Condensation of this acid with benzyl lecanorate (**276**) gave the

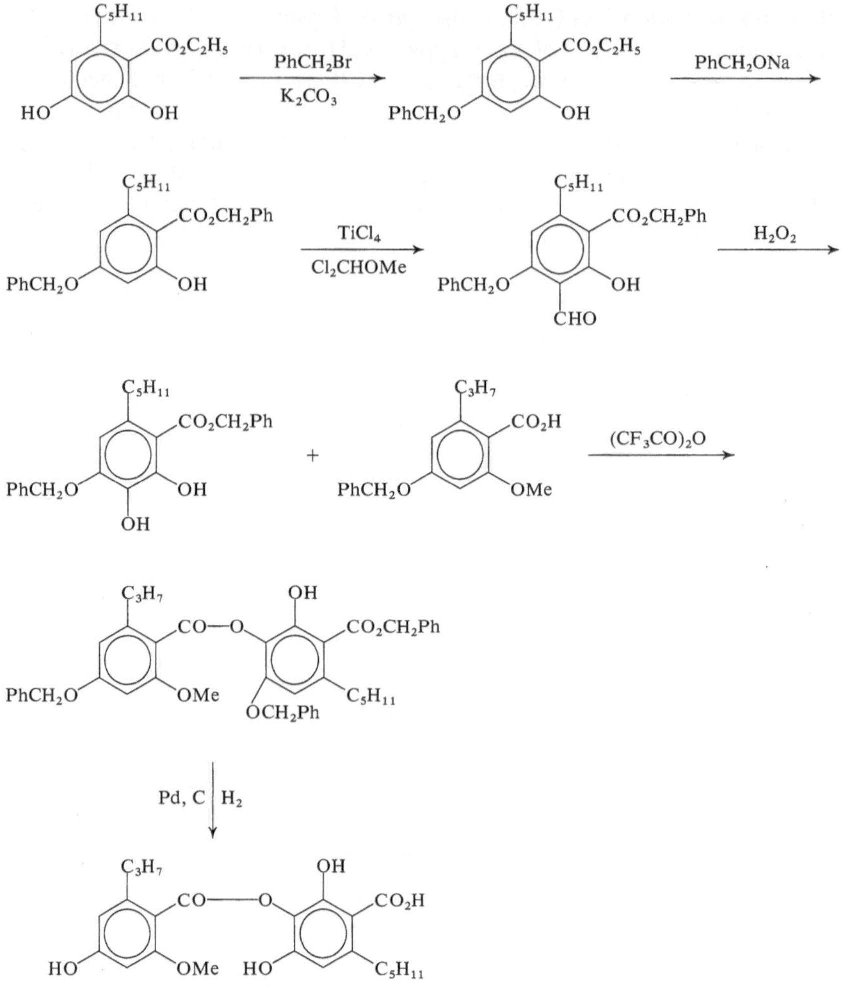

Scheme 34. Synthesis of paludosic acid

tridepside ester (277). Subsequent hydrogenolysis over palladised carbon gave 5-O-methylhiascic acid (227), identical with the natural material.

Finally aphthosin (278), an orcinol tetradepside was obtained through condensing evernic acid (279) and methyl lecanorate (280) (Scheme 36) (37). The properties of this synthetic material were identical with those of the natural product isolated by BACHELOR and KING (10), although a reinvestigation of the constituents of *Peltigera aphthosa* failed to verify the natural occurrence of this tetradepside (234).

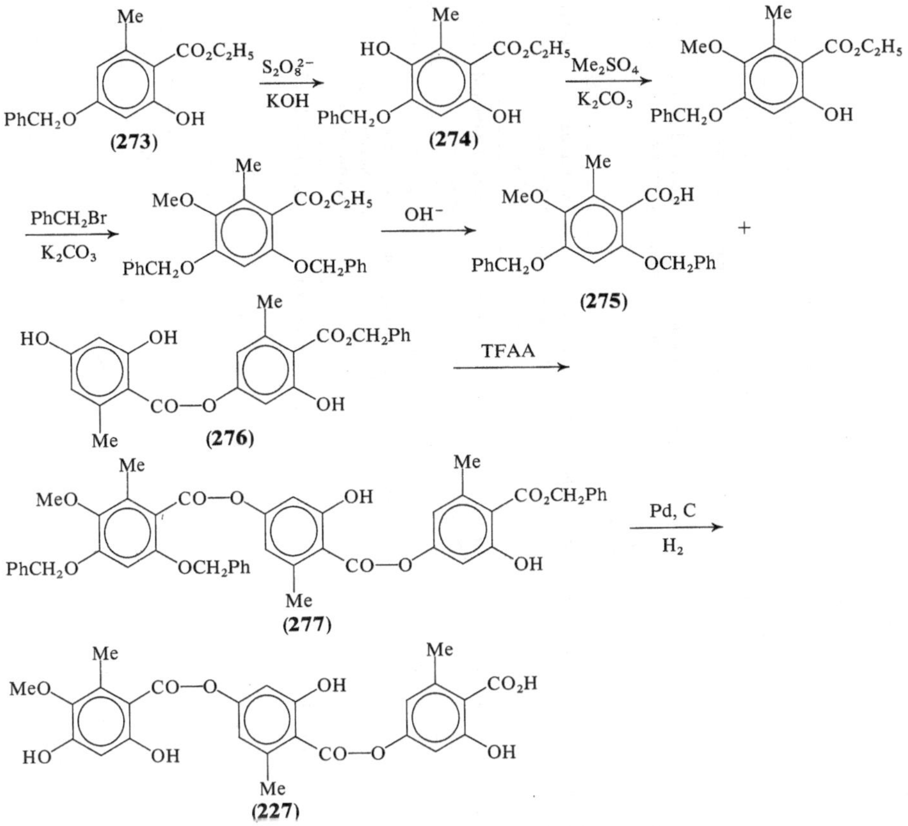

Scheme 35. Synthesis of 5-O-methylhiascic acid

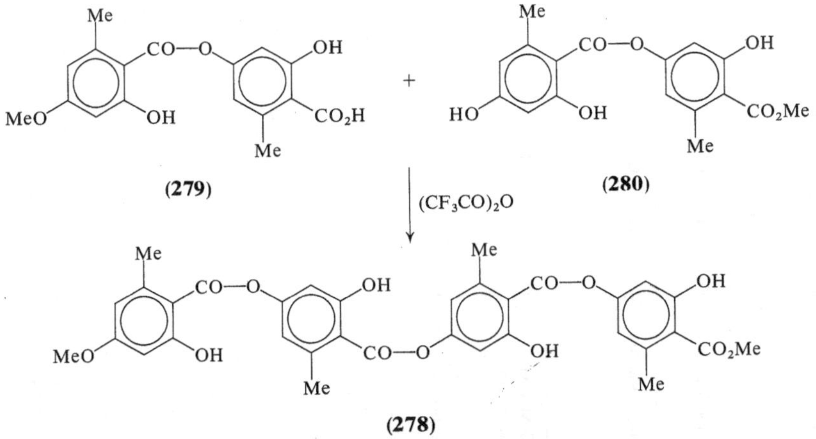

Scheme 36. Synthesis of aphthosin

Table 4. *New Lichen Depsides*

Orcinol *para*-depsides   Structure (**204**)

	C_2	C_3	C_4	C_5	C_6	$C_{3'}$	$C_{2'}$	$C_{1'}$	$C_{6'}$	$C_{5'}$	Source	Separation	Structure determination	Ref.
Methyl lecanorate	OH	H	OH	H	Me	H	OH	CO_2Me	Me	H	*Pseudocyphellaria crocata*	Column chromatography	Spectral data and synthesis	(232)
Methyl evernate	OH	H	OMe	H	Me	H	OH	CO_2Me	Me	H	*Pseudocyphellaria crocata* *Lobaria linita* *Peltigera aphthosa*	Column chromatography	Spectral data and synthesis	(232, 233, 234)
2-O-Methyl-evernic acid	OH	H	OMe	H	Me	H	OMe	CO_2H	Me	H	*Evernia prunastri*	Column chromatography	Spectral data (^{13}C-n.m.r.) synthesis	(252)
Nordivaricatic acid	OH	H	OH	H	Pr	H	OH	CO_2H	Pr	H	*Heterodea beugleholei*	Benzylation/hydrogenolysis	Spectral data and synthesis	(132)
3-Chloro-divaricatic acid	OH	Cl	OMe	H	Pr	H	OH	CO_2H	Pr	H	*Thelomma mammosum*	Column chromatography	Spectral data (^{13}C-n.m.r.) comparison with 5-chloro isomer	(197)
2-O-Methyl divaricatic acid	OMe	H	OMe	H	Pr	H	OH	CO_2H	Pr	H	*Ramalina sayreana*	Not isolated	Microhydrolysis	(76)
4-O-Demethyl-stenosporic acid	OH	H	OH	H	Pr	H	OH	CO_2H	C_5H_{11}	H	brown *Parmelia* sp.	Not isolated	Tentative from Rf correlation	(77)
4-O-Demethyl-imbricaric acid	OH	H	OH	H	C_5H_{11}	H	OH	CO_2H	Pr	H	*Cetraria* sp.	Not isolated	Tentative from Rf correlation	(72)
2-O-Methyl-stenosporic acid	OMe	H	OMe	H	Pr	H	OMe	CO_2H	C_5H_{11}	H	*Ramalina sayreana*	Not isolated	Microhydrolysis	(76)
2'-O-Methyl-anziaic acid	OH	H	OH	H	C_5H_{11}	H	OMe	CO_2H	C_5H_{11}	H	*Lecidea dilucens*	Not isolated	Microhydrolysis, synthesis	(83, 111)

Compound									Source	Isolation	Characterization	Ref.
3,5-Dichloro-2'-O-methyl anziaic acid	OH	Cl	C_5H_{11}	H	OMe	CO_2H	C_5H_{11}	H	*Lecanora sulphurella*	Crystallization	Spectral data, hydrolysis, partial synthesis	(183)
2-O-Methyl-perlatolic acid	OMe	H	C_5H_{11}	H	OH	CO_2H	C_5H_{11}	H	*Pertusaria tuberculifera*	Not isolated	Microhydrolysis, synthesis	(80, 111)
2'-O-Methyl-perlatolic acid	OH	H	C_5H_{11}	H	OMe	CO_2H	C_5H_{11}	H	*Pertusaria globularis* / *Lecidea* sp.	Not isolated / Crystallization	Microhydrolysis, synthesis / Spectral data, hydrolysis	(80, 111, 191)
4-O-Demethyl-planaic acid	OMe	H	C_5H_{11}	H	OMe	CO_2H	C_5H_{11}	H	*Lecidea lithophila*	Crystallization	Microhydrolysis, synthesis	(84, 111)
Oxostenosporic acid	OH	H	Pr	H	OMe	CO_2H	$CH_2\text{-}CO\text{-}C_3H_7$	H	*Parmelia verruciifera*	Preparative TLC	Microhydrolysis	(77)
4-O-Demethyl-loxodellic acid	OH	H	$CH_2\text{-}CO\text{-}C_3H_7$	H	OH	CO_2H	Pr	H	brown *Parmelia* sp.	Not isolated	Tentative from Rf correlation	(85)
Loxodellic acid	OH	H	$CH_2\text{-}CO\text{-}C_3H_7$	H	OMe	CO_2H	Pr	H	*Parmelia loxodella*	Preparative TLC	Microhydrolysis	(81)
4-O-Demethyl-glomelliferic acid	OH	H	$CH_2\text{-}CO\text{-}C_3H_7$	H	OH	CO_2H	C_5H_{11}	H	brown *Parmelia* sp.	Not isolated	Tentative from Rf correlation	(85)

Table 4 (continued)

Compound	C_2	C_3	C_4	C_5	C_6	$C_{3'}$	$C_{6'}$	$C_{1'}$	$C_{2'}$	$C_{5'}$	Source	Separation	Structure determination	Ref.
4-O-Demethyl-glomellic acid	OH	H	OH	H	CH_2—CO—C_3H_7	H	CH_2—CO—C_3H_7	CO_2H	OH	H	brown *Parmelia* sp.	Not isolated	Tentative from Rf correlation	(85)
Glomellic acid	OH	H	OMe	H	CH_2—CO—C_3H_7	H	CH_2—CO—C_3H_7	CO_2H	OH	H	*Parmelia glomellifera*	Crystallization	Spectral data, hydrolysis	(180)
Miriquidic acid	OH	H	OMe	H	CH_2—CO—C_2H_5	H	C_5H_{11}	CO_2H	OH	H	*Lecidea lilienstroemii*, *L. leucophaea*	Crystallization	Spectral data, hydrolysis	(193)
4-O-Methyl-olivetoric acid	OH	H	OMe	H	CH_2—CO—C_5H_{11}	H	C_5H_{11}	CO_2H	OH	H	*Parmelia brattii*	Not isolated	Microhydrolysis, synthesis	(81, 120)
2-O-Methyl-confluentic acid	OMe	H	OMe	H	CH_2—CO—C_5H_{11}	H	C_5H_{11}	CO_2H	OMe	H	*Lecidea fuscoatra*	Crystallization	Spectroscopic data, hydrolysis and synthesis	(190, 119)
4-O-Demethyl-microphyllinic acid	OH	H	OH	H	CH_2—CO—C_5H_{11}	H	C_5H_{11}	CO_2H	OH	H	*Cetraria* sp.	Not isolated	Tentative from Rf correlation	(72)

	C2	C3	C4	C5	C6 (CH2–CO–C5H11)	OMe	CO2H	OMe	CO–C5H11	H	Structure determination	Ref.
2′-O-Methyl-microphyllinic acid	OH	H	OMe	H	CH2–CO–C5H11	OMe	CO2H	OMe	CO–C5H11	H	Spectral data, hydrolysis	(49)

Orcinol β-orcinol para-depsides

	C2	C3	C4	C5	C6	C2′	C1′	C6′	C3′	Source	Separation	Structure determination	Ref.
2-O-Methyl-obtusatic acid	OMe	Me	OMe	H	Me	OH	CO2H	Me	H	*Xanthoparmelia tucsonensis*	Not isolated	Synthesis, comparative TLC	(45)
Evernine	OH	H	OMe	H	Me	OH	CO2Me	Me	H	*Evernia prunastri*	Column chromatography	Spectral data (13C-n.m.r.), hydrolysis, synthesis	(251, 253)
3′-Methylevernic acid	OH	H	OMe	H	Me	OH	CO2H	Me	Me	*Evernia prunastri*	Column chromatography	Spectral data (13C-n.m.r.), synthesis	(252)
Methyl 3′-methyl-lecanorate	OH	H	OH	H	Me	OH	CO2Me	Me	Me	*Evernia prunastri*	Column chromatography	Spectral data (13C-n.m.r.), synthesis	(252)

β-Orcinol para-depsides

	C2	C3	C4	C5	C6	C2′	C1′	C6′	C3′	Source	Separation	Structure determination	Ref.
3-α-Hydroxy-barbatic acid	OH	CH2OH	OMe	H	Me	OH	CO2H	Me	Me	*Xanthoparmelia moctezumensis*	Not isolated	Synthesis, comparative TLC, HPLC	(87)
Methyl-4-O-demethyl-barbatate	OH	Me	OH	H	Me	OH	CO2Me	Me	Me	*Oropogon loxensis*	Not isolated	Synthesis, comparative TLC, HPLC	(73)

Table 4 (continued)

	C_2	C_3	C_4	C_5	C_6	$C_{3'}$	$C_{2'}$	$C_{1'}$	$C_{6'}$	$C_{5'}$	Source	Separation	Structure determination	Ref.
2'-O-Methyl-atranorin	OH	CHO	OH	H	Me	Me	OMe	CO_2Me	Me	H	Oropogon loxensis	Preparative TLC	Microhydrolysis	(73)
Methyl 3-α-hydroxy-4-O-demethyl-barbatate	OH	CH_2OH	OH	H	Me	Me	OH	CO_2Me	Me	H	Oropogon loxensis	Preparative TLC	Synthesis, microhydrolysis	(73)

Orcinol meta-depsides　Structure (232)

	C_2	C_3	C_4	C_5	C_6	$C_{2'}$	$C_{1'}$	$C_{6'}$	$C_{5'}$	$C_{4'}$	Source	Separation	Structure determination	Ref.
2-O-Methyl-sekikaic acid	OMe	H	OMe	H	Pr	OH	CO_2H	Pr	H	OMe	Ramalina asahinae	Not isolated	Rf correlation, synthesis	(75, 44)
2,4-Di-O-methyl-norsekikaic acid	OMe	H	OMe	H	Pr	OH	CO_2H	Pr	H	OH	Ramalina asahinae	Not isolated	Synthesis, comparative TLC	(44)
4-O-Methyl-norsekikaic acid	OH	H	OH	H	Pr	OH	CO_2H	Pr	H	OMe	Ramalina luciae	Preparative TLC	Spectral properties, microhydrolysis, partial synthesis	(241, 242)
4-O-Methyl-norhomosekikaic acid	OH	H	OH	H	Pr	OH	CO_2H	C_5H_{11}	H	OMe	Ramalina cf. pacifica	Not isolated	Tentative from Rf correlation	(241)
4-O-Methyl paludosic acid	OMe	H	OH	H	Pr	OH	CO_2H	C_5H_{11}	H	OMe	Ramalina asahinae	Not isolated	Synthesis, comparative TLC	(44)
4-O-Methyl-cryptochloro-phaeic acid	OMe	H	OH	H	C_5H_{11}	OH	CO_2H	C_5H_{11}	H	OMe	Ramalina asahinae	Not isolated	Tentative from Rf correlation	(76)

| 4,4'-Di-O-methylcrypto-chlorophaeic acid | OMe | H | OMe | H | C_5H_{11} | OH | CO_2H | C_5H_{11} | H | OMe | Ramalina asahinae | Not isolated | Synthesis, comparative TLC | (44) |

Orcinol tridepsides   Structure (226)

	C_2	C_3	C_4	C_5	$C_{2'}$	$C_{1'}$	$C_{2''}$	Source	Separation	Structure determination	Ref.
Orcinyl lecanorate	OH	H	OH	H	OH	H	OH	Parmelia borreri	Preparative TLC	Spectral data, synthesis	(49)
Ovoic acid	OH	H	OH	H	OMe	CO_2H	OH	Parmelia substygia	Crystallization	Spectral data (^{13}C-n.m.r.), hydrolysis	(194)
2'',4-Di-O-methylgyrophoric acid	OH	H	OMe	H	OH	CO_2H	OMe	Evernia prunastri	Column chromatography	Spectral properties (^{13}C-n.m.r.), synthesis	(252)
2,4-Di-O-methylgyrophoric acid	OMe	H	OMe	H	OH	CO_2H	OH	Parmelia damaziana	Vacuum liquid chromatography	Spectral data, synthesis	(124)
2-O-Acetyl-tenuiorin	OAc	H	OMe	H	OH	CO_2Me	OH	Pseudocyphellaria australiensis	Preparative TLC	Spectral data, synthesis	(36)
2'-O-Methyl-tenuiorin	OH	H	OMe	H	OMe	CO_2Me	OH	Pseudocyphellaria faveolata	Preparative TLC	Spectral data, synthesis	(126)
2''-O-Methyl-tenuiorin	OH	H	OMe	H	OH	CO_2Me	OMe	Pseudocyphellaria faveolata	Preparative TLC	Spectral data, synthesis	(126)
2',2'-Di-O-methyltenuiorin	OH	H	OMe	H	OMe	CO_2Me	OMe	Pseudocyphellaria faveolata	Preparative TLC	Spectral data, synthesis	(126)
5-O-Methyl-hiascic acid	OH	H	OH	OMe	OH	CO_2H	OH	Parmelia sp.	Benzylation/ hydrogenolysis	Spectral data, synthesis	(122)

Table 4 (continued)

	C_2	C_3	C_4	C_5	$C_{2'}$	$C_{1''}$	$C_{2''}$	Source	Separation	Structure determination	Ref.
4,5-Di-O-methylhiascic acid	OH	H	OMe	OMe	OH	CO_2H	OH	*Parmelia pseudofatiscens*	Benzylation/ hydrogenolysis	Spectral data, synthesis	(117)
3-Methoxy-2,4-di-O-methyl-gyrophoric acid	OMe	OMe	OMe	H	OH	CO_2H	OH	*Parmelia subfatiscens*	Benzylation/ hydrogenolysis	Spectral data, synthesis	(123)
2,4,5-Tri-O-methylhiascic acid	OMe	H	OMe	OMe	OH	CO_2H	OH	*Parmelia damaziana*	Vacuum liquid chromatography	Spectral data, synthesis	(124)

11. Depsidones and Related Diphenyl Ethers

11.1 Introduction

The depsidones are based on the 11*H*-dibenzo[*b,e*][1,4]dioxepin-11-one ring system and are numbered systematically as in formula (**281**) which will be adhered to in this discussion. For convenience the benzenoid rings are designated as A and B. Many workers in this field use the numbering system in formula (**282**) which emphasizes the relationship of depsidones to the *para*-depsides.

The depsidones may be classified into compounds in which both benzenoid rings are derived from the polyketide orsellinic acid; chlorination and O-methylation, as in lecideoidin (**283**) or decarboxylation as in diploicin (**284**) being common secondary modifications. Homologues of orsellinic acid, as in colensoic acid (**285**) or analogues with oxygenated side chains as in vittatolic acid (**286**) are also encountered. Somewhat rarer types belonging to this group involve oxygenation at a nuclear position not demanded by their polyketide derivation as in hydroxycolensoic acid (**287**). The second class of depsidones contains those compounds in which both rings are derived from the *C*-methylated polyketide β-orcinol carboxylic acid; the simplest member being hypoprotocetraric acid (**288**). Side chain oxidation is very common in this group as exemplified by the series: virensic acid (**289**), protocetraric acid (**290**), and salazinic acid (**291**) in which the degree of side chain oxidation increases. Chlorination, O-methylation, and decarboxylation are also common secondary modifications in this group as in vicanicin (**292**). The third class of depsidones contain those compounds in which one ring is derived from orsellinic acid and the other from β-orcinol carboxylic acid as in the simplest member nornotatic acid (**293**). A group of fungal depsidones also belongs to this class.

(281)

(282)

(283)

(284)

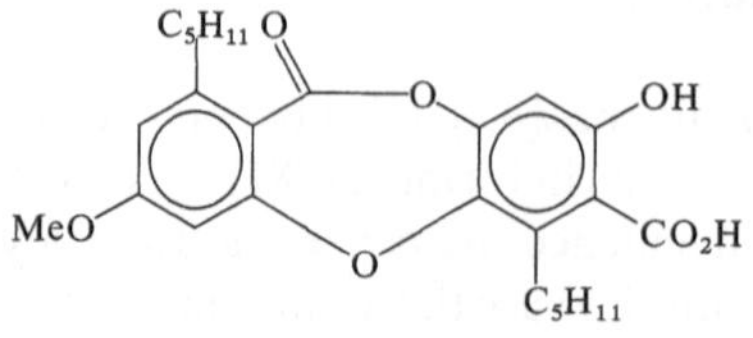

(285)

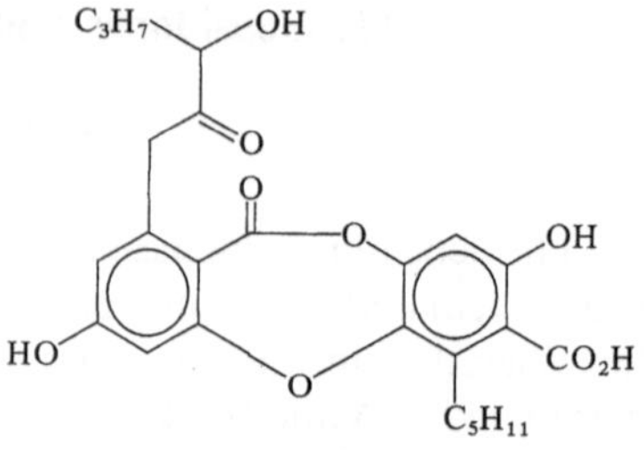

(286)

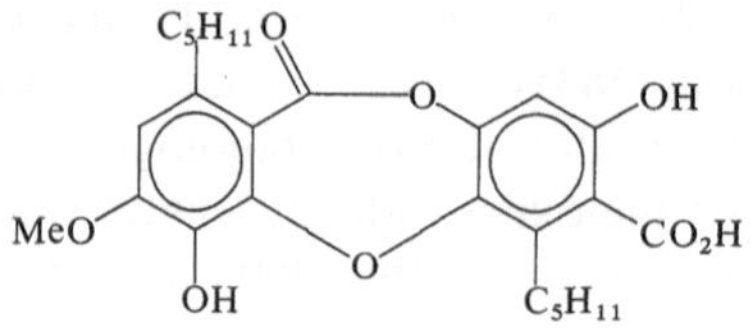

(287)

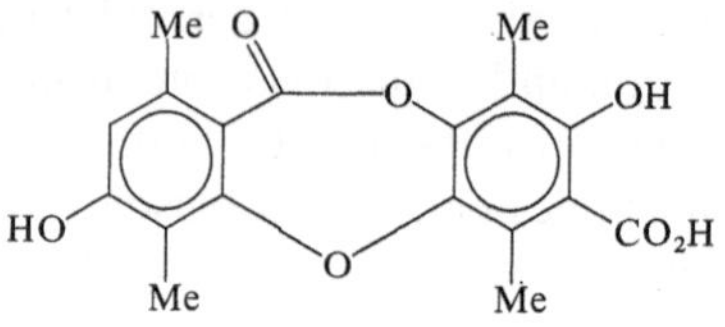

(288)

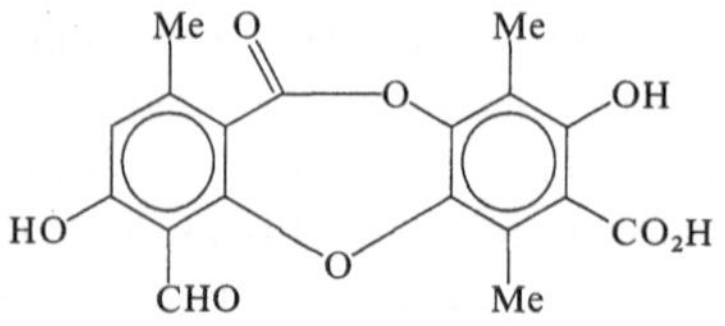

(289)

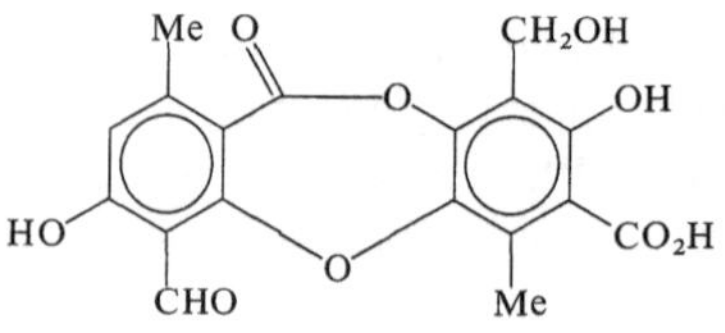

(290)

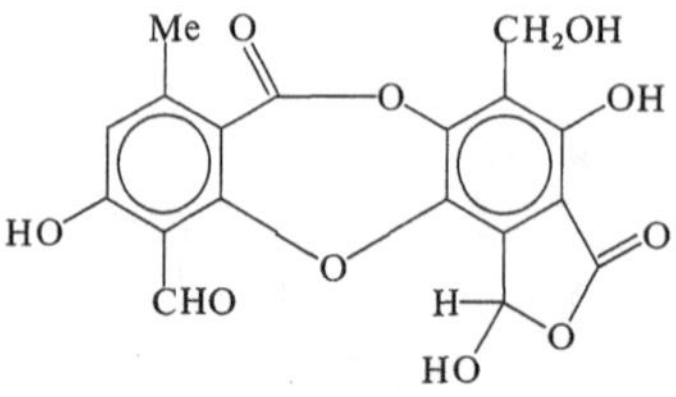

(291)

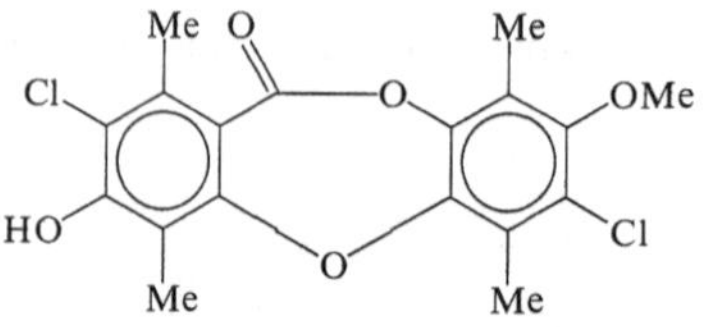

(292)

The biogenesis of depsidones was first explained (*12, 135*) by the attractive theory that they arose from intramolecular phenolic oxidative coupling of *para*-depsides, and although at least nine *para*-depside-depsidone pairs of corresponding structures exist they usually occur in widely separated lichen genera. However in several lichens (*62, 63*) the pair olivetoric acid (**294**) and physodic acid (**295**) actually co-occur.

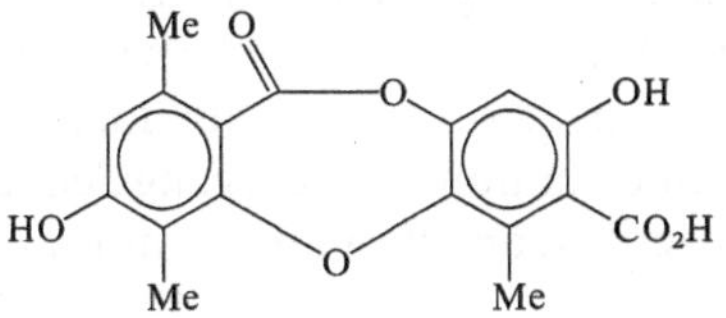

(293)

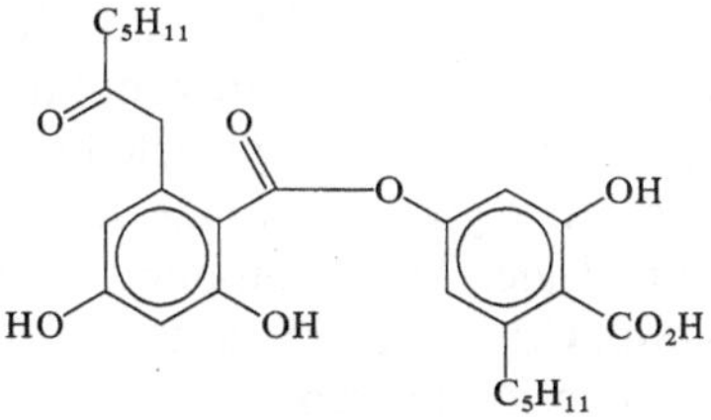

(294)

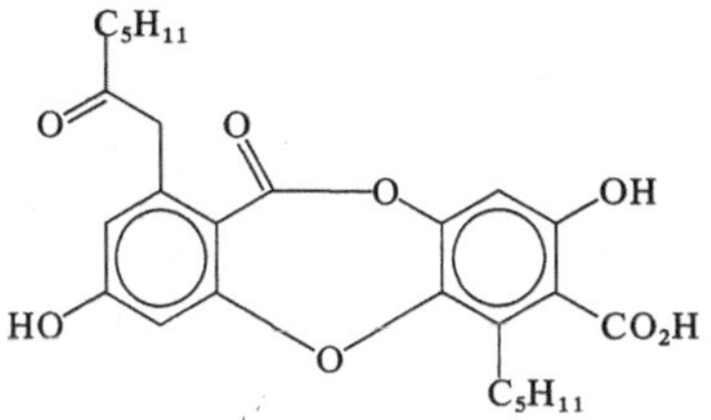

(295)

In the ten years since the last review there has been a resurgence of interest in the chemistry of depsidones in terms of the detection (*79*), the isolation of new compounds and the total synthesis of these substances. The new work is discussed in the order of the above classification and of increasing structural complexity of the compounds. A methodical study and interpretation of the ^{13}C-n.m.r. spectra of orcinol (*311*) and β-orcinol depsidones (*312*) has been reported by SUNDHOLM and HUNECK.

11.2 Orcinol Depsidones

Gangaleoidin (**296**) (Scheme 37), first isolated by NOLAN and his coworkers who advanced an erroneous structure (*94*) may be regarded as one of the biogenetically simpler of the depsidones despite its three chlorosubstituents. Unsuccessful attempts were made by HENDRICKSON and his coworkers (*160*) to establish structure (**296**) but this was achieved by SARGENT, ELIX, and VOGEL(*287*) in a manner which well illustrates modern methods of depsidone structural elucidation. The mass spectrum of gangaleoidin (**296**) showed prominent ring A ions (**297**) and (**298**), which were shifted by 14 a.m.u. on methylation of gangaleoidin thus locating the phenolic group and the chloro-substituents. The methylated methanolysis product (**299**) of O-methylgangaleoidin now exhibited ions (**300**) and (**301**) due to ring B. Hydrogenolytic dechlorination of compound (**299**) gave the diphenyl ether (**302**) which was identified by Ullmann ether synthesis using the components (303) and (304). DJURA, SARGENT, and VOGEL (*107*) completed the structure proof by a synthesis of gangaleoidin. Thus the acid (305), available in high yield by Ullmann ether synthesis using benzyl as the protective group, gave the depsidone (306) on lactonization with hot acetic anhydride. This compound on partial chlorination yielded gangaleoidin (296). CULLEN and SARGENT (*88*) have described an alternative synthesis of the depsidone (306).

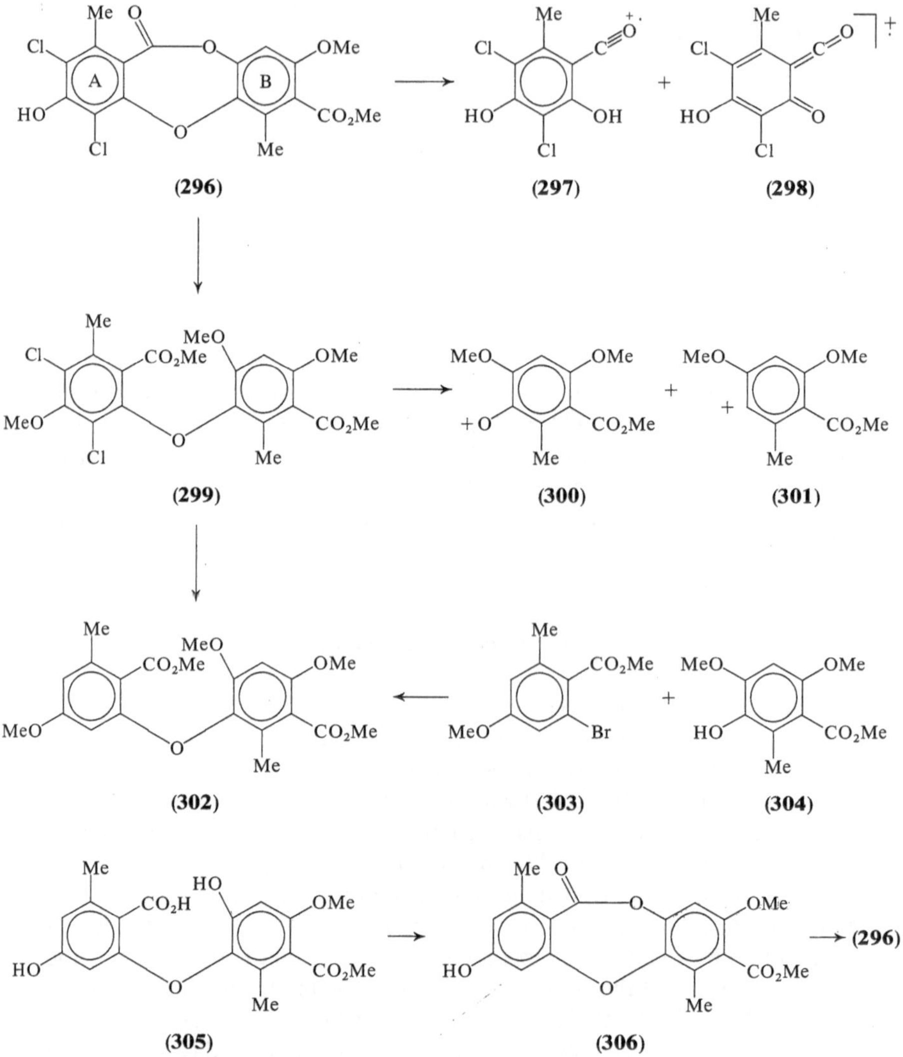

Scheme 37. The degradation and synthesis of gangaleoidin

CHESTER, ELIX, and JONES (*50*) described the isolation and structural determination of two close relatives of gangaleoidin: lecideoidin (**283**) and dechlorolecideoidin (**307**) (Scheme 38) from an Australian *Lecidea* species. Lecideoidin (**283**), shown by high resolution mass spectrometry to have the molecular formula $C_{17}H_{12}Cl_2O_7$, had one less CH_3 unit than gangaleoidin as indicated by its ^{1}H-n.m.r. spectrum. Exhaustive methylation of lecideoidin (**283**) gave a derivative (**308**) different from O-methylgangaleoidin and the mass spectrum of lecideoidin indicated that only one chloro-substituent was present in its A-ring. The degradation product (**309**) of lecideoidin exhibited a high field aromatic proton in its ^{1}H-n.m.r. spectrum characteristic of a tri-*o*-substituted diphenyl ether, and this fixed the location of the A-ring chlorosubstituent. Such evidence is particularly useful in depsidone structural elucidation. The structural assignment (**283**) was also confirmed by the ^{13}C-n.m.r. spectrum. Chlorination of compound (**308**) gave the same product (**310**) as that obtained by chlorination of O-methylgangaleoidin so that structure (**283**) follows for lecideoidin. Similar considerations allowed structure (**307**) to be proposed for dechlorolecideoidin.

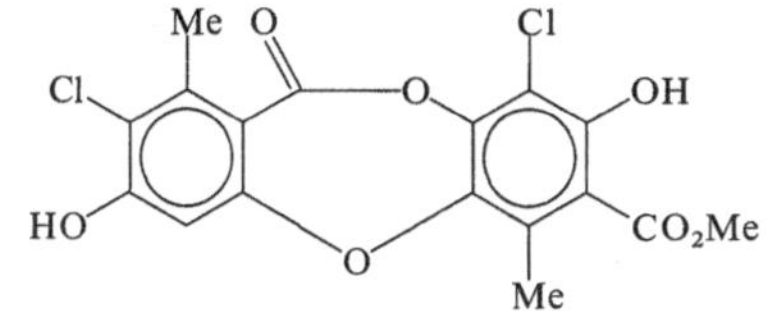

(**283**) 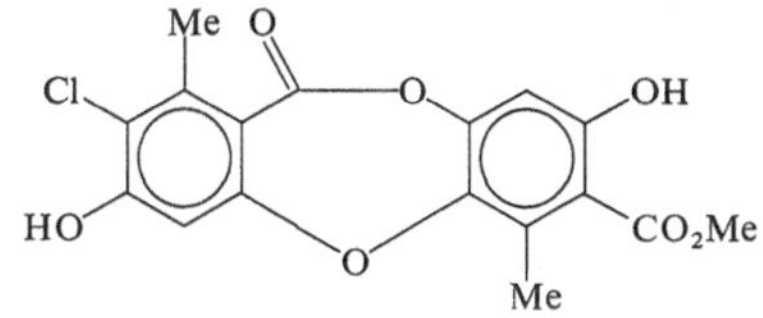   (**307**)

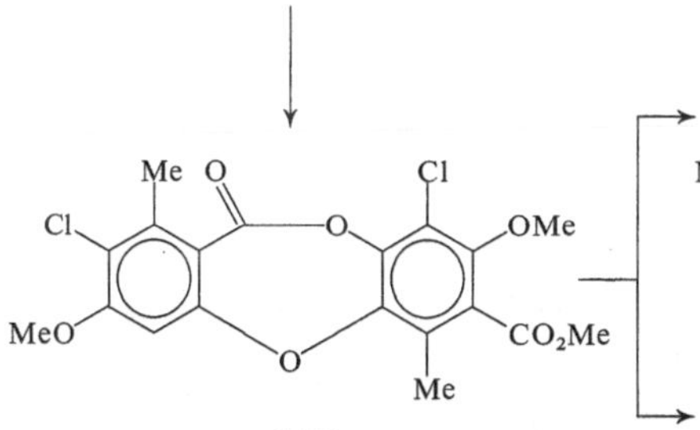

(**308**)

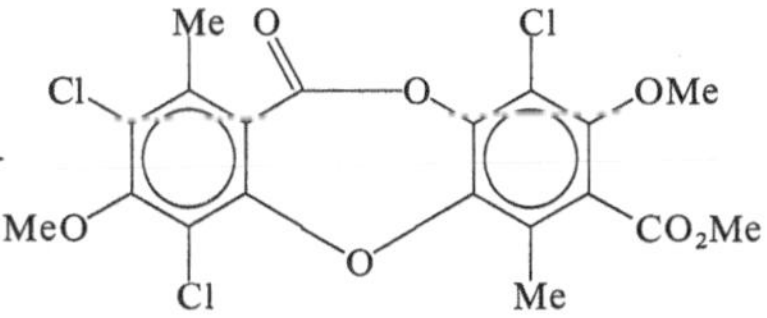

(**310**)

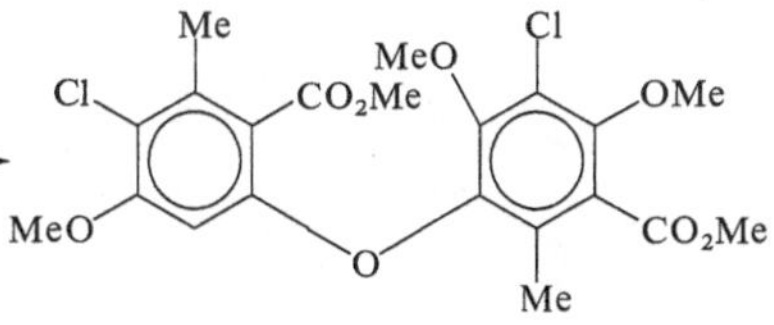

(**309**)

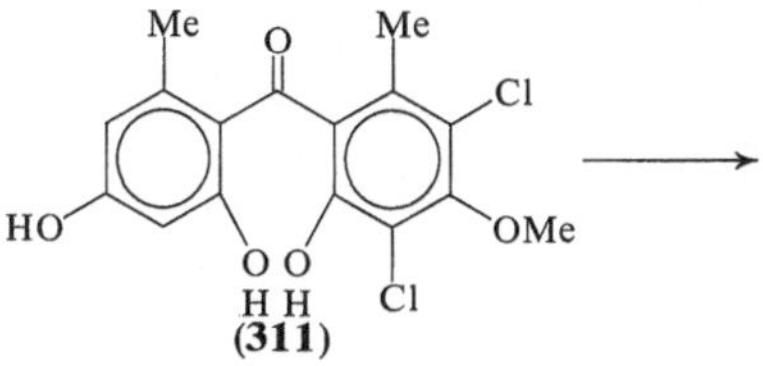

(**311**) 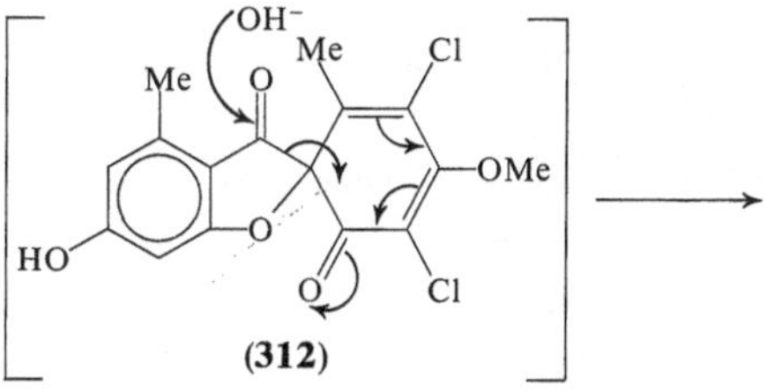   (**312**)

(313)   (314)

(315)   (284)

(316)   (317)

(318)

(319)   (320)

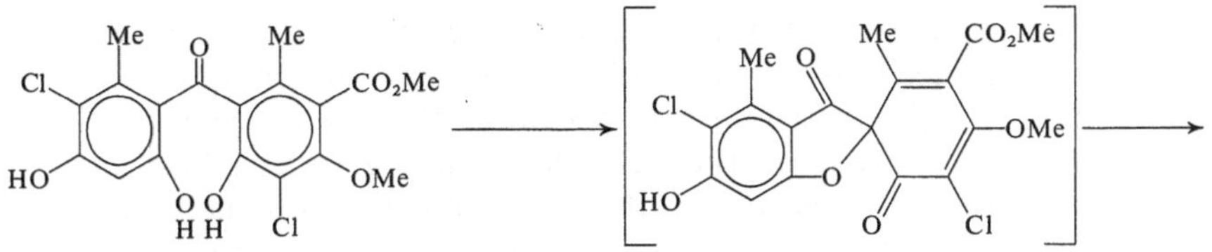

Scheme 38. Degradation and synthesis of lecideoidon and related depsidones

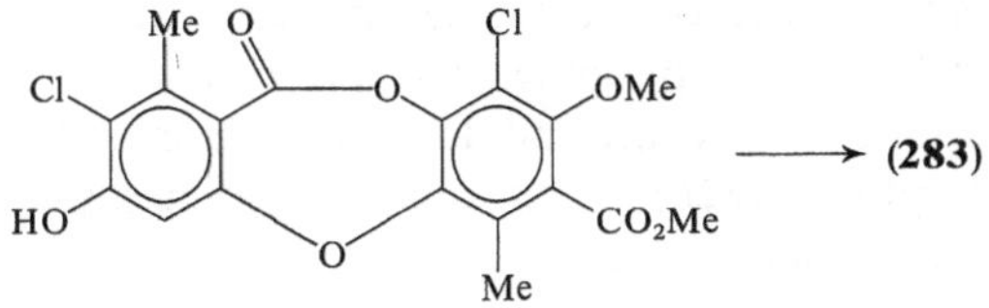

McEwen and Sargent (237) have confirmed both structures by synthesis. Their method is adapted from that of Hendrickson and his co-workers (160) and is of considerable generality as will be seen in the sequel. Hendrickson's group (160) subjected the benzophenone (311) to oxidation with alkaline ferricyanide and assumed that a grisadienedione intermediate (312) underwent hydrolytic vinylogous β-diketonic fission to yield a diphenyl ether carboxylic acid (313) which was not characterized but treated with hot acetic anhydride and yielded the depsidone acetate (314). This was subjected to mild hydrolysis with hot aqueous pyridine yielding the depsidone (315) which was convertible by chlorination into diploicin (284). Sala and Sargent (276) found, however, that the closely related benzophenone (316) on very brief alkaline ferricyanide oxidation gave the grisadienedione (317). They rationalized the direction of oxidative coupling in terms of the rate of oxidation being higher at the A-ring in the benzophenone (316). When the grisadienedione (317) was exposed to the basic reaction conditions for a longer period the depsidone dechlorodiploicin (318) resulted. A ketene was suggested as the intermediate in this rearrangement since the grisadienedione (319), in which phenolate ion formation is blocked, was stable under mild basic conditions. Both the grisadienediones (317) and (319) underwent thermal rearrangement to the depsidones (318) and (320) (O-methyl-dechlorodiploicin) which can be formulated as sigmatropic reactions. This general method has been used for lecideoidin (283). The benzophenone (321) on alkaline ferricyanide oxidation gave directly the depsidone (323), presumably *via* the grisadienedione (322). The depsidone (323) on demethylation with boron trichloride furnished lecideoidin (283). A similar route gave dechlorolecideoidin (307).

Diploicin (**284**) (*258*) is very similar in structure to gangaleoidin (**296**). The ring B carboxyl has been lost and all the nuclear positions not occupied by carbon or oxygen substitutents are chlorinated. It was one of the first chlorine containing natural products to be isolated. Ollis's synthesis (*30*) is of interest since it depends on the intramolecular phenolic oxidative coupling of a *para*-depside and may be closely related to the natural process. Attempts to effect similar oxidative couplings, however, have been fruitless (*60*). Seshadri and his co-workers (*248*) heated trichlorolecanoric acid (**324**) (*249*) (Scheme 39) with copper bronze in pyridine. The product (**325**), formed in poor yield, had undergone ring closure and decarboxylation and it was converted into O-methyldiploicin (**326**) by chlorination and methylation. Allusion has already been made to the diploicin synthesis of Hendrickson and his co-workers (*160*). Djura, Sargent, and Vogel (*107*) described a synthesis of diploicin (**284**) in which the diphenyl ether linkage was constructed by an Ullmann reaction; thus ring closure of the acid (**327**) with hot acetic anhydride gave the depsidone (**328**) which on chlorination with *N*-chlorosuccinimide in presence of toluene-*p*-sulfonic acid in boiling dioxan gave a trichlorocompound assigned structure (**329**) (*276*). Further chlorination of compound (**329**) was effected with chlorine in acetic acid and this afforded diploicin (**284**).

(**324**)   (**325**)

(**326**)

(**327**)   (**328**)

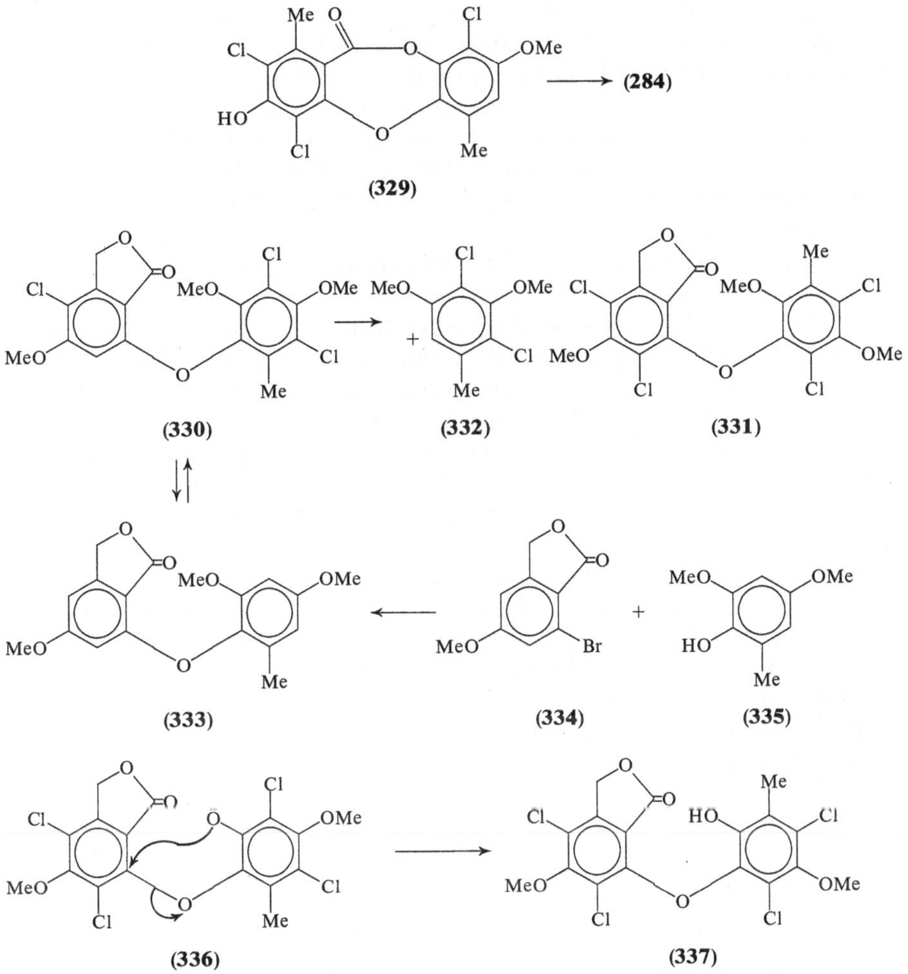

(329) ⟶ **(284)**

(330) → **(332)** + **(331)**

(333) ← **(334)** + **(335)**

(336) ⟶ **(337)**

Scheme 39. Chemistry of *Buellia canescens* metabolites

Recently SALA, SARGENT, and ELIX (*278*) have re-examined the lichen *Buellia canescens* from which NOLAN originally isolated diploicin (**284**). In addition to diploicin they isolated dechlorodiploicin (**318**) and its methyl ether (**320**) the structures of which followed by degradation and the synthesis mentioned previously. Other metabolites isolated from this source were the phenoxyphthalides buellolide (**330**) and canesolide (**331**). The molecular formula $C_{18}H_{15}Cl_3O_5$ of buellolide (**330**) was established by high resolution mass spectrometry and the base peak of the spectrum was attributed to the ion (**322**). The ^{1}H-n.m.r. spectrum of buellolide indicated the presence of an aromatic C-Me group, three methoxy groups, the methylene protons of a phthalide and a high field aromatic proton indicative of a tri-o-substituted diphenyl ether. Hydrogenolytic dechlorination of buellolide (330) yielded the diphenyl ether (333) which was synthesized in an Ullmann reaction from the components (334) and (335). Chlorination of the diphenyl ether (333) with sulfuryl chloride gave back buellolide (330). The structure of canesolide (331) followed from similar considerations.

Buellolide (330) and canesolide (331) probably arise biogenetically by catabolism of their congeneric depsidones. The name "pseudodepsidones" has been coined for this type of diphenyl ether (139). Thus dechlorodiploicin (318) or its methyl ether (320) would undergo fission of the depside linkage, oxidation of the methyl group at the 1-position, and O-methylation thus yielding buellolide (330). Canesolide (331), however, would arise from diploicin (284) by a similar sequence, but the fission of the depside linkage, in this case, must be accompanied by a Smiles rearrangement [(336)→(337)].

Mahandru and Gilbert (238) have isolated from Fulgensia fulgida a further member of the diploicin group of depsidones, fulgidin, for which they propose structure (340) (Scheme 40). No evidence was presented for the position of the chlorine on ring B so structure (329), synthesized by Sala and Sargent (276), is equally tenable. Sala and Sargent (276) also synthesized compound (340) by oxidative coupling of the benzophenone (338) and thermal rearrangement of the resultant grisadienedione (339). A direct comparison between the synthetic and natural material was not possible so that the structure of fulgidin is still in doubt.

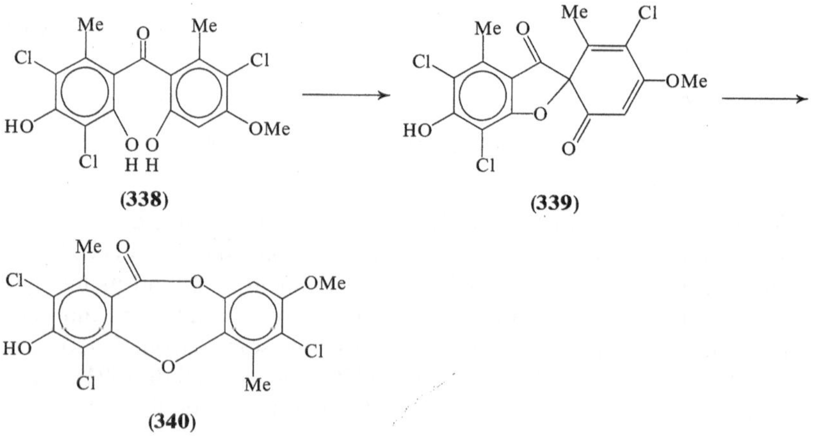

Scheme 40. Synthesis of fulgidin

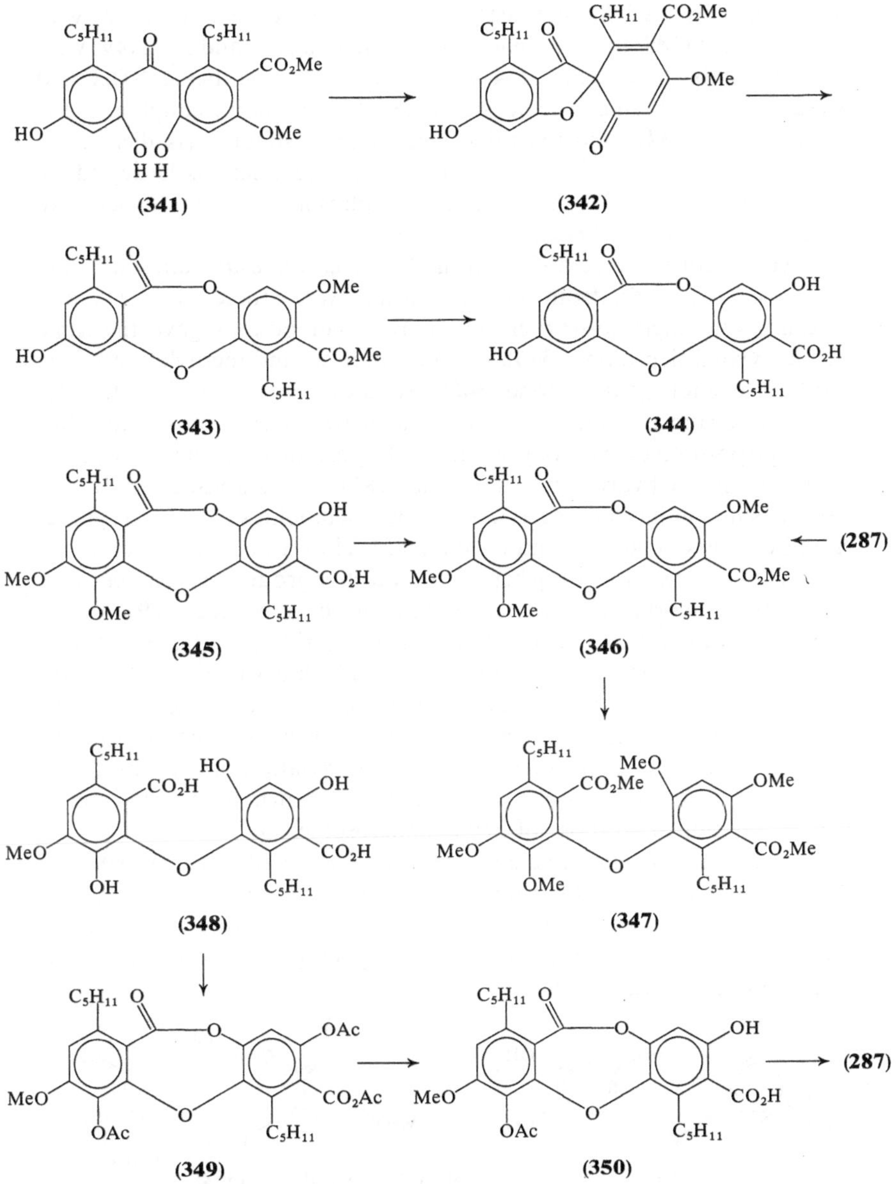

Scheme 41. Chemistry of the colensoic acids

Colensoic acid, from *Stereocaulon colensoi,* was assigned structure (285) by FOX, KLEIN, and HUNECK (*141*). This has been confirmed by DJURA and SARGENT (*102*) by a synthesis in which the diphenyl ether linkage was constructed by an Ullmann reaction. Norcolensoic acid co-occurs with colensoic acid (285) in a *Lecanora* species and CHESTER and ELIX (49) were able to establish structure (344) (Scheme 41) by synthesis. SALA and SARGENT (276) obtained the depsidone (343) by oxidative coupling of the benzophenone (341) and rearrangement of the resultant grisadienedione (342) under thermal, acidic, or basic conditions. Treatment of the depsidone (343) with lithium iodide in hot hexamethylphosphoric triamide then gave norcolensoic acid (344) (49).

Hydroxycolensoic acid (287) from *Parmelia formosana* and methoxy-colensoic acid (345), from *Parmelia livida* were studied by ELIX and SARGENT and their co-workers (106). Both compounds gave the same permethylated depsidone (346). On methanolysis and methylation of the resultant phenol the depsidone (346) gave a diphenyl ether (347) the ^{1}H-n.m.r. spectrum of which suggested that it was tetra-*o*-substituted thus fixing the position of extra oxygenation. The position of the phenolic group in the A-ring of hydroxycolensoic acid (287) was established by n.m.r. spectral shifts induced on acetylation. Both metabolites have been synthe-sized by DJURA, SARGENT, and CLARK (105). Thus the acid (348), obtained by an Ullmann reaction using benzyl as protective group, on treatment with hot acetic anhydride gave the crystalline mixed anhydride (349). Partial hydrolysis was achieved using hot aqueous pyridine and the resultant acetate (350) was cleaved with boron trichloride thus yielding hydroxycol-ensoic acid (287). A similar sequence gave methoxycolensoic acid (345). DJURA (100) has suggested, since methoxycolensoic acid co-occurs with colensoic acid, that the former arises by hydroxylation of the latter. This hypothesis appears plausible since fungi are known to possess mono-oxygenases similar to those of hepatic microsomes (9).

The structure of grayanic acid (351) has been confirmed by synthesis (101). It co-occurs with the ring-opened diphenyl ether congrayanic acid (352) in the lichen *Gymnoderma melacarpum* (47). Interestingly the dibenzo-furan melacarpic acid (353), a product of carbon-carbon oxidative coupl-ing, also occurs in this lichen.

(351)   (352)

(353)

Variolaric acid was assigned the unique structure (**354**) (Scheme 42), in which the diphenyl ether is *ortho* to the phenolic group of ring B, by NOLAN and his co-workers (*245*). This structure was confirmed by RANA, SARGENT, and ELIX (*270*) by synthesis of the degradation product (**355**). More recently JONGEN, SALA, and SARGENT (*200*) have synthesized variolaric acid from the synthetic depsidone (**356**). This was brominated under radical conditions and the crude product was subjected to hydrolysis by boiling in aqueous dioxan. The major product was assigned structure (**357**) on the grounds of its mass spectrum. Boron tribromide cleaved only one of the O-methyl groups yielding compound (**358**), which on treatment with lithium iodide in hot hexamethylphosphoric triamide gave variolaric acid (**354**) (Scheme 42).

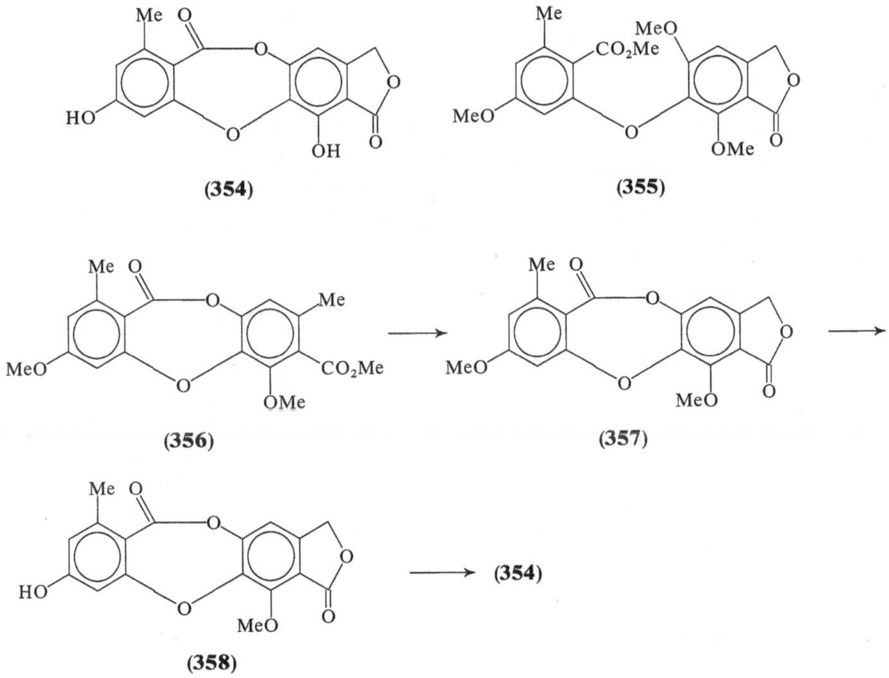

Scheme 42. Synthesis of variolaric acid

ELIX(*112*) and YOSIOKA and his co-workers (*163*) have reported the isolation and structural elucidation of the O-methylphysodic acid (**359**) from *Hypogymnia billardieri* and *H. vittata*, where it co-occurs with physodic acid (**295**). In this connection ELIX developed the useful technique of treating the mixture of lichen acids with phenyldiazomethane, which

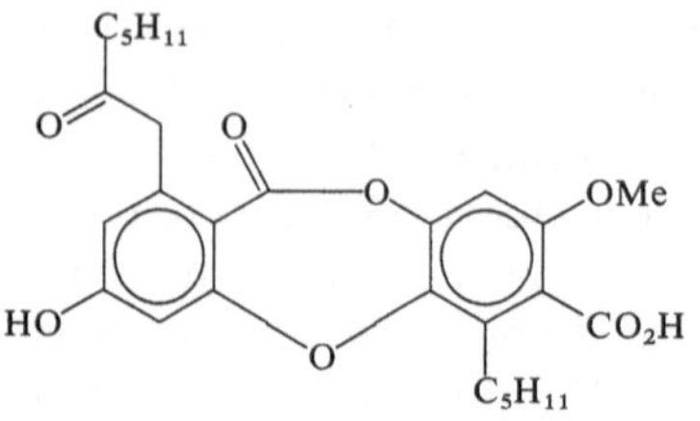

(359) 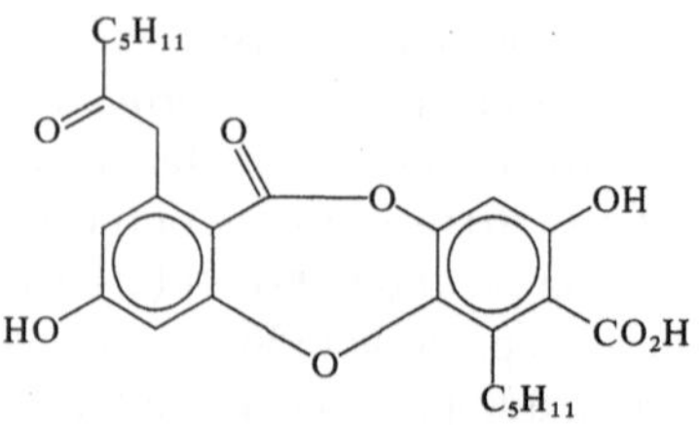   (295)

reacts with carboxy groups but not phenols, and then separating the benzyl
esters by chromatography. The purified esters can then be converted back
into the acids by hydrogenolysis.

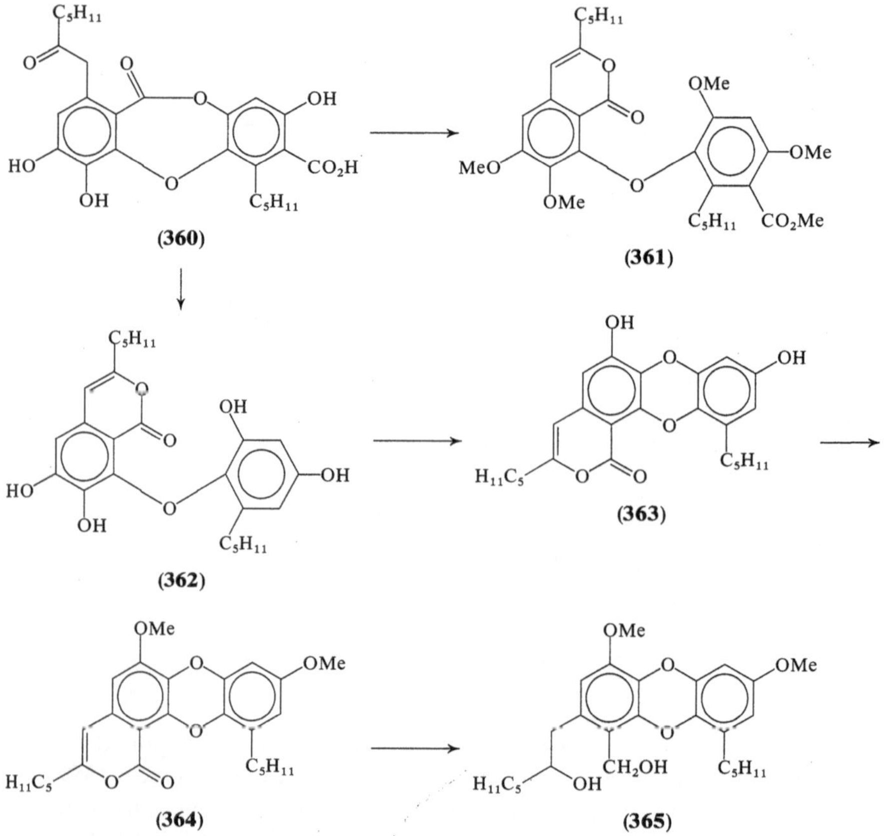

(360)   (361)

(362)   (363)

(364)   (365)

Scheme 43. Physodic acid derivatives

Hydroxyphysodic acid (**360**) (Scheme 43) occurs in the lichens *Hypogymnia enteromorpha* (*162*), *H. billardieri* (*112*), and *H. physodes* (*116, 302*). ELIX and his co-workers have produced convincing evidence for structure (**360**) although MOLHO and co-workers advanced an alternative structure (*158*). The mass spectrum of the new metabolite indicated that the site of extra hydroxylation was on ring A (*112*). On treatment with an excess of ethereal diazomethane hydroxyphysodic acid gave the diphenyl ether (**361**) which must be tetra-*o*-substituted from its ^{1}H-n.m.r. spectrum (*116*). With potassium hydrogen carbonate in boiling acetone hydroxyphysodic acid yielded the phenoxyisocoumarin (**362**) which was converted into the dioxin lactone (**363**) with boiling aqueous formic acid, a reaction mechanistically similar to the ring closure of 2,2′-dihydroxybiphenyls to dibenzofurans (*267*). Compound (**363**) was also formed on treatment of hydroxyphysodic acid under similar conditions. Structure (**363**) follows from an analysis of its long range ^{13}C-^{1}H coupling interactions (*199*) and by the conversion of its di-O-methyl ether (**364**) into the diol (**365**). The formation of the dioxin lactone (**363**) lends support to the location of the extra hydroxy group at the 4-position. YOSIOKA and his co-workers (*162*) also assigned structure (**360**) to hydroxyphysodic acid on the grounds of the observation of a nuclear Overhauser effect (8%) between the α-methylene protons of the side chain and the ring A aromatic proton in the ^{1}H-n.m.r. spectrum of the tri-O-acetyl derivative.

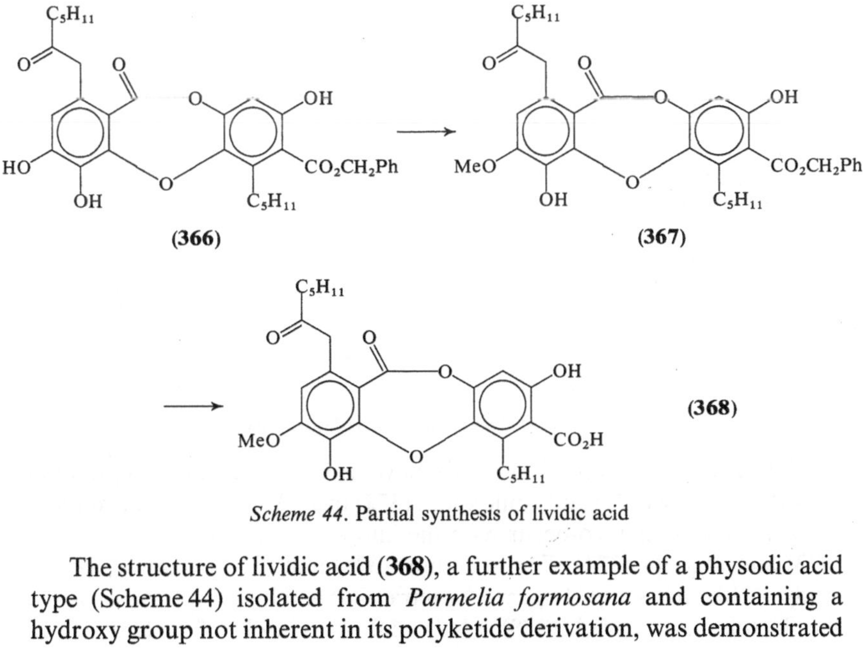

(**366**)   (**367**)

(**368**)

Scheme 44. Partial synthesis of lividic acid

The structure of lividic acid (**368**), a further example of a physodic acid type (Scheme 44) isolated from *Parmelia formosana* and containing a hydroxy group not inherent in its polyketide derivation, was demonstrated by Elix and Engkaninan (115) by inter-relation of its derivatives with those of hydroxyphysodic acid. The methylation pattern was demonstrated by judicious use of acetylation induced shifts in the n.m.r. spectra of its derivatives. Treatment of benzyl hydroxyphysodate (366) with diazomethane at room temperature gave benzyl lividate (367) which was converted into lividic acid (368) by hydrogenolysis.

Yosioka and his co-workers (163) isolated the interesting depsidone vittatolic acid (369) (Scheme 45) from *Hypogymnia vittata*. It is the first optically active depsidone, and its similarity to physodic acid (295) was easily recognized from its ^{1}H-n.m.r. spectrum. On reduction with zinc and hydrochloric acid in acetic acid vittatolic acid (369) gave physodic acid (295). The location of the side chain hydroxy group was fixed by double irradiation experiments and is at a position expected from the polyketide derivation of the metabolite. The absolute configuration of the chiral centre in vittatolic acid was determined by application of Horeau's method and Brewster's benzoate rule to the degradation product (370) obtained on treatment of vittatolic acid with diazomethane.

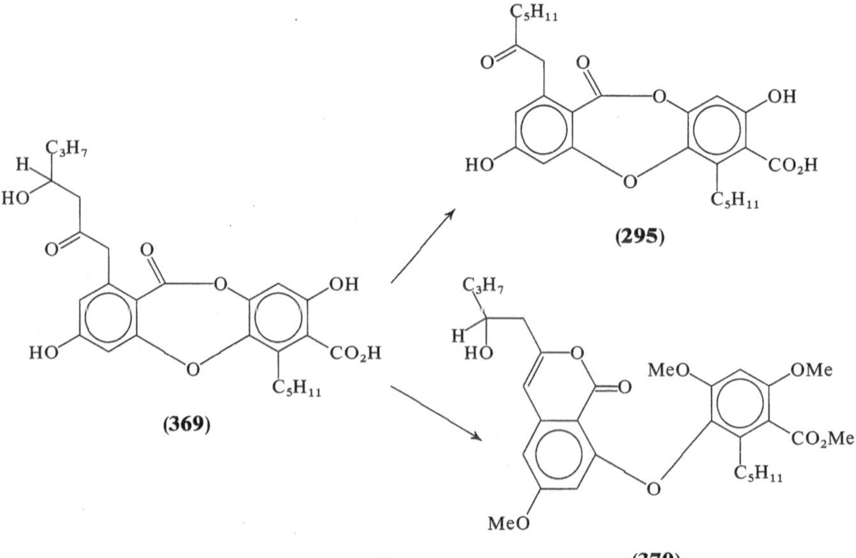

Scheme 45. Degradation of vittatolic acid

Begg, Chester, and Elix (13) have isolated the new depsidones conorlobaridone (371) and conloxodin (372) from *Xanthoparmelia xanthosorediata* where they co-occur with the known depsidones norlobaridone (373) and loxodin (374). The structures (371) and (372) followed from a study of their mass and ^{1}H-n.m.r. spectra. Foo and his co-workers isolated the lactols norlobariol (375) (140) and loxodinol (376) (139) from

X. scabrosa where they co-occur with norlobaridone (**373**) and loxodin (**374**). These pseudodepsidones, presumably derived by catabolism of their congeneric depsidones, were identified by their spectral properties. Norlobariol (**375**) was obtained from norlobaridone (**373**) by treatment with hot aqueous sodium hydroxide (*149*). Loxodinol (**376**) was similarly obtained from loxodin (**374**).

(371) (372) (373) (374) (375) (376)

ELIX, FERGUSON, and SARGENT (*121*) have studied the tautomerism of alectoronic acid (**377**), α-collatolic acid (**382**) and related depsides. The ^{1}H-n.m.r. spectrum of alectoronic acid (Scheme 48) at ambient temperature exhibited a sharp two proton signal ($\delta\,4.06$) due to the benzylic methylene protons of the A-ring side chain and a broad two proton signal (4.05 – 3.10) due to the protons H_A and H_B. At $-40°$ C the signal for H_A and H_B appeared as an AB pattern at $\delta\,3.33$ and 3.70 ($J\,16.0$ Hz). This behaviour is explained by rapid tautomeric exchange of H_A and H_B at ambient temperature, but at $-40°$ C the rate is slower. α-Collatollic acid (**382**) parallels this behaviour. Brief treatment of alectoronic acid at room temperature with diazomethane trapped the more acidic normal tautomer affording the normal esters methyl alectoronate (**380**) and methyl α-collatolate (**381**), but treatment of alectoronic acid with methanol containing a catalytic amount of sulfuric acid trapped the lactol tautomer and gave predominantly the pseudo-esters (**378**) and (**379**).

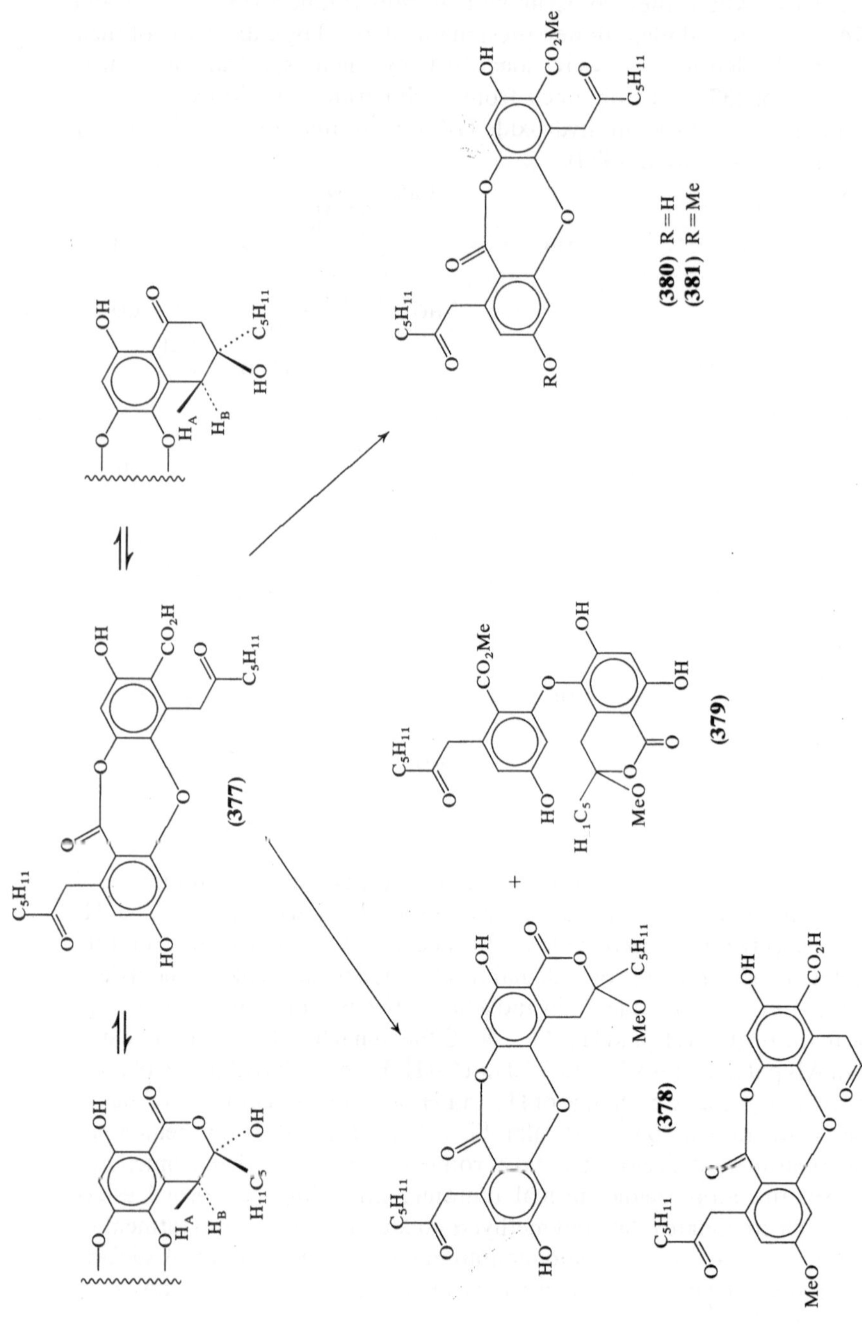

Scheme 46. Tautomerism of alectoronic acid

11.3 β-Orcinol Depsidones

The simplest of the β-orcinol depsidones, hypoprotocetraric acid (288), has been synthesized by SALA and SARGENT (277) (Scheme 47). Oxidative coupling of the benzophenone (383) with alkaline ferricyanide gave the depsidone (384). Demethylation of this with boron trichloride gave methyl hypoprotocetrarate (385) which was de-esterified using lithium iodide in hot hexamethylphosphoric triamide. CRESP et al. (61) isolated O-methyl-hypoprotocetraric acid (386) from Parmelia notata; the methylation pattern followed from the mass spectrum and the structure was confirmed by the conversion of both O-methylhypoprotocetraric acid (386) and hypoprotocetraric acid (288) into methyl O-methylhypoprotocetrarate (387). The di-O-methyl ether (388), obtained by methylation of the synthetic depsidone (384), was treated with boron trichloride and the resultant methyl O-methylhypoprotocetrarate (387) on de-esterification gave SALA and SARGENT (277) synthetic O-methylhypoprotocetraric acid (386). An earlier synthesis (60) gave a poorer overall yield.

YOSIOKA and co-workers (338) isolated the known depsidone vicanicin (292) from a Caloplaca species. They were unable to obtain an authentic sample so they confirmed structure (292) by spectroscopy and by subjecting the ethyl ether (389) to the nitric acid oxidation technique of DEAN et al. (98) which afforded the identifiable fragments (392) and (393) (Scheme 48). SARGENT and ELIX and their co-workers (288) isolated vicanicin (292) from Psoroma sphinctrinum; another chemical strain produced norvicanicin (391), also available by boron tribromide induced demethylation of vicanicin (292). Vicanicin was synthesized (276) from the benzophenone (394) which on oxidation afforded the grisadienedione converted by exposure to base or acid into the depsidone (395) and thence by chlorination into vicanicin (292).

Virensic acid (289) has been synthesized by SALA and SARGENT (277) (Scheme 49). Radical bromination of the depsidone (396), obtained by acetylation of the synthetic depsidone (387), and mild hydrolysis of the product gave the hydroxymethyl compound (397) identified by its charac-teristic mass spectral ring A ion. Oxidation with pyridinium chloro-chromate converted this product into the aldehyde (398), which on treatment with lithium iodide in hot hexamethylphosphoric triamide gave synthetic virensic acid (289). Methyl virensate (399) has been isolated from the lichens Pseudocyphellaria granulata and P. faveolata by GOH and WILKINS (146), and has been obtained from natural virensic acid (289) by methylation (1). Physciosporin (chlorogranulatin) (400) also occurs in Pseudocyphellaria species, and is formed on chlorination of methyl virensate (399) (146). Its structure is also supported by its mass spectral fragmentation and by its conversion by hydrogenation into methyl hypoprotocetrarate (385) (235).

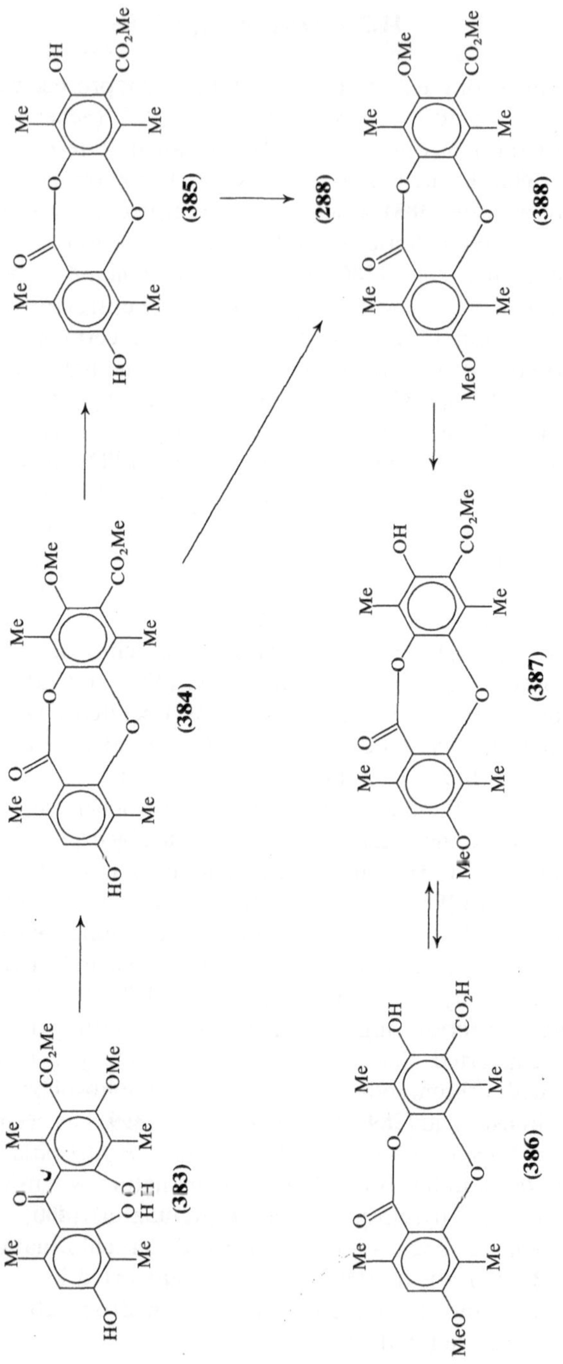

Scheme 47. Synthesis of hypoprotocetraric acid and O-methylhypoprotocetraric acid

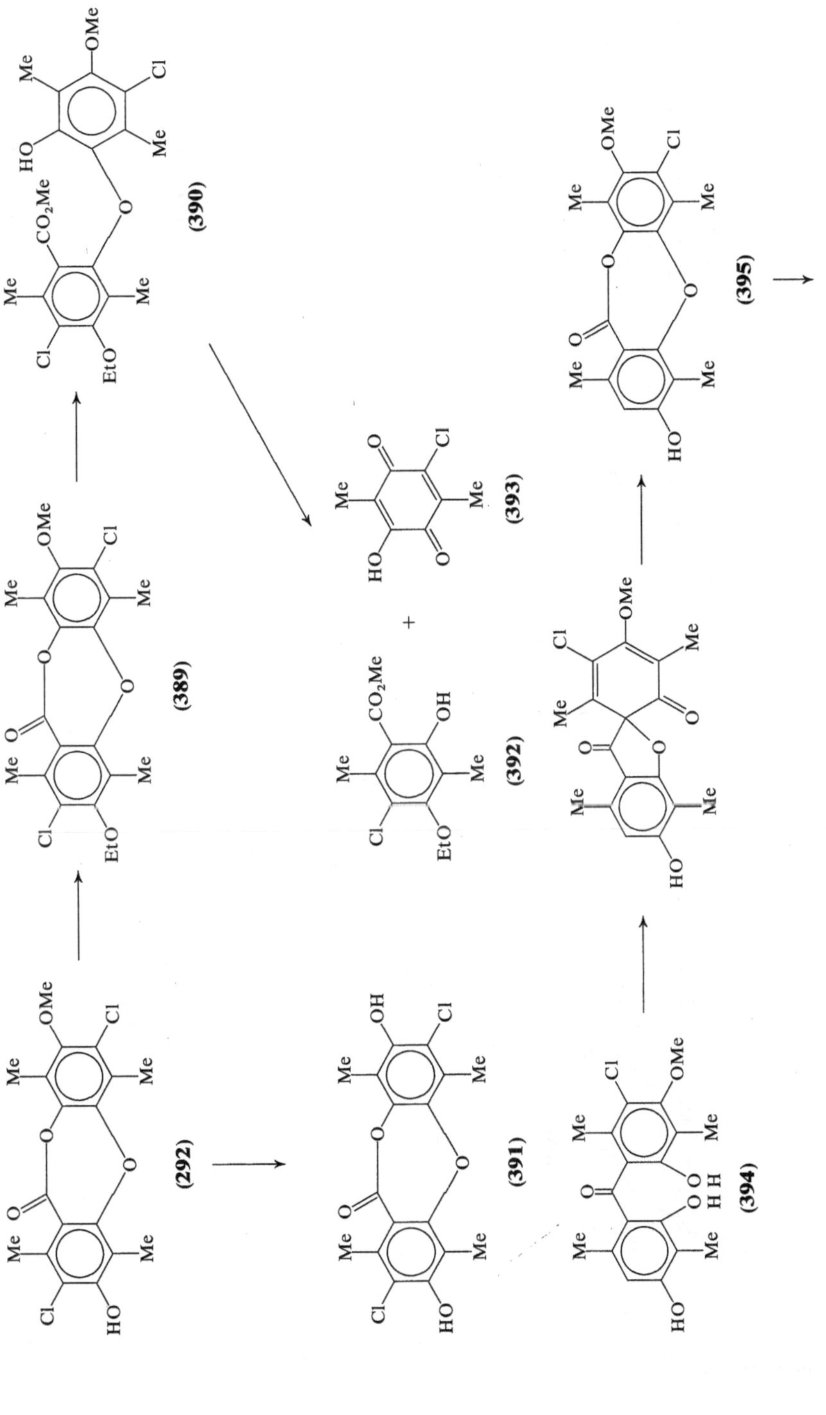

Scheme 48. Degradation and synthesis of vicanicin

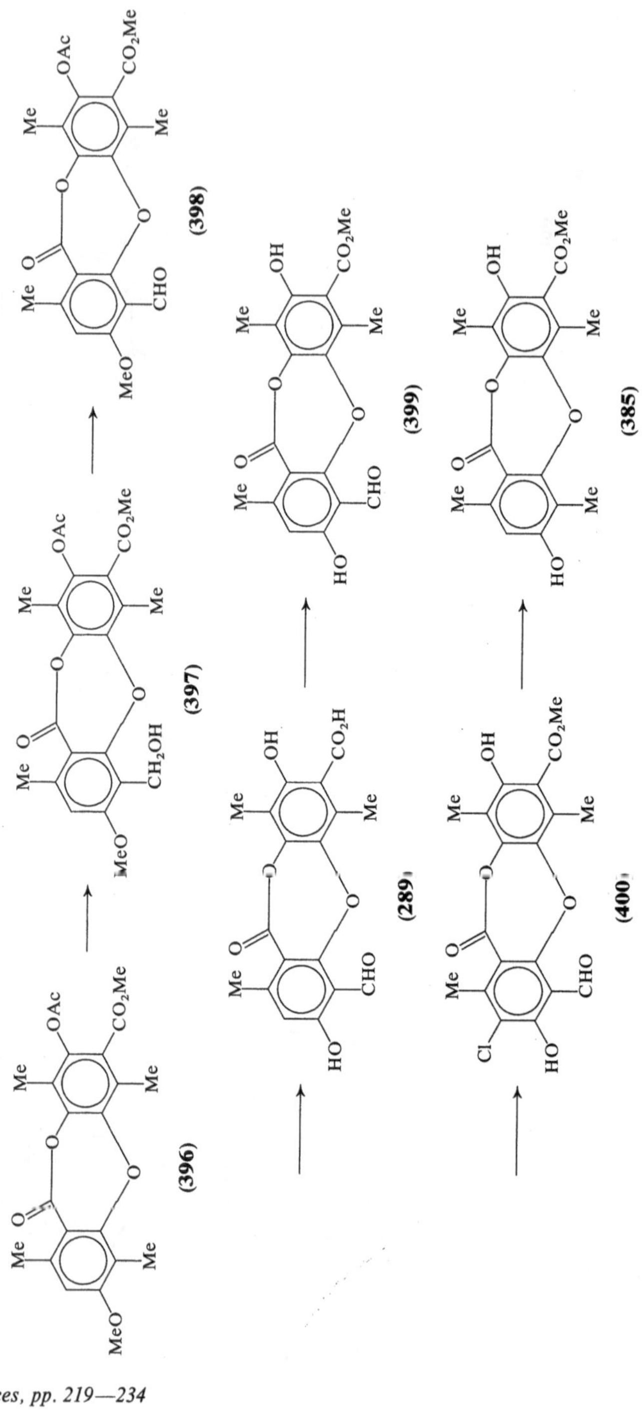

Scheme 49. Virensic acid and related depsidones

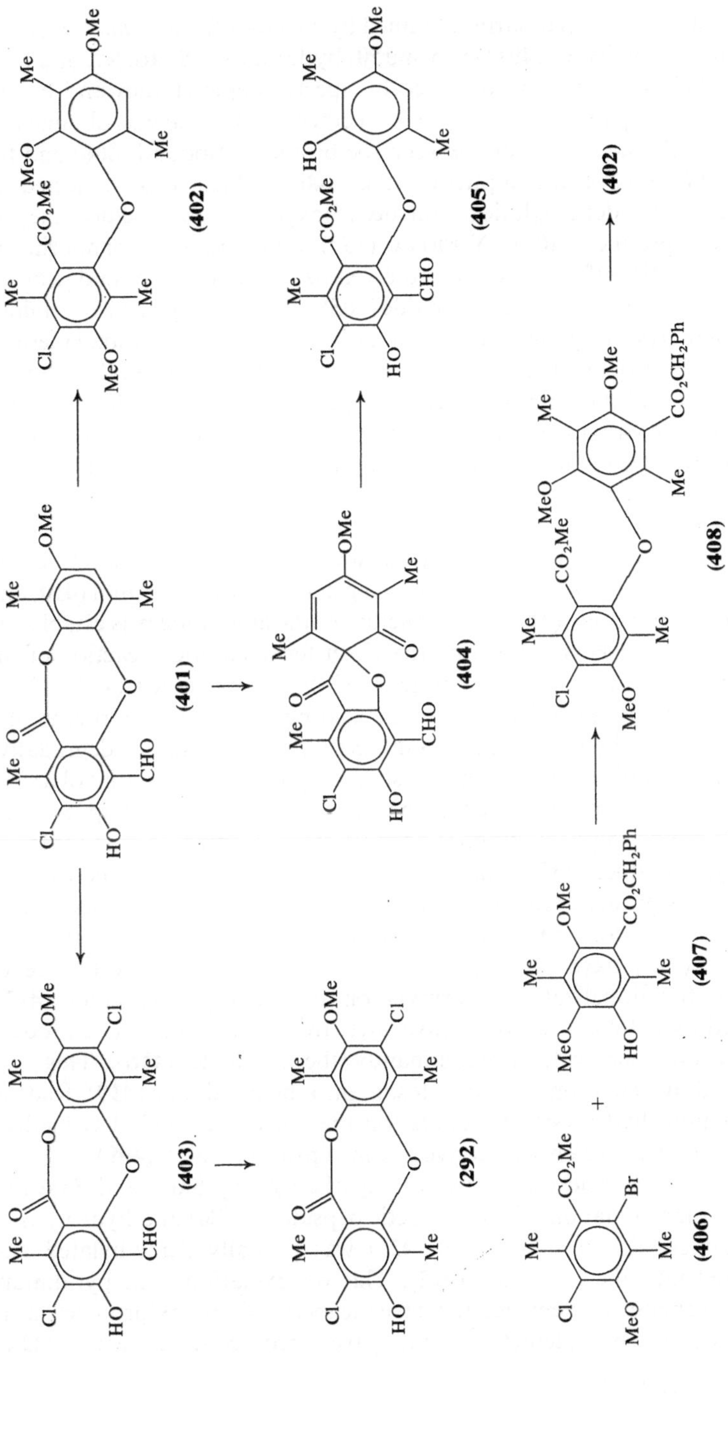

Scheme 50. Chemistry of pannarin

The structure of pannarin, obtained by Yosioka from *Pannaria* species (*337*), was amended to (**401**) (Scheme 50) by Jackman, Sargent, and Elix (*198*). Their structural proof relied on spectroscopic studies and on the degradation of pannarin to the diphenyl ether (**402**) which was obtained by synthesis. Ullmann reaction between the bromo-compound (**406**) and the phenol (**407**) gave the diphenyl ether (**408**) which was subjected to hydrogenolytic debenzylation and decarboxylation and yielded the degradation product (**402**). Yosioka (*337*) observed that pannarin on treatment with sulfuric acid yielded a yellow isomeric compound, isopannarin, which gave pannarin methoxide (**405**) on methanolysis, also obtained by methanolysis of pannarin. Yosioka (*337*) proposed a quinone methide structure for isopannarin, but the grisadienedione structure (**404**) is more likely in view of the reported electronic spectrum. Argopsin (**403**), isolated by Bodo and Molho (*19*), and by Huneck and Lamb (*184*) from *Argopsis friesiana* was also obtained by chlorination of pannarin (*184*). The French workers also converted the new depsidone into vicanicin (**292**) by Clemmensen reduction.

Dean (*95*) cast doubt on structure (**409**) (Scheme 51) for psoromic acid since pyrolysis of this compound in soda glass (*264*) yields atranol (**410**) and the phthalic anydride (**411**). The structure of the anhydride was established by Asahina and Hayashi (*6*) by the violet ferric chloride reaction of the derived ester (**412**) and by synthesis of its methyl ether (**413**). Dean proposed structure (**414**) for psoromic acid but the original structure has been vindicated by Huneck and Sargent (*187*) as a result of degradative and spectroscopic studies. Doubtlessly the pyrolysis reaction involves the intermediacy of the grisadienedione (**414**). Methyl O-methylpsoromate (**415**) on methanolysis undergoes a Smiles rearrangement yielding (**416**) as the major product (*187*). The reaction is favoured by the three electron withdrawing groups at the positions *ortho* to the diphenyl ether linkage and produces the more resonance stabilized phenolate anion. Sala and Sargent (*275*) have synthesized psoromic acid by their technique of selective functionalization. Formylation of the diphenyl ether (**417**), obtained by Ullmann reaction, gave, after treatment of the crude product with boron trichloride, predominantly the aldehyde (**418**). This was converted by oxidation and ring closure into the depsidone (**419**). Radical bromination, hydrolysis and oxidation then afforded methyl O-methyl-psoromate (**415**) which was converted into psoromic acid (**409**).

Protocetraric acid (**424**) has been synthesized by Sala and Sargent (*277*) by dibromination of the synthetic depsidone (**420**) and hydrolysis of the product. The resultant diol (**421**) was partially demethylated and converted into the acetonide (**422**). This on oxidation with pyridinium chlorochromate and deprotection gave the methyl O-methylprotocetrarate (**423**) which on demethylation gave protocetraric acid (**424**).

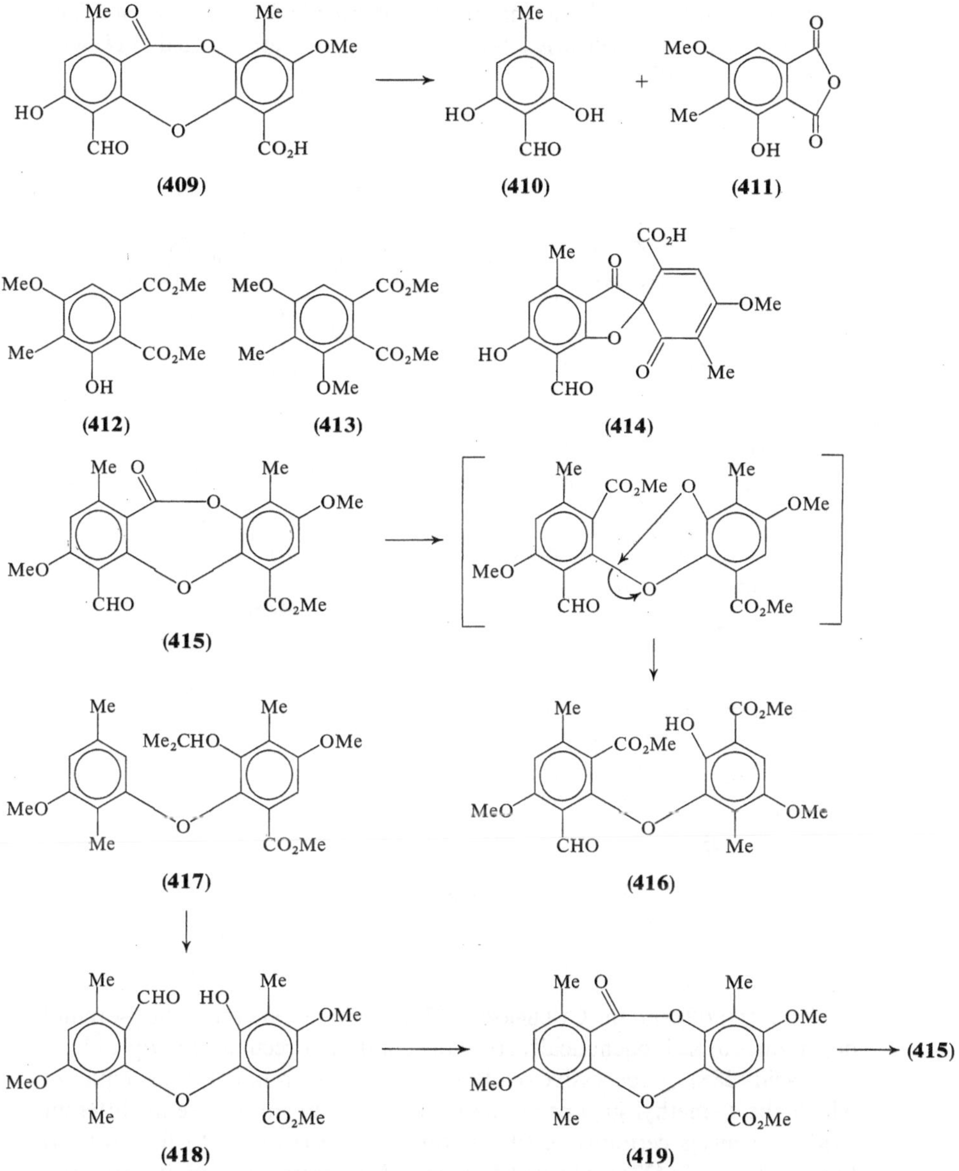

(409) (410) (411)

(412) (413) (414)

(415) (416)

(417)

(418) (419) (415)

Scheme 51. Degradation and synthesis of psoromic acid

Succinprotocetraric acid, from *Parmelia reptans* and *Chondropsis semi-viridis* (*11*), and malonprotocetraric acid from *Parmotrema conformatum* (*208*), are derivatives of protocetraric acid (**424**) in which the benzyl alcohol function is esterified with succinic and malonic acids. Keogh (*208*) obtained synthetic malonprotocetraric acid by boiling protocetraric acid (**424**) with malonic acid in dioxan.

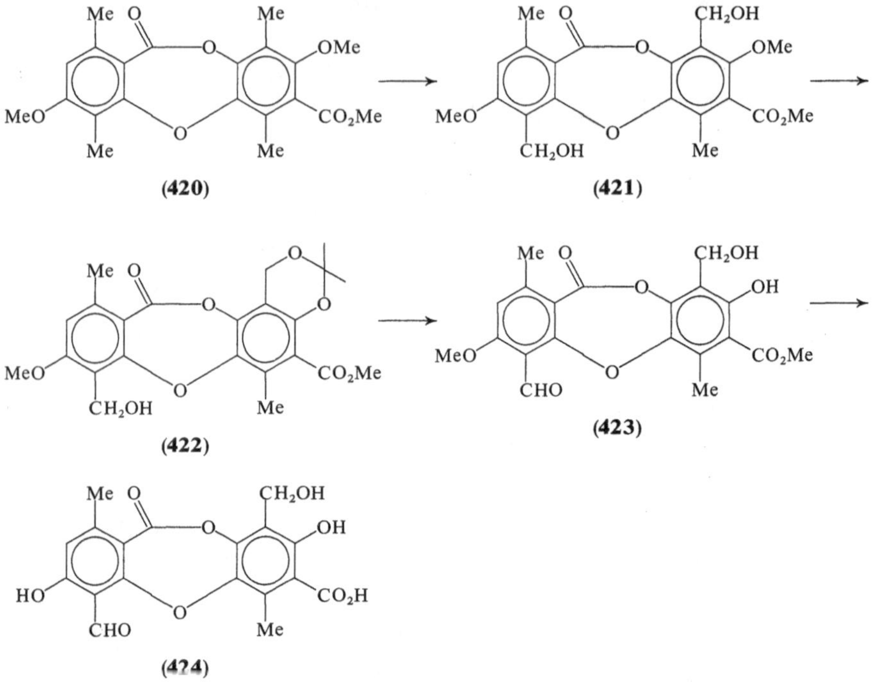

Scheme 52. Synthesis of protocetraric acid

Keogh (*209*) and the Culbersons (*73*) have suggested that the β-orcinol depsidones arise biogenetically from the common precursor hypoprotocetraric acid (**288**) by successive oxidations. The large number of examples in which the *C*-methyl groups at the 4-, 6- and 9-positions are at different oxidation levels certainly testify to this view. Keogh (*209*) has isolated hypostictic acid (**427**) (Scheme 53) from a *Thelometra* species. Its structure followed by reduction of natural stictic acid (**426**) (*8*). In *Xanthoparmelia quintaria* hypostictic acid (**427**) co-occurs with hyposalazinic acid (**431**), identified by cochromatography with a sample obtained by reduction of natural salazinic acid (**291**). Shibata and his co-workers (*295*) isolated cryptostictic acid (**425**) and methylstictic acid (**428**) from *Lobaria oregana*.

Again the structure of cryptostictic acid (**425**) was confirmed by partial reduction of stictic acid (**426**). KEOGH and co-workers (*263*) have isolated two further depsidones with a CH_2OH function in the 4-position. Thus connorstictic acid (**430**) was obtained from *Pertusaria pseudocorallina* and consalazinic acid (**433**) from *Parmotrema subisidiosum*. The structure of these compounds was deduced from the spectroscopic properties and confirmed by catalytic reduction of norstictic acid (**429**) to connorstictic acid (**430**), and salazinic acid (**291**) to consalazinic acid (**433**) over platinum. Connorstictic acid (**430**) was also isolated from *Lecidea aspidula* and the structure confirmed independently by ELIX and LAJIDE (*125*). Hypoconstictic acid (**434**) has been isolated from *Nephroma antarcticum* by WILKINS *et al.* (*243*) where it co-occurs with hypostictic acid (**427**). This compound was separated as the corresponding triacetate (**435**) and the structure verified by catalytic reduction of constictic acid triacetate (**436**) whereupon (**435**) was obtained.

Galbinic acid (**432**) from *Usnea undulata* is an acetyl derivative of salazinic acid (**291**) from which it can be prepared by heating it in acetic acid containing *N,N*-dimethylformamide (*114*).

(**437**)

Menegazziaic acid (**437**) was isolated from *Menegazzia asahinae* and *M. terebrata* by YOSIOKA and his co-workers (*164*). The similarity of menegazziaic acid (**437**) to stictic acid (**426**) was apparent from its ^{1}H-n.m.r. spectrum except that a formyl proton signal was absent. Menegazziaic acid possessed one carbon atom less than stictic acid and the mass spectrum supported the location of a hydroxy group on ring A. Structure (**437**) was convincingly confirmed by Baeyer-Villiger oxidation of stictic acid diacetate (**438**) which yielded the phenol (**439**) readily converted by acetylation into menegazziaic acid triacetate (**440**) (Scheme 54). DJURA (*100*) has suggested, in an extension of the theory of KEOGH mentioned above, that since menegazziaic acid (**437**) co-occurs with stictic acid (**426**) the latter undergoes the biological equivalent of a Baeyer-Villiger oxidation and is therefore the precursor of menegazziaic acid.

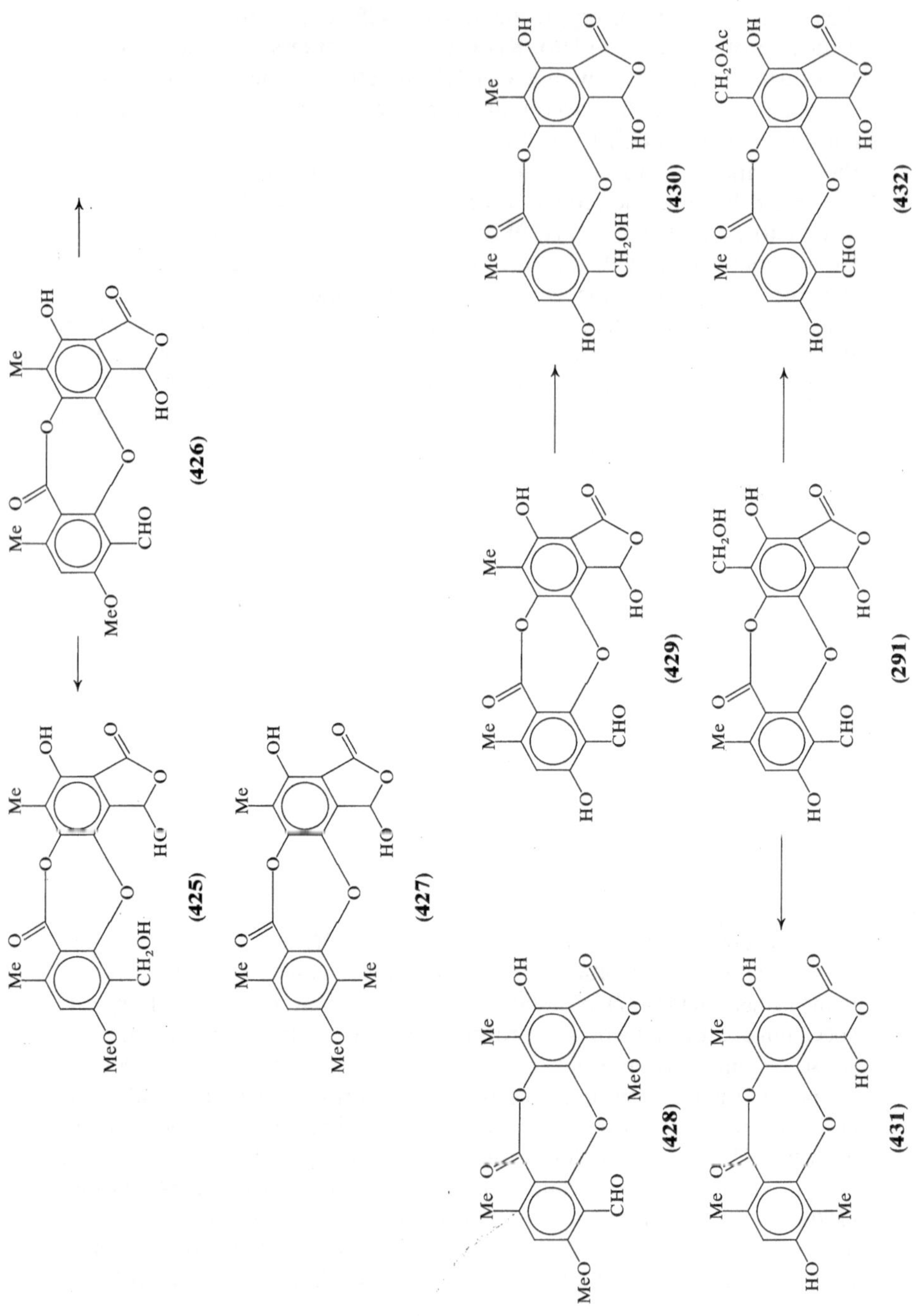

(426)

(425)

(427)

(430)

(432)

(429)

(291)

(428)

(431)

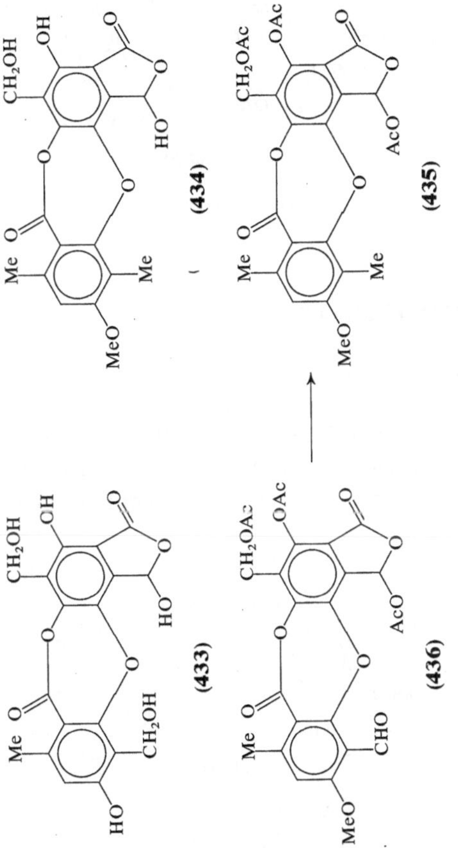

Scheme 53. Reactions of stictic acid and related depsidones

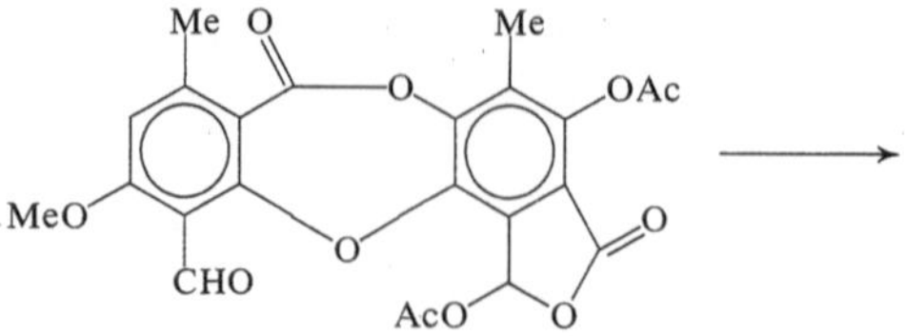

(438)

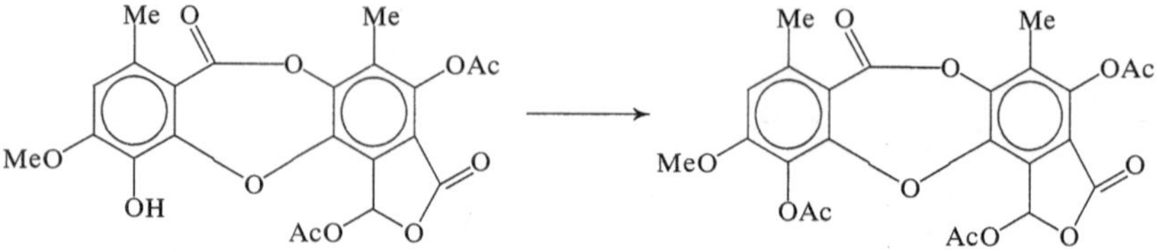

(439)   (440)

Scheme 54. Synthesis of menegazziaic acid triacetate

11.4 Mixed Depsidones

The biogenetically simplest of the third class of depsidones, those in which one benzenoid ring is derived from orsellinic acid and one from β-orcinol carboxylic acid, is nornotatic acid (293) (Scheme 55). T.l.c. evidence was presented by CULBERSON and HALE (82) for its occurrence in many lichens. They advanced structure (293) on the basis of the chromatographic

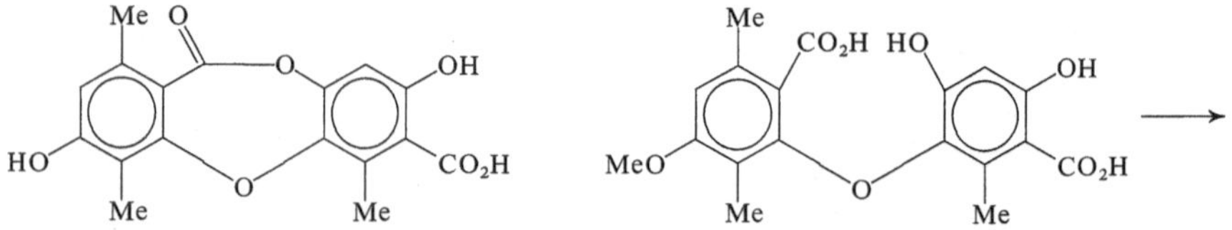

(293)   (441)

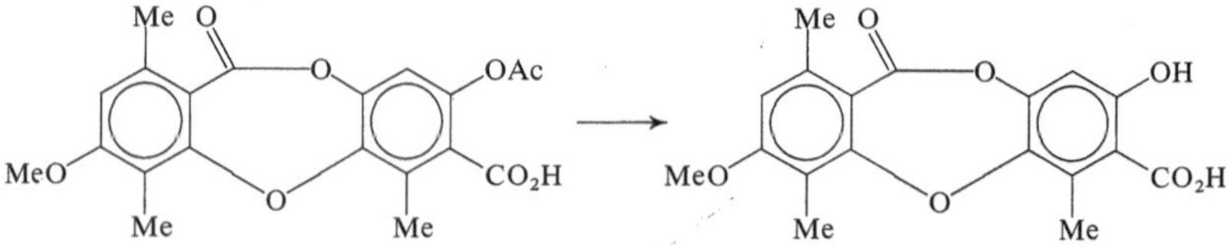

(442)   (443)

Scheme 55. Synthesis of notatic acid

behaviour of its diazomethane methylation products. DJURA and SARGENT (*103*) synthesized the compound of structure (**293**) by a route in which the diphenyl ether linkage was formed by an Ullmann reaction and proved its identity with nornotatic acid isolated from *Parmelia weberi*. Notatic acid (**443**) (*61*) often co-occurs with nornotatic acid (**293**) (*82*). Of the syntheses that have been described (*60, 103*) the best is that of DJURA and SARGENT (*103*) in which the synthetic diphenyl ether (**441**) was cyclized with hot acetic anhydride and the resultant depsidone acetate (**442**) was hydrolyzed to notatic acid (**443**).

Caloploicin was assigned structure (**444**) (*338*) which has been confirmed by synthesis (*276, 286*). Fulgoicin (**445**) is a dechlorocaloploicin (*238*).

(444)   (445)

11.5 Fungal Depsidones

In contrast to the lichen depsidones those of fungi are not widely distributed and as yet they are confined to *Aspergillus nidulans, A. unguis* which is morphologically indistinguishable from *A. nidulans* (*307*), and *Chaetomium mollicellum*. They all belong to the "mixed" series of depsidones and are briefly mentioned here because of some structural features of biosynthetic interest. *A. nidulans* grown on a Czapek-Dox medium produces nidulin (**446**) as a major metabolite (*96*) as well as the minor metabolites O-nornidulin (**447**) (*98*) and dechloro-O-nornidulin (**448**) (*97*). The structures of these metabolites were established by classical methods whereby fragments due to both rings A and B and their pendant substituents were identified (*39*). When grown on a Czapek-Dox medium free from chloride *A. nidulans* produces tridechloro-O-nornidulin (*202, 297, 307*), the structure (**449**) of which followed from degradative experiments (*202*), and from its general similarity to nidulin (*297, 307*). *A. unguis* also produces four other minor depsidone metabolites (*203*). SARGENT and his co-workers (*104, 137*) have confirmed the structures of nidulin (**446**) and tridechloro-O-nornidulin (**449**) by the synthesis of their dihydro derivatives by methods based on the oxidative coupling of appropriate benzophenones.

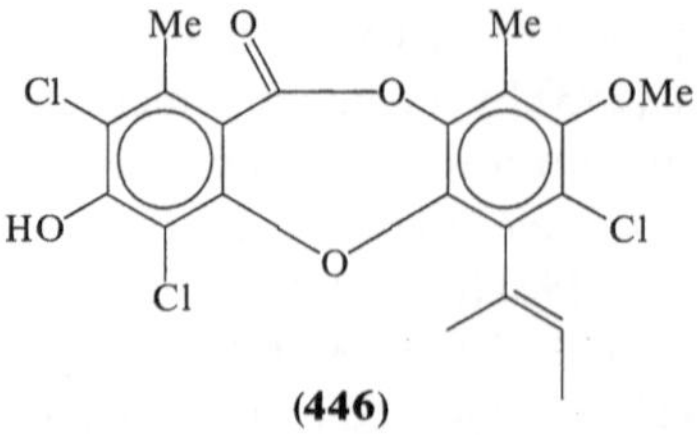

(446)

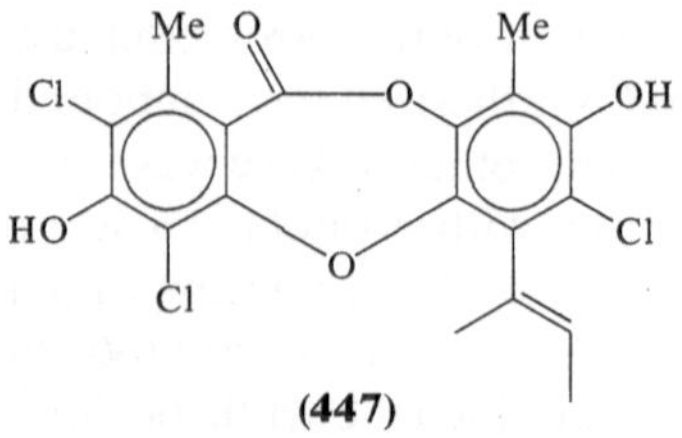

(447)

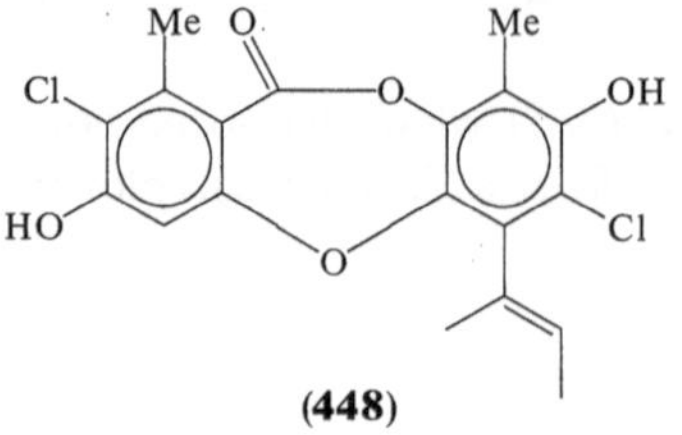

(448)

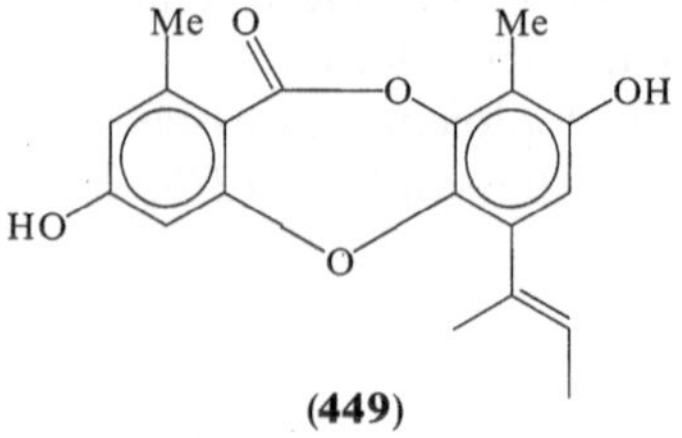

(449)

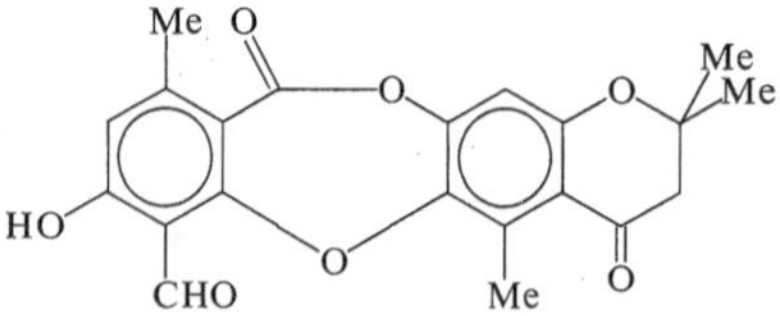

(450)

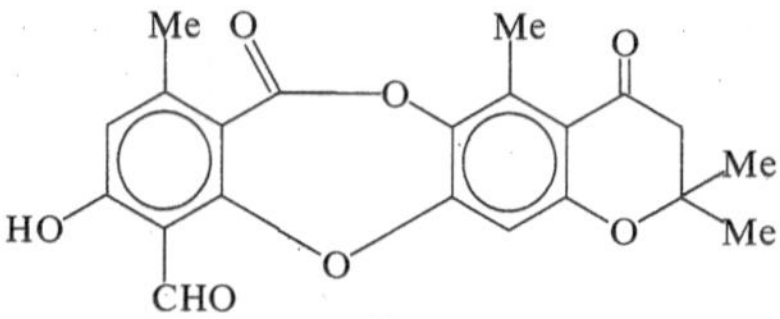

(451)

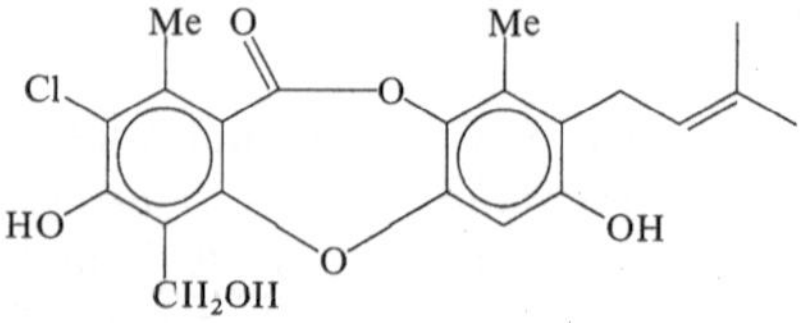

(452)

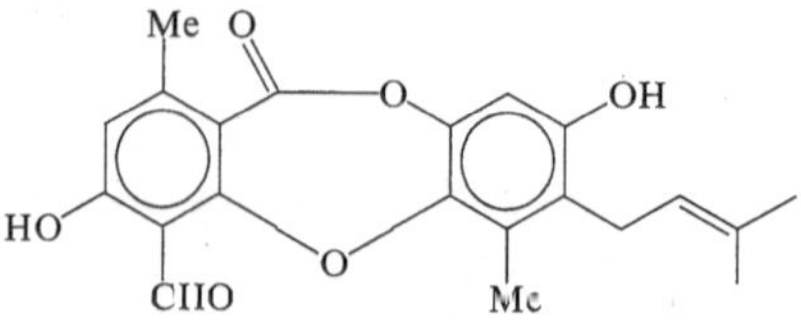

(453)

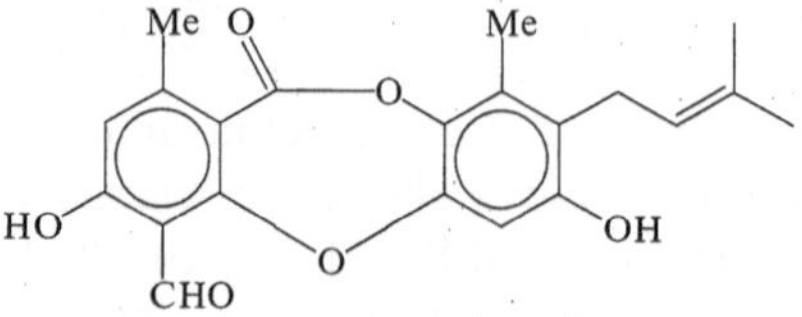

(454)

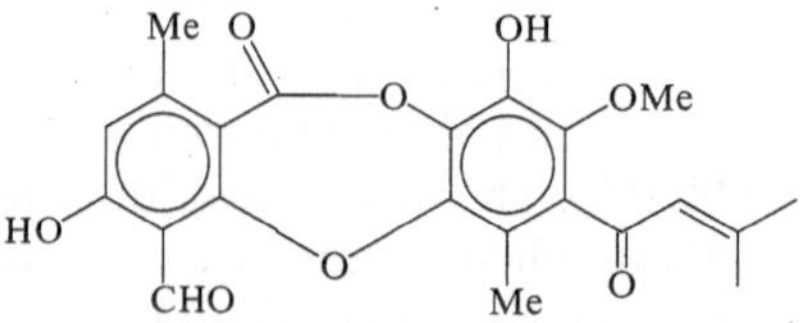

(455)

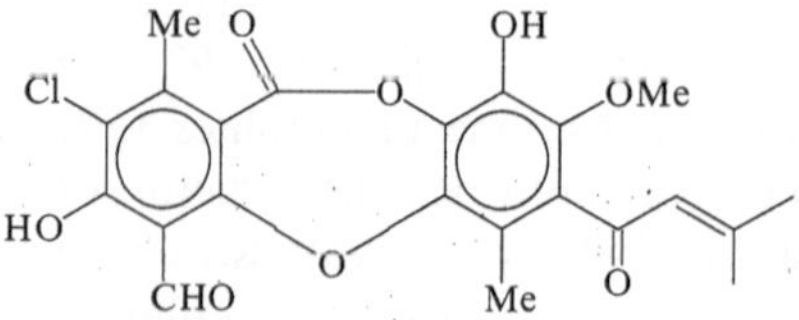

(456)

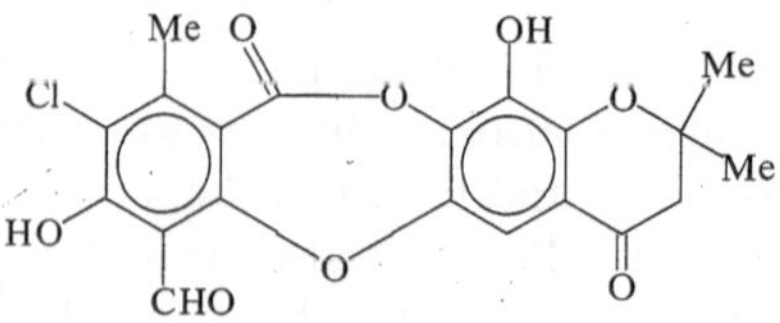

(457)

SIERANKIEWICZ and GATENBECK (*298*) have studied the biosynthesis of tridechloro-O-nornidulin (**449**) using ^{13}C-n.m.r. spectroscopy, and have shown that it is formed, as expected, from two polyketide moieties. The ring A orsellinic acid moiety is derived from one acetate unit and two malonate units. The ring B moiety is formed from one acetate unit and four malonate units, and in addition two one carbon units located as the methyl group at the 9-position and in the butenyl side chain adjacent to the ring.

BÜCHI and WILLIARD (*38*) have isolated eight toxic new depsidones from *Chaetomium mollicellum*. The structures of mollicellin A (**450**) and B (**451**) were established by X-ray crystallography and the structures of three others were demonstrated by chemical correlations of D (**452**), G (**453**), and H (**454**) with A and B. The remaining metabolites are tentatively regarded as possessing structures C (**455**), E (**456**), and F (**457**). Mollicellins A and B yield the same methanolysis product so that methanolysis of B proceeds with Smiles rearrangement. Mollicellin A has been synthesized. These metabolites encompass many of the secondary modifications encountered in the lichen depsidones: the C_1 substituent at the 4-position occurs as formyl or hydroxymethyl in D (**452**), chlorination in D (**452**), E (**456**), and F (**457**) possess extra hydroxyl groups in ring B (in the lichen depsidones these are always in ring A). In addition mevalonoid side chains are present in place of the usual ring B carboxyl group — the dihydropyrone ring of mollicellin A may well be derived by cyclization of the 8-hydroxy group of a precursor on to the terminus of the double bond of a senecioyl side chain at the 7-position. Mollicellin G may be a precursor of mollicellin H since lactone opening of G, Smiles rearrangement, and ring closure would yield H (*276*).

11.6 Biosynthesis of Depsidones

SALA and SARGENT (*276*), in view of their *in vitro* experiments where the oxidative coupling of benzophenones readily yielded grisadienediones which rearranged to depsidones under mild acidic or basic conditions, have suggested that this pathway might occur *in vivo*. The lichen or mould would presumably use benzoylbenzoic acids such as (**458**) (Scheme 56) which is envisaged to be involved in the biosynthesis of diploicin (**284**). The rearrangement of the grisadiendione, *e.g.* (**459**), would presumably occur on an enzyme surface such that the ring-C carbonyl group is protonated, or only the ring-A phenol is ionized since it would be expected that ionization of the phenol and carboxylic acid on ring C would cause this ring to be a poor leaving group. Further transformations of the resultant depsidones such as decarboxylation, chlorination, O-methylation, C-prenylation, hydroxylation, or oxidation of methyl substituents as in the β-orcinol depsidones are unexceptional. No benzophenone has ever been isolated from a lichen but they are obviously involved in the biosynthesis of the fungal grisans and the xanthones and may be regarded as reactive and short lived intermediates.

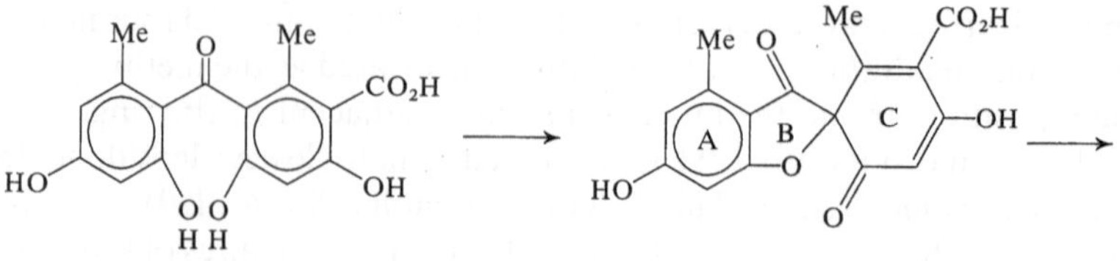

(458)   (459)

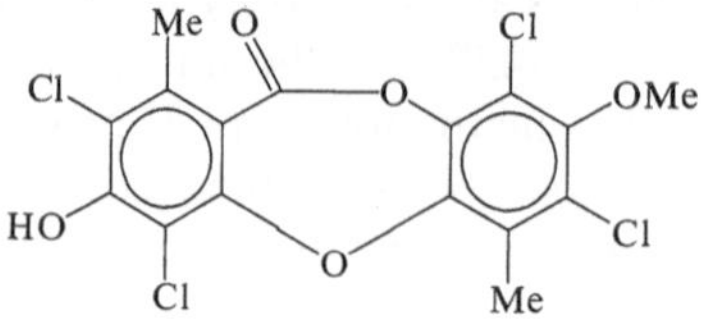

(284)

Scheme 56. Proposed biosynthesis of diploicin

Sala and Sargent (276) postulate that the benzophenone precursors to depsidones would arise by acylation of one orsellinic acid type by another. Thus p-depside and depsidone biosynthesis would diverge at the mononuclear stage. A decision between the Barton-Erdtman theory (12, 135) and that of Sala and Sargent would require the feeding of appropriate advanced precursors. This is probably the most challenging experiment yet to be carried out in the area of depsidone chemistry.

12. Dibenzofurans and Biogenetically Related Compounds

12.1 Dibenzofurans

Only two new dibenzofurans have been isolated from lichens in the past decade. Melacarpic acid (460) was isolated from Gymnoderma melacarpum and the structure established from the ^{1}H-n.m.r. and mass spectra (47). The

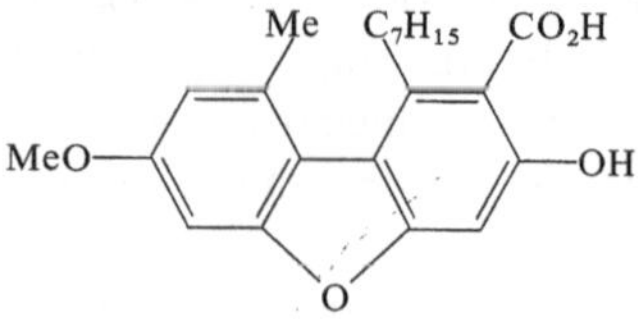

(460)

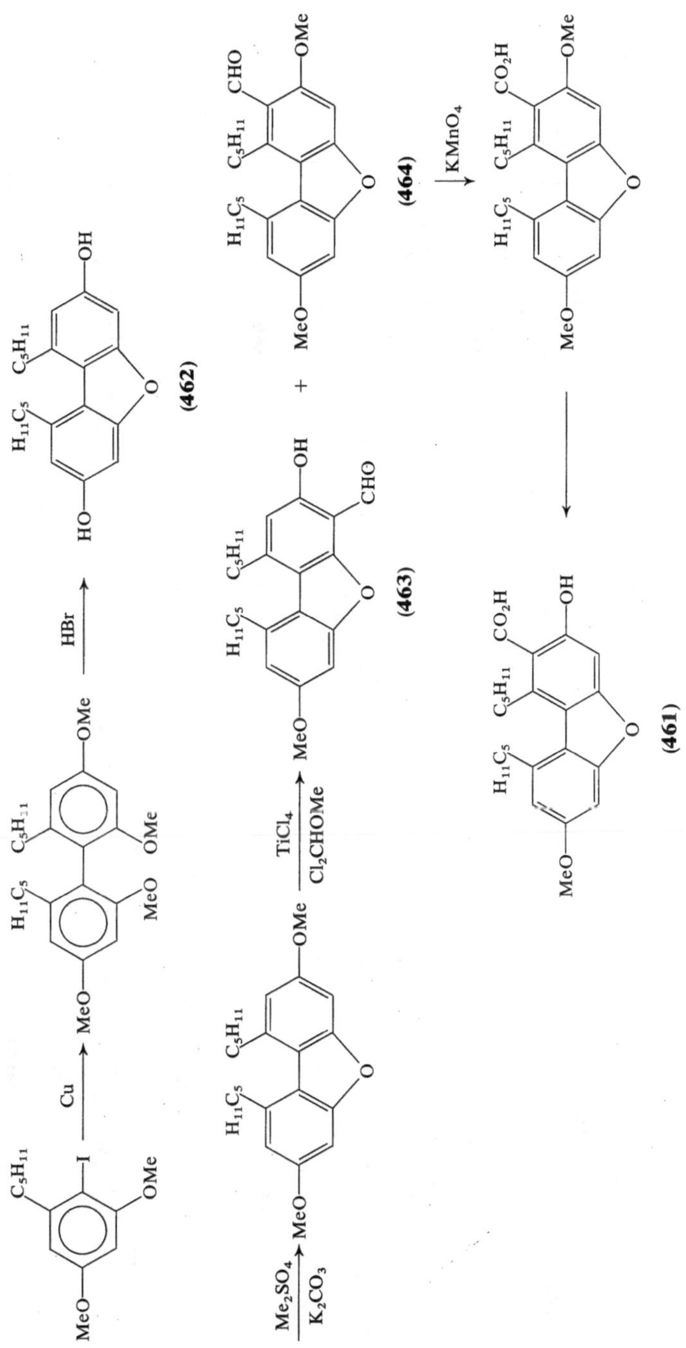

Scheme 57. The synthesis of condidymic acid

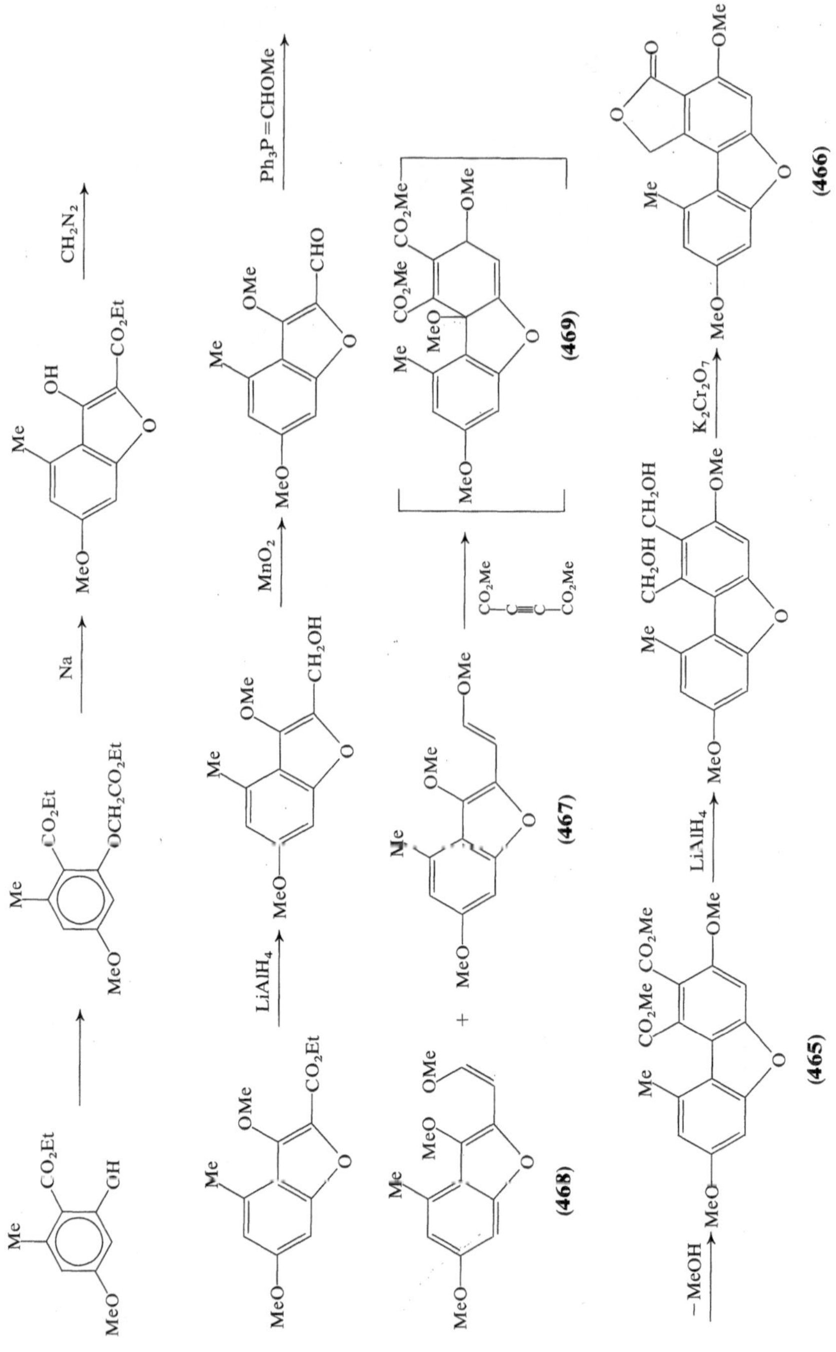

Scheme 58. Synthesis of di-O-methylstrepsilin

structure of condidymic acid (**461**), isolated from *Cladonia squamosula* (*48*), was confirmed by total synthesis (*48*) (Scheme 57). Ullmann reaction of 2-iodo-di-O-methylolivetol and subsequent cyclodehydration/demethylation of the product with hydrobromic acid gave the symmetrical dibenzofuran intermediate (**462**). Methylation followed by formylation yielded a mixture of the 4-formyl and 2-formyl dibenzofurans (**463**) and (**464**), which were separated chromatographically. 3,7-Dimethoxy-1,9-dipentyl-2-formyldibenzofuran (**464**) was then oxidized and selectively demethylated to yield condidymic acid (**461**).

Following a preliminary publication (*27*), BREWER and ELIX (*28*) reported a second synthetic route to dimethyl 3,7-dimethoxy-9-methyl-dibenzofuran-1,2-dicarboxylate (**465**) (Scheme 58), an intermediate in the synthesis of di-O-methylstrepsilin (**466**).

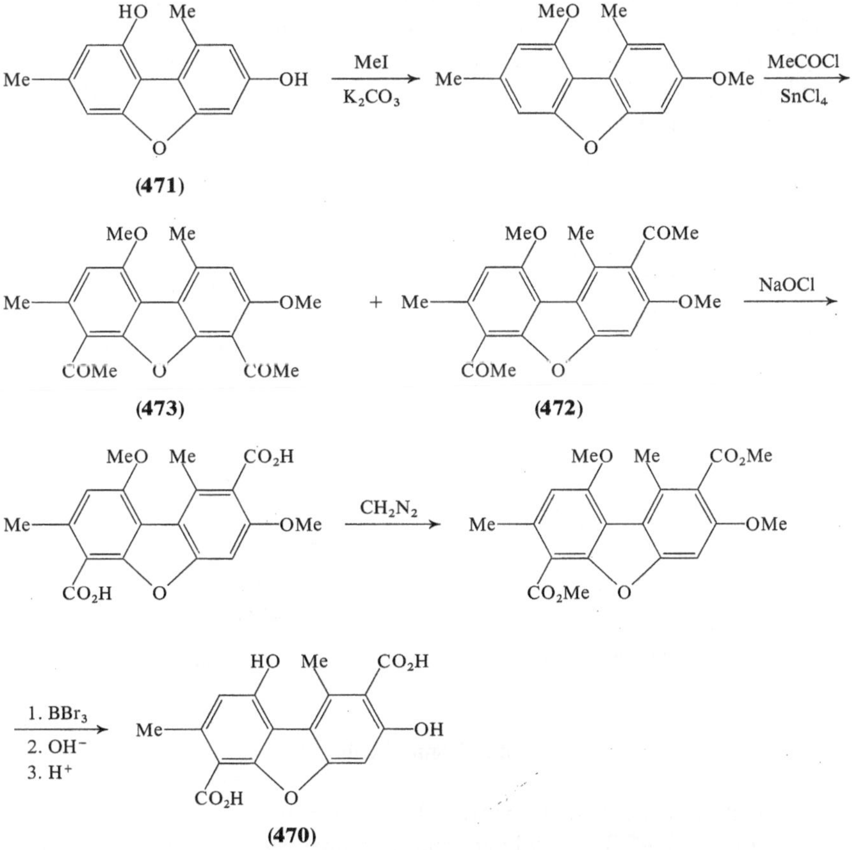

Scheme 59. Synthesis of pannaric acid

As in the previous route (27), the key step in this synthesis was the Diels-Alder reaction between the diene (467), chromatographically separated from the Z-isomer (468), and dimethyl acetylenedicarboxylate. The presence of the 3-methoxy group in (467) allows elimination of methanol from the Diels-Alder adduct (469) to give the aromatized dibenzofuran (465) as the isolated product. This eliminates the bromination/dehydrobromination steps of the original synthesis (27).

The synthesis of the lichen dibenzofuran pannaric acid (470) from pannarol (471) was reported by ELIX (110) (Scheme 59). Pannarol has been synthesized via a mixed Ullmann reaction and subsequent hydrobromic acid induced cyclisation of the unsymmetrical biphenyl separated from the mixture of three products (3). Pannarol (471) was methylated and then acetylated to yield a mixture of 2,6-diacetyl-3,9-dimethoxy-1,7-dimethyl-dibenzofuran (472) and 4,6-diacetyl-3,9-dimethoxy-1,7-dimethyldibenzo-furan (473), which were separated by preparative t.l.c. The major product (472) was then oxidized and demethylated to yield pannaric acid (470).

LELE and HOSANGADI (227) obtained dibenzofuran (474) from depsidone (281) (11H-dibenzo[b,e][1,4]dioxepin-11-one) by irradiation (Scheme 60). These authors postulated that lichen dibenzofurans could originate biosynthetically from depsidones by a similar photodecarboxylation reaction. However such a photodecarboxylation would not give the substitution pattern of the natural dibenzofurans, hence biosynthesis through phenolic coupling and cyclodehydration is a more plausible route. ELIX and MURPHY (127) undertook similar work, photocyclizing 2-methoxyaryl aryl ethers obtained from lichen depsidones, but again the dibenzofurans produced contained an "unnatural" substitution pattern.

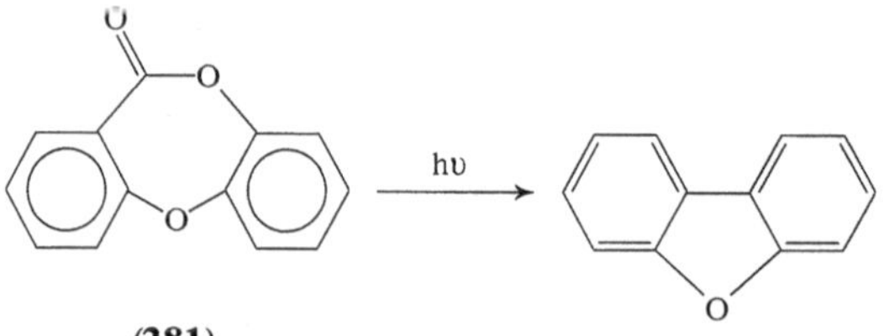

(281)

(474)

Scheme 60. Synthesis of dibenzofuran from depsidone

12.2 Usnic Acids

Two new usnic acid derivatives have recently been described. (−)-Placodiolic acid (475) was isolated from Lecanora rubina by HUNECK (173). The structure was established spectroscopically (u. v., m.s. and ^{1}H-n.m.r.)

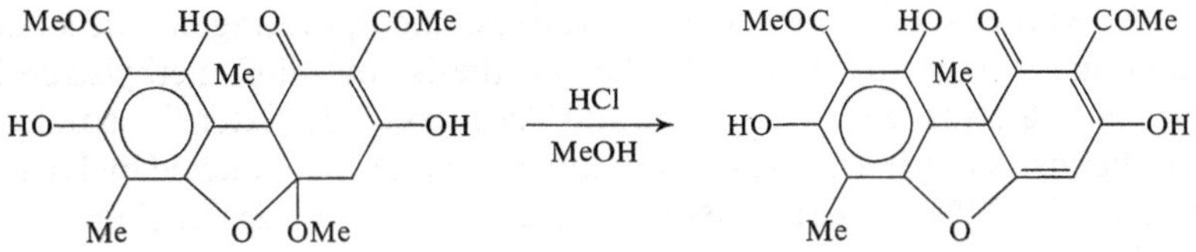

(475)   **(476)**

Scheme 61. Conversion of (−)-placodiolic acid to (−)-isousnic acid

and confirmed by conversion of **(475)** to (−)-isousnic acid **(476)** (Scheme 61). Similarly (−)-pseudoplacodiolic acid **(447)**, isolated from *Rhizoplaca chrysoleuca* was structurally identified and converted to (−)-usnic acid **(478)** upon treatment with acid (*178*) (Scheme 62). The absolute configuration of **(477)** was determined by X-ray analysis of the (−)-α-phenylethylimine derivative.

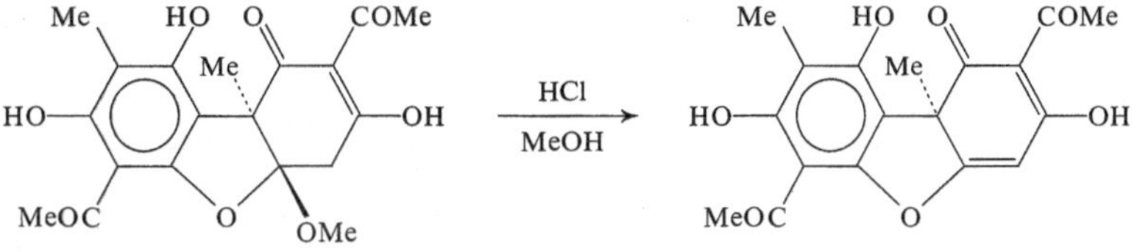

(477)   **(478)**

Scheme 62. Conversion of (−)-pseudoplacodiolic acid to (−)-usnic acid

Three-dimensional crystal structures of usnic acid and 2-desacylusnic acid were determined by X-ray diffraction (*236, 259*), but it was not until 1981 that HUNECK *et al.* (*178*) established the absolute configurations of the usnic acids. X-Ray diffraction of the (−)-α-phenylethylimine of (+)-usnic acid established the configuration 4a(*R*) for this isomer. By utilising the relationship of (+)-usnic acid and (+)-isousnic acid to (−)-dihydrousnic acid (Scheme 63), (+)-isousnic acid must also possess a 4a(*R*) configuration.

$$(+)\text{-isousnic acid} \xrightarrow{\text{H}_2} (+)\text{-isodihydrousnic acid}$$

$$\downarrow \Delta$$

$$(+)\text{-usnic acid} \xrightarrow{\text{H}_2} (-)\text{-dihydrousnic acid}$$

Scheme 63. Interrelation of (+)-usnic and (+)-isousnic acids

A synthesis of usnic acid derivatives, incorporating a Diels-Alder reaction similar to that of the synthesis of di-O-methylstrepsilin (Scheme 58), was reported by Elix and Tronson (*133*). Extensive studies on the chemistry of usnic acid and derivatives have been conducted by Kutney *et al.* (*217—225*) and Takahashi *et al.* (*315—321, 324, 326*). Included in the work of Kutney *et al.* was a base-catalysed usnic acid-isousnic acid rearrangement (*224, 225*), which constituted the first synthesis of isousnic acid (**476**) (Scheme 64). A mechanism for this rearrangement was proposed (*221*).

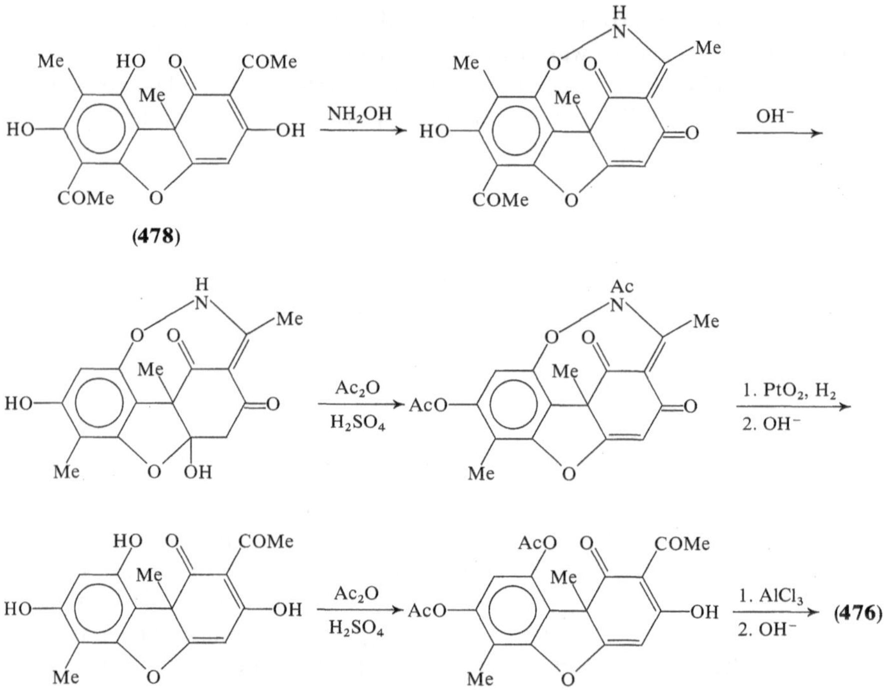

Scheme 64. Base catalysed usnic acid-isousnic acid rearrangement

Kutney, Sanchez, and Yee (*222*) recorded and reported the mass spectra of usnic acid and some derivatives and included a detailed interpretation of these spectra.

Chance and Tibbetts (*42*) reported a specific method for identifying usnic acid by t.l.c., and Fahselt (*136*) successfully identified usnic acid from the gas-liquid chromatogram of a crude lichen extract without purification or derivatization.

12.3 The Diphenyl Ether Leprolomin (479)

Leprolomin (**479**), isolated from *Psoroma leprolomum* (*118*), is a unique lichen diphenyl ether. Unlike other lichen diphenyl ethers which appear to be derived by hydrolysis or further modification of depsidones [cf. norlobariol (**375**)], leprolomin is apparently derived from phenolic coupling of two methylphloroacetophenone units and hence would appear to be biogenetically related to the usnic acids. From spectroscopic data of leprolomin and derivatives, six possible structures were considered, the unique structure being determined by X-ray diffraction of leprolomin triacetate (*118*).

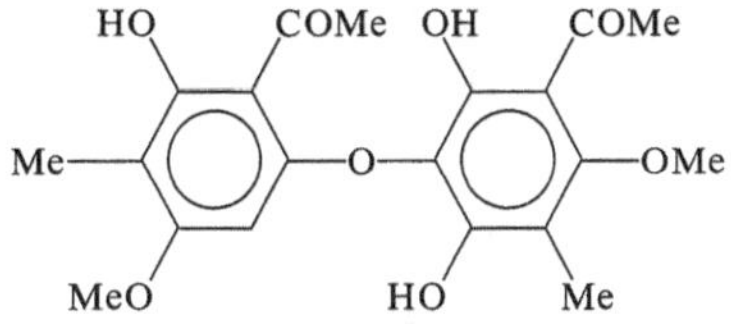

(479)

13. Mevalonic Acid Derivatives

13.1 A Sesterterpene

Retigeranic acid (**480**), isolated from the *Lobaria retigera* and related species, is the first example of a lichen sesterterpene. As sufficient information could not be obtained from routine spectral data (*205*), the structure (and absolute configuration) of (**480**) was determined by X-ray diffraction of the *p*-bromoanilide derivative of retigeranic acid (*204*). A possible biosynthetic route to (**480**) has also been outlined (*205*).

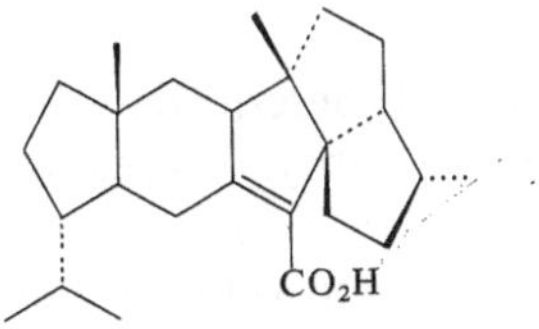

(480)

13.2 Triterpenoids

13.2.1 Hopanes

YOSIOKA et al. (342) have published a concluding statement concerning the stereostructure of the widely distributed lichen metabolite, zeorin (481). From X-ray analysis of 6-O-p-bromobenzoylzeorin the configuration of C-21 has been established as C-21βH (246). This confirmed that zeorin, like leucotylin (482), was a hopane derivative with C-21βH configuration.

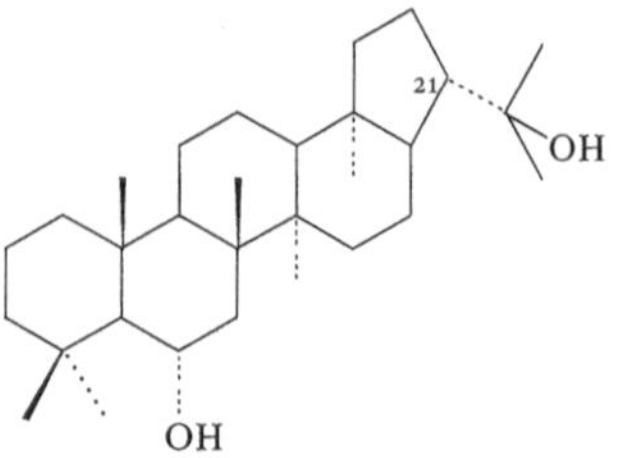

(481)

The acid catalysed epimerization of the C-21 group of leucotylin (482), to give isoleucotylin (483) in addition to leucotylidiene (484) has been reported (342) (Scheme 65). Methyl leucotylate is known to react in a similar manner (343, 341).

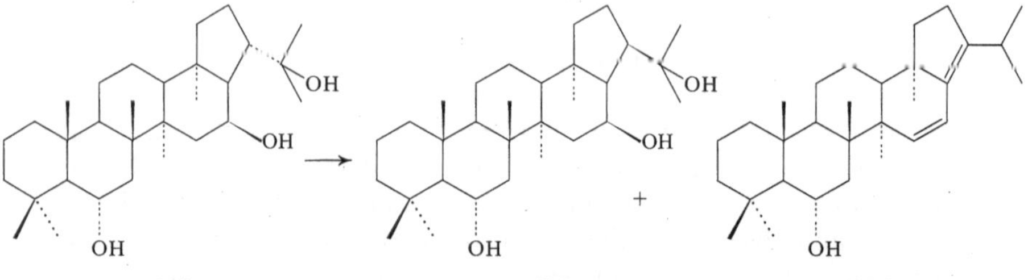

(482)　　　　　　(483)　　　　　　(484)

Scheme 65. Action of acid on leucotylin

In addition to the previously described pyxinic acid (485) (339), YOSIOKA et al. (345) have isolated methyl pyxinate (486) and methyl 3-O-acetylpyxinate (487) from Pyxine endochrysina. The structure of these esters was verified by direct comparison with the synthetic derivatives obtained from pyxinic acid.

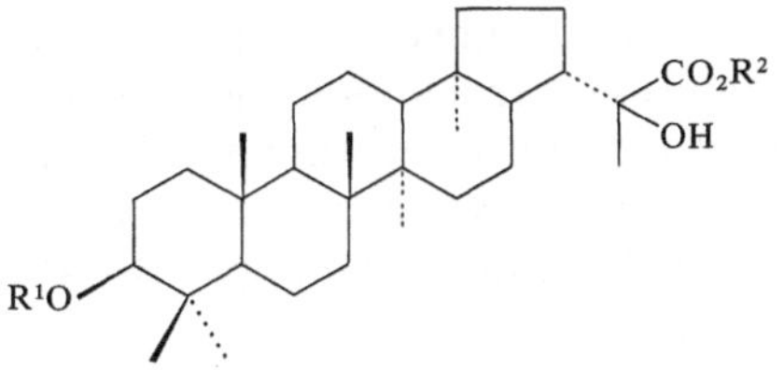

CO_2R^2
OH

(485) R^1=H, R^2=H
(486) R^1=H, R^2=Me
(487) R^1=Ac, R^2=Me

R^1O

A new hopane derivative, amphistictinic acid (488), has been isolated from *Pseudocyphellaria amphisticta* (273). The structural assignment followed principally from ^{1}H-n.m.r. and mass spectral data and was confirmed by chemical conversion to hopane-15α,22-diol.

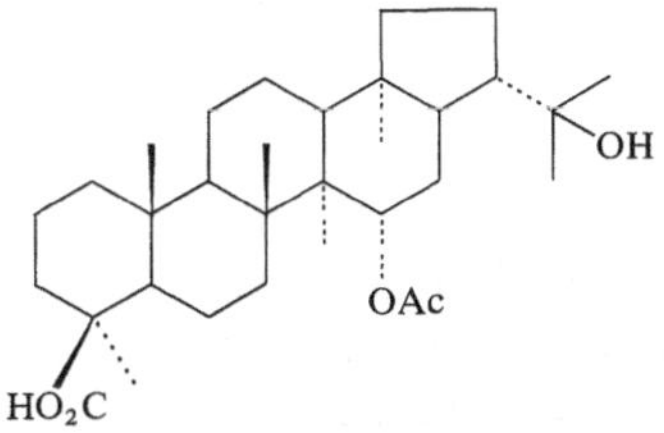

OH

OAc

HO_2C

(488)

16-O-Acetylleucotylin (489) and 20α-acetoxy-6α,22-dihydroxyhopane (490) were isolated from chemical strains of *Physcia aipolia* (134). The structures of these products were determined primarily from their ^{13}C-n.m.r. spectra. The compound with the C-20 acetoxy group is of particular interest as it is the first example of a hopane triterpenoid with an oxygen substituent in this position.

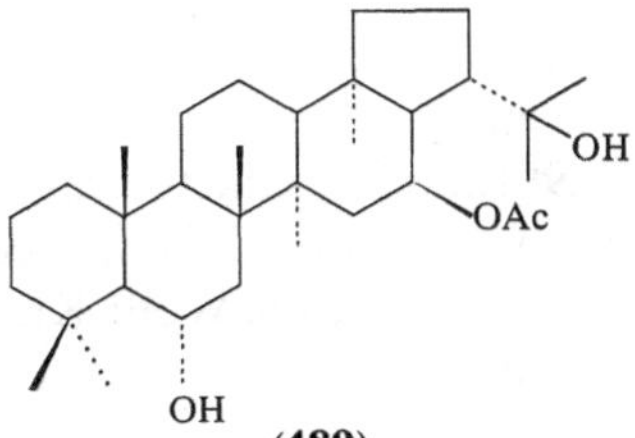

OH
OAc

OH
(489)

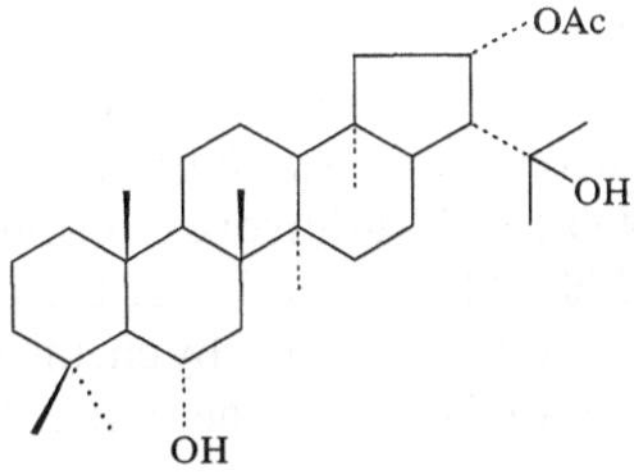

OAc
OH

OH
(490)

CORBETT and WILKINS (59) have revised the structures of the tri-terpenoids obtained from *Pseudocyphellaria mougeotiana* (52). Following the discovery of the presence of the known triterpenoids, 7β,22-di-hydroxyhopane (491) and 15α,22-dihydroxyhopane (492), the structures of the remaining triterpenoids present in this lichen were reinvestigated. It was found that the previously reported 11β,22-dihydroxyhopane was in fact a mixture of the above diols, (**491**) and (**492**). In addition, the configuration of C-7 in the previously reported triol and derivatives was revised from C-7βH to C-7αH, such that the now accepted structures are 6α,7β,22-trihydroxyhopane (**493**), 6α-acetoxy-7β,22-dihydroxyhopane (**494**) and 7β-acetoxy-6α,22-dihydroxyhopane (**495**).

(**491**) R^1=H, R^2=OH, R^3=H
(**492**) R^1=H, R^2=H, R^3=OH
(**493**) R^1=OH, R^2=OH, R^3=H
(**494**) R^1=OAc, R^2=OH, R^3=H
(**495**) R^1=OH, R^2=OAc, R^3=H

Two fern-9(11)-ene derivatives, retigeric acid A (**496**) and retigeric acid B (**497**), have been isolated from lichens of the *Lobaria retigera* group (*313*). The structures were deduced from ^{1}H-n.m.r. and mass spectral properties of the acids and derivatives and were confirmed by X-ray analysis of *p*-bromophenacyl retigerate A (*314*).

(**496**) R=Me
(**497**) R=COOH

Gonzalez, Martin, and Perez (*148*) identified three new fernene derivatives from *Xanthoria resendei:* 12α-acetoxy-3β-hydroxyfern-9(11)-ene (**498**), 3β,12α-dihydroxyfern-9(11)-ene (**499**) and 3,12-diketofern-9(11)-ene (**500**). The structures of these triterpenes were deduced from spectral evidence and confirmed by chemical interconversion and the Wolff-Kishner reduction of (**500**) to the parent fern-9(11)-ene.

(**498**) R^1=aH, βOH, R^2=aOAc, βH
(**499**) R^1=aH, βOH, R^2=aOH, βH
(**500**) R^1=O, R^2=O

13.2.2 Dammaranes

In addition to diacetylpyxinol (**501**), YOSIOKA, YAMAUCHI, and KITAGAWA (*345*) have identified three new dammarane triterpenoids from *Pyxine endochrysina:* pyxinol (**502**), 3-O-acetylpyxinol (**503**) and 12-desoxydiacetylpyxinol (**504**).

More recently HUNECK (*176*) identified two further dammarane triterpenoids from the related species *Pyxine coccifera,* namely 25-acetoxy-20(*S*),24(*R*)-epoxy-3-oxodammarane (**505**) and 25-acetoxy-3β-hydroxy-20(*S*),24(*R*)-epoxy-dammarane (**506**).

(**501**) $R^1 = aH$, βOAc, $R^2 = OH$, $R^3 = Ac$
(**502**) $R^1 = aH$, βOH, $R^2 = OH$, $R^3 = H$
(**503**) $R^1 = aH$, βOAc, $R^2 = OH$, $R^3 = H$
(**504**) $R^1 = aH$, βOAc, $R^2 = H$, $R^3 = Ac$
(**505**) $R^1 = O$, $R^2 = H$, $R^3 = Ac$
(**506**) $R^1 = aH$, βOH, $R^2 = H$, $R^3 = Ac$

13.2.3 Stictanes

Ten new metabolites (**507—516**), isolated from *Pseudocyphellaria* species by CORBETT, WILKINS and co-workers (*51*), provide the first examples of this new group of triterpenoids. The structures of these metabolites followed primarily from a detailed comparison of the ^{1}H-n.m.r. spectra with those of other triterpenoid groups. CORBETT and co-workers had previously synthesized a number of the parent triterpanes, 18α-oleanane (*54*), 17αH-hopane (*55*), 17αH-moretane (*55*) and 14α-taraxerane (*53*), by methods similar to those exemplified for stictane (Scheme 66). Thus oxidation of the alcohol (**516**) gave the diketone (**517**) which upon Wolff-Kishner reduction yielded the parent stictane (**518**). The physical and spectroscopic properties of this triterpane confirmed the novelty of this group of compounds.

(**507**) $R^1 = OH$, $R^2 = aH$, βOH, $R^3 = aOH$, βH
(**508**) $R^1 = OAc$, $R^2 = aH$, βOAc, $R^3 = aOAc$, βH

(**509**) $R^1 = OAc$, $R^2 = aH$, βOAc, $R^3 = aOH$, βH
(**510**) $R^1 = OAc$, $R^2 = aH$, βOH, $R^3 = aOH$, βH
(**511**) $R^1 = OH$, $R^2 = aH$, βOAc, $R^3 = aOH$, βH
(**512**) $R^1 = OAc$, $R^2 = aH$, βOAc, $R^3 = O$
(**513**) $R^1 = H$, $R^2 = aH$, βOH, $R^3 = aOH$, βH
(**514**) $R^1 = H$, $R^2 = aH$, βOAc, $R^3 = aOAc$, βH
(**515**) $R^1 = H$, $R^2 = aH$, βOAc, $R^3 = aOH$, βH
(**516**) $R^1 = H$, $R^2 = O$, $R^3 = aOH$, βH

14*

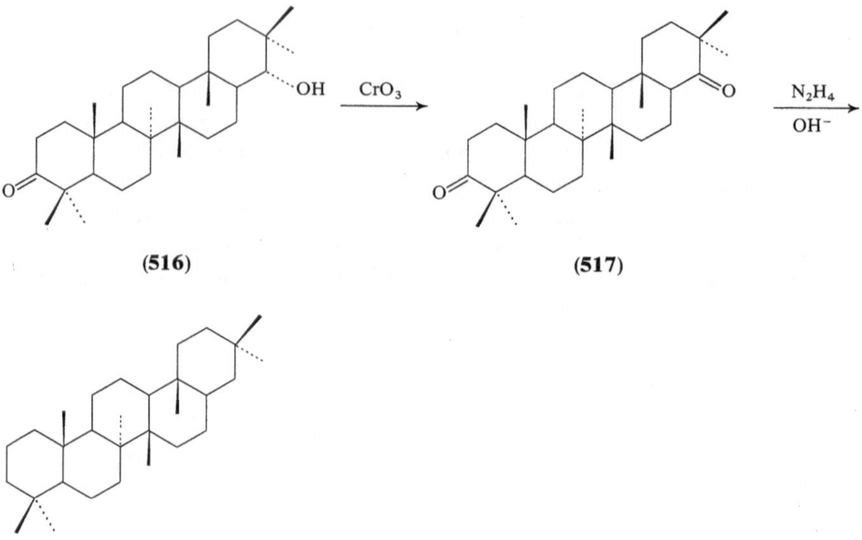

(516)   **(517)**

(518)

Scheme 66. Synthesis of stictane

Elimination of the C-22 hydroxy group from such compounds led to ring contraction and the production of a new triterpenoid structure, for which the name flavicane was proposed. Hence **(509)** on treatment with phosphoryl chloride in pyridine gave the flavicane **(519)** (Scheme 67).

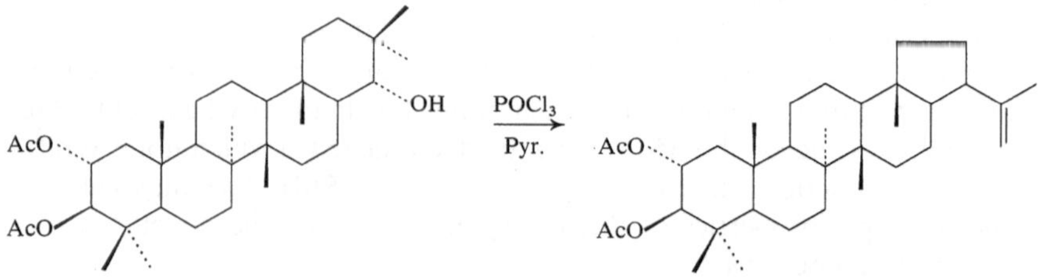

(509)   **(519)**

Scheme 67. Stictane-flavicane rearrangement

The 8α-Me configuration and boat structure for ring B of the stictanes (and the related flavicanes) rather than the usual 8β-Me and chair ring B of pentacyclic triterpenoids was established by a detailed analysis of the ^{1}H-n.m.r. spectra of a number of stictane derivatives (*57*). Further evidence for the proposed structure of flavicane was obtained from a comparison of the ^{1}H-n.m.r. and mass spectra of 17,21-secohopane and 17,21-secofla-vicane derivatives (*58*).

RAO and SESHADRI (*271*) isolated a triterpenoid, retigeradiol, from lichens of the genus *Lobaria*. This compound was identified as a saturated pentacyclic disecondary diol, and on evidence obtained from the spectra of the parent compound and derivatives, retigeradiol was tentatively assigned the structure 3β,19β-dihydroxytaraxerane (**520**). CORBETT, HENG and WILKINS (*56*) compared the spectral data for retigeradiol with that of 3β,22α-dihydroxystictane (**513**) and concluded that the structure of re-tigeradiol should be revised to 3β,22α-dihydroxystictane. The latter compound was also shown to occur in *Lobaria retigera*.

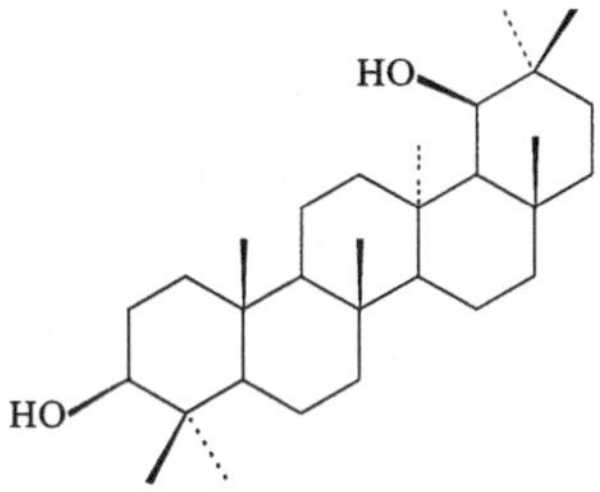

(**520**)

A new group of secostictanes was isolated from *Pseudocyphellaria degelii* (*147*). Three triterpenoids, 22α-hydroxy-3,4-secostict-4(23)-en-3-oic acid (**521**), 22α-hydroxy-3,4-secostict-4(23)-en-3-al (**522**) and 3-acetoxy-22α-hydroxy-3,4-secostict-4(23)-ene (**523**), were identified principally from spectroscopic data.

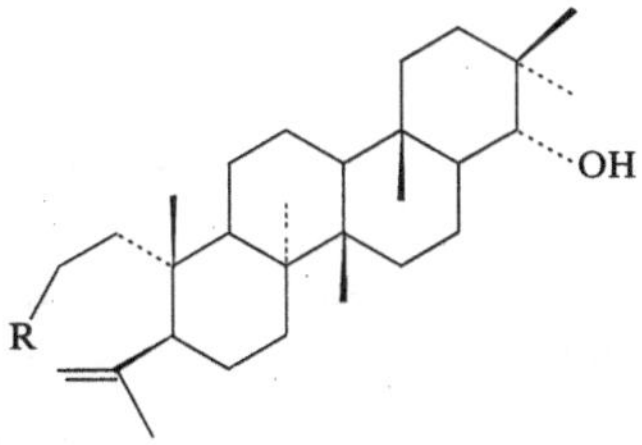

(**521**) R=COOH
(**522**) R=CHO
(**523**) R=CH$_2$OAc

13.3 Steroids

The following fungal and higher plant steroids have now also been isolated from lichen species: ergosterol-5,8-peroxide (*325*), cholestan-3β-ol, 24-methylcholestan-3β-ol, 24-ethylcholestan-3β-ol, 24-methylcholest-7-en-3β-ol, 24-ethylcholest-7-en-3β-ol, cholest-5-en-3β-ol, 24-methylcholest-5-

en-3β-ol, 24-ethylcholest-5-en-3β-ol, 24-methylcholesta-5,22-dien-3β-ol, 24-ethylcholesta-5,22-dien-3β-ol, 24-methylcholesta-7,24(28)-dien-3β-ol (*332*), episterol and tecosterol (*274*).

Lichesterol (ergosta-5,8,22-trien-3β-ol) (**524**), is a novel sterol isolated from *Xanthoria parietina* (*228*).

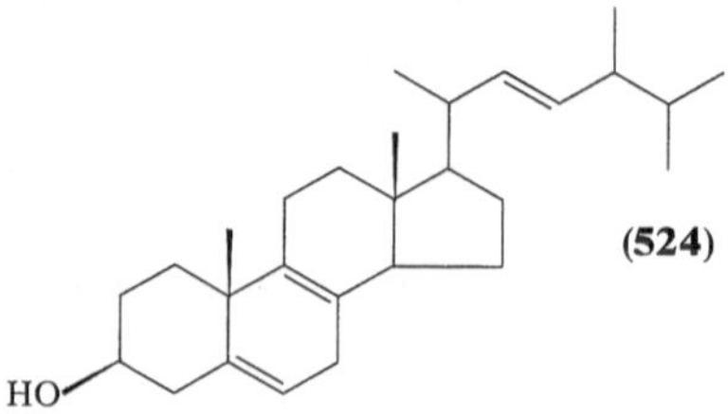

(**524**)

Evidence for two sterol "pools" in lichens has been presented (*228, 229*). The first pool, associated with the mycobiont, is readily extracted from the lichen and consists mainly of C_{28} sterols while the second pool is tightly bound and can only be extracted after hydroxide digestion of the lichen. This second group consists largely of C_{29} sterols which are dominant in the phycobiont.

BRUUN and MOTZFELD (*35*) isolated the novel peroxyergosteryl divaricatinate (**525**) from the lichen *Haematomma ventosum*. This product is the only known example of a steryl ester of an aromatic polyketide derived acid. The depside divaricatic acid (**234**) also present in this lichen contains the same A-ring component and it is possible that the ester formation in both of these metabolites is effected by the same enzyme.

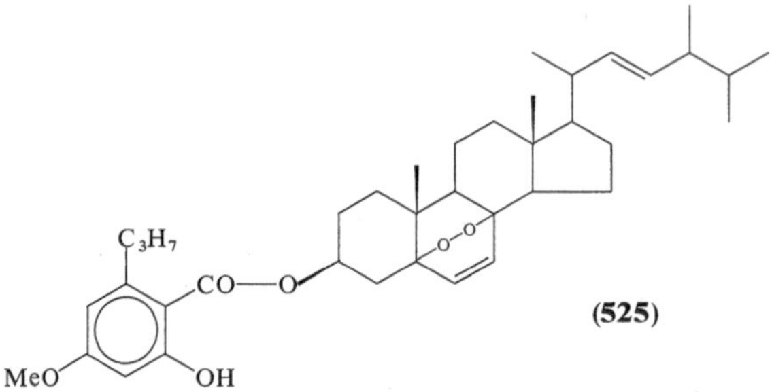

(**525**)

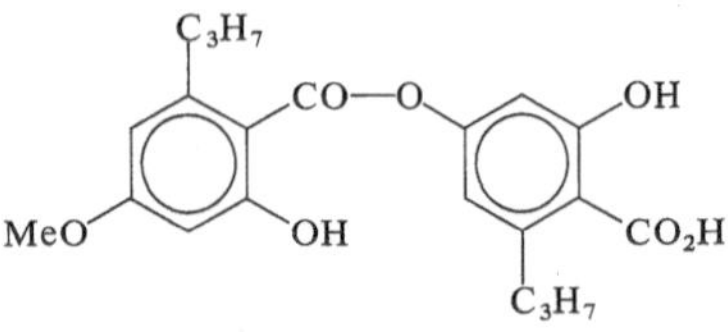

(**234**)

13.4 Carotenoids

In the past decade the following carotenoids have been detected in lichens: α-carotene, β-carotene, γ-carotene, δ-carotene, lycopene, neurosporene, 3,4-dehydrolycopene, α-cryptoxanthin, β-cryptoxanthin, canthax-

anthin, lutein, lutein epoxide, β-carotene epoxide, zeaxanthin, iso-
zeaxanthin, rubixanthin, astaxanthin, astaxanthin ester, violaxanthin,
aurochrome, mutatochrome, mutatoxanthin, lycopene-5,6-epoxide,
dihydro-ζ-carotene, neoxanthin, canaxanthin, flavochrome, capsanthin,
capsornbin, neurosporoxanthin, lycoxanthin, gazaniaxanthin, auroxanthin
and torularhodin (89—92). All are known to occur in higher plants.

14. Lichen Chemotaxonomy

BRODO (29) has recently summarized the application of chemical criteria
in lichen taxonomy in the following terms "chemical investigations now
form an integral part of all serious taxonomic studies on lichen forming
fungi. In fact within the past decade or so, more and more evidence has
accumulated to show that taxonomic units at all levels can be characterized
to a greater or lesser extent by their chemical products. One can argue *ad
nauseam* whether these chemically characterised populations represent
species, subspecies or whatever, but almost no-one will deny that chemical
products do reflect broad or narrow taxonomic relationships."

Hence it is the taxonomic interpretation of the observed chemical
variations that is controversial although HAWKSWORTH (155) has attempted
to forward some guidelines for such interpretation.

Most morphologically defined species have a constant chemistry,
usually one cortical substance (e. g. usnic acid, atranorin) and one or more
medullary substances, and justify the use of this criterion in lichen
taxonomy. The three common patterns of chemical variation are those of
replacement type compounds, accessory type compounds and chemosyndromic variation. These are discussed in turn.

14.1 Replacement Type Compounds and Chemosyndromic Variation

Here congeneric chemotypes show simple replacement of one or a few
substances, a classical example being that of the chemical strains of
Pseudevernia furfuracea. From all appearances these lichen populations are
morphologically indistinguishable but have variable chemical composition
(150, 151) (Scheme 68). Biogenetically the first two strains appear very
closely related [metabolites are biosequential (113)] but the third is quite
remote. It is now generally accepted that when there is such a geographical
and biogenetic demarkation, such taxa should be recognised as species and
the North American taxon is distinguished as *Pseudevernia consocians*.

Strain 1, containing olivetoric acid

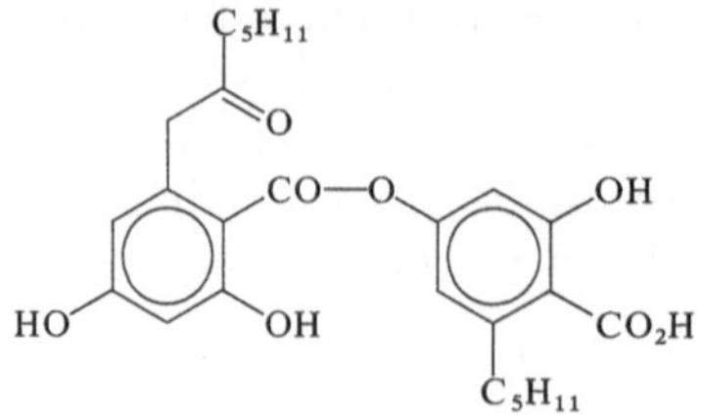

occur in Europe

Strain 2, containing physodic acid

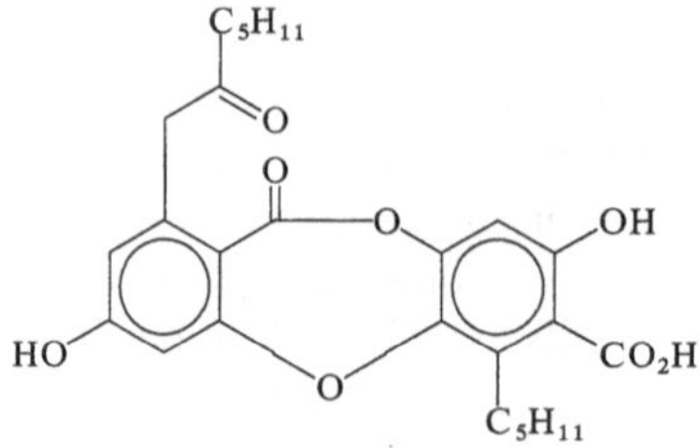

Strain 3, containing lecanoric acid

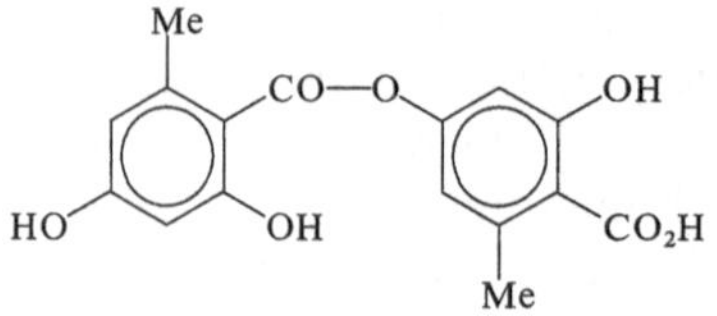

occurs in North America

Scheme 68. Chemical strains of *Pseudevernia furfuracea*

In fact the geographical distribution of chemical strains (1) and (2) differed significantly within Europe (*150, 151, 156*) and it was suggested by HAWKSWORTH (*155*) that these two strains should be recognised as varieties. However the discovery (*63*) that approximately 0.5% of a single population of specimens from Spain exhibited joint occurrence of both acids [i.e. a chemical intermediate or chemical combinant (*113*)] convinced some lichenologists that these strains represent a single species which shows some genetic variation (*93*).

In summary, most lichenologists who have chosen to recognise chemically distinct races as species have supported their decision primarily on the basis of the different geographic distributions that such races usually show. However it has been suggested by the CULBERSONS (*68, 69, 74*) that the best evidence that chemical variation is under genetic control rather than being environmentally determined is the fact that the chemical races, where sympatric, maintain their integrity even when growing side by side. Alternatively the occurrence of intermediates (chemical combinants) in areas of sympatry confirms that such races belong to a single species. However the existence of chemosyndromic variation in some groups of lichens may make the recognition of intermediates more difficult (*72, 77, 73, 87*).

A chemosyndrome refers to a group of biogenetically related metabolites and in this pattern of chemical variation the major metabolite (or metabolites) in any one taxon is *invariably* accompanied by minor quantities of several biosequentially related substances. Further, the compounds that are major constituents of some species may become minor constituents of other related taxa and vice versa. For example a group of North American *Xanthoparmelia* produce the β-orcinol depsides barbatic acid (**221**), 3-α-hydroxybarbatic acid (**222**), baeomycesic acid (**223**), squamatic acid (**224**), 4-O-demethylbarbatic acid (**526**), diffractaic acid (**527**) and 2-O-methylobtusatic acid (**212**) in varying quantities (*87*). These medullary depsides differ from one another by the level of oxidation of the C_1 substituent at position 3 as well as the degree of O- and C-methylation.

(**526**) R = H
(**527**) R = Me

Table 5 illustrates the overlapping chemosyndromes of these species. Hence a true intermediate or chemical combinant cannot be simply defined as containing both of two replacement compounds, but would have to contain both chemical constellations in *comparable* concentrations.

Table 5. *Chemosyndromic Variation in some Xanthoparmeliae*

Depside Sp:	X. ajoensis	X. moctezumensis	X. tucsonensis	X sp A	X sp B
(**221**)	minor	minor	minor	major	trace
(**222**)	trace	major	trace	major	
(**223**)	trace	trace		trace	
(**224**)	minor	trace	trace	trace	
(**526**)	trace	trace	trace	major	
(**527**)	major		major		major
(**212**)					minor

A further important feature of chemical races is their ecology. In several cases that have received detailed study, different chemical races are ecologically sorted into distinct habitats in their regions of sympatry (70, 71, 74). Although the underlying physiological causes of this sorting or the related phytogeographically significant distributions remain unknown, they do indicate that the chemical races have a more than superficial genetic basis.

In summary, most chemotypes (i.e. disregarding accessory chemical variations) have subtle morphological, ecological or distributional tendencies and consequently should be accorded some taxonomic recognition (29, 113, 155).

14.2 Accessory Metabolites

These substances are ones which occur sporadically in a species, usually in addition to the constant constituents, and have no correlation with any morphological or distributional variations and are accorded no taxonomic significance. Such substances commonly occur as accessory compounds in more than one species and often vary in quantity from deficiency to abundance. Accessory substances have been further subdivided into two groups (113):

(a) Biosequential accessory substances, which are present in trace quantities and are biogenetically closely related to the major metabolites, e.g.

methylstictic acid ⎫
norstictic acid   ⎪
cryptostictic acid ⎬ in stictic acid containing species
connorstictic acid ⎪
constictic acid   ⎪
menegazziaic acid ⎭

In many respects this chemical variation resembles that of the chemosyndromes, except that these minor metabolites may be present or absent (but this may reflect the limits of detection by t.l.c.).

(b) Biogenetically distant compounds which are unrelated to the invariant lichen substances and may be present in significant quantities. For instance, the most common accessory compounds of this nature detected in the genus *Xanthoparmelia* are the aliphatic acids, constipatic acid (**60**), protoconstipatic acid (**58**), dehydroconstipatic acid (**59**) and dehydroprotoconstipatic acid (**528**) (113). Although they occur sporadically in a number of species in addition to the invariant usnic acid, depsides or depsidones, these aliphatic acids appear more common in specimens collected in arid areas, e.g. northern South Australia, western New South Wales and the Sonoran Desert (Mexico). It appears possible that such environmental conditions stimulate the biosynthesis of these compounds, and that the same species growing in more temperate areas simply produces the invariant depsides (or depsidones).

(58) R = CHOHMe
(528) R = COMe

Chemically aberrant specimens which do not fit either of the above categories have been observed on rare occassions in several genera (*29, 113*) but the origin and taxonomic significance (if any) of these variations has yet to be clarified.

References

1. AGHORAMURTHY, K., K. G. SARMA, and T. R. SESHADRI: Chemical Investigation of Indian Lichens. XXIV. The Chemical Components of *Alectoria virens* Tayl. Constitution of a New Depsidione, Virensic Acid. Tetrahedron **12**, 173 (1961).
2. AHMAD, S., and G. HUSSAIN: Derivatives of Phloroacetophenone from the Lichen *Pseudevernia furfuraceae*. Pakistan J. Sci. Ind. Res. **19**, 126 (1976).
3. ÅKERMARK, B., H. ERDTMAN, and C. A. WACHTMEISTER: Chemistry of Lichens. XIII. Structure of Pannaric Acid. Acta Chem. Scand. **13**, 1855 (1959).
4. ARSHAD, M., J. P. DEVLIN, and W. D. OLLIS: Synthesis of Sordidone and Thiophanic Acid, Two Chlorine-containing Lichen Metabolites. J. Chem. Soc. (London) C **1971**, 1324.
5. ARSHAD, M., J. P. DEVLIN, W. D. OLLIS, and R. E. WHEELER: The Constitution of Sordidone and its Relation to Thiophanic Acid. Chem. Commun. **1968**, 154.
6. ASAHINA, Y., and H. HAYASHI: Untersuchungen über Flechtenstoffe, XXVI. Mitteil.: Über Psoromsäure. Ber. dtsch. chem. Ges. **66**, 1023 (1933).
7. ASAHINA, Y., and S. SHIBATA: Chemistry of Lichen Substances. Japan Society for Promotion of Science, Tokyo, 1954.
8. ASAHINA, Y., M. YANAGITA, and T.OMAKI: Untersuchungen über Flechtenstoffe, XXV. Mitteil.: Über Stictinsäure. Ber. dtsch. chem. Ges. **66**, 943 (1933).
9. AURET, B. J., D. R. BOYD, P. M. ROBINSON, C. G. WATSON, J. W. DALY, and D. M. DERINA: The NIH Shift During the Hydroxylation of Aromatic Substrates by Fungi. Chem. Commun. **1971**, 1585.
10. BACHELOR, F. W., and G. G. KING: Chemical Constituents of Lichens: Aphthosin, a Homologue of Peltigerin. Phytochem. **9**, 2587 (1970).
11. BAKER, C., J. A. ELIX, D. P. H. MURPHY, S. KUROKAWA, and M. V. SARGENT: *Parmelia reptans*, a New Lichen Species Producing the Depsidone, Succinprotocetraric Acid. Austral. J. Bot. **21**, 137 (1973).
12. BARTON, D. H. R., and T. COHEN: Some Biogenetic Aspects of Phenol Oxidation. Festschrift A. Stoll, p. 117. Basel: Birkhäuser 1957.
13. BEGG, W. R., D. O. CHESTER, and J. A. ELIX: The Structure of Conorlobaridone and Conloxodin. New Depsidones from the Lichen *Xanthoparmelia xanthosorediata*. Austral. J. Chem. **32**, 927 (1979).

14. BEGG, W. R., J. A. ELIX, and A. J. JONES: Nonacyclic Amides from Lichens of the Genus *Xanthoparmelia*. Tetrahedron Letters **1978**, 1047.
15. BERNARD, T., M. JOUCLA, G. GOAS, and J.HAMELIN: Characterisation de la Sticticine chez le Lichen *Lobaria laetevirens*. Phytochem. **19**, 1967 (1980).
16. BIRKINSHAW, J. H., J. C. ROBERTS, and P. ROFFEY: Studies in Mycological Chemistry. Part XIX. "Product B" (Averantin) [1,3,6,8-Tetrahydroxy-2-(1-hydroxyhexyl)-anthraquinone], a Pigment from *Aspergillus versicolor* (Vuillemin) Tiraboschi. J. Chem. Soc. (London) C **1966**, 855.
17. BODO, B.: L'Acide Bourgeanique: Nouveau Metabolite des Lichens. Structure, Synthése et Biosynthése. Bull. Museum National D'Histoire Naturelle, 3rd series, No. 349, 23 (1975).
18. BODO, B., P. HEBRARD, L. MOLHO, and D. MOLHO: Un Nouvel Acide Aliphatique des Lichens, *Desmaziera evernioides* et *Ramalina bourgeana*. Tetrahedron Letters **1973**, 1631.
19. BODO, B., and D. MOLHO: Structure de l'Argopsine, Nouvelle Chlorodepsidone du Lichen *Agropsis megalospora*. C. R. hebd. séances Acad. Sci., Ser. C **278**, 625 (1974).
20. — — Structure des Acides Isomuronique et Neuropogolique, Nouveaux Acides Aliphatiques du Lichen *Neuropogon trachycarpus*. Phytochem. **19**, 1117 (1980).
21. BODO, B., D. MOLHO, and J. POLONSKY: Sur la Biosynthese de l'Acide Bourgeanique. Tetrahedron Letters **1974**, 1443.
22. BOHMAN-LINDGREN, G.: Chemical Studies on Lichens XXXIII. Roccanin, a New Cyclic Tetrapeptide from *Roccella canariensis*. Tetrahedron **28**, 4625 (1972).
23. BOHMAN-LINDGREN, G., and U. RAGNARSSON: Chemical Studies on Lichens — XXXIV. The Synthesis of cyclo-(R-β-phenyl-β-alanyl-L-protyl)$_2$, a Peptide Isolated from *Roccella canariensis*. Tetrahedron **28**, 4631 (1972).
24. BOLOGNESE, A., F. CHIOCCARA, and G. SCHERILLO: Isolation and Characterisation of Atranorin and 4,6-Dihydroxy-2-methoxy-3-methylacetophenone from *Stereocaulon vesuvianum*. Phytochem. **13**, 1989 (1974).
25. BRANDÄNGE, S., L. MÖRCH, and S. VALTEN: The Structure of Caperatic Acid. Acta. Chem. Scand. **29 B**, 889 (1975).
26. BRAUN, M.: Regioselective Synthesis of the Anthraquinones Digitopurpone and Islandicin. Angew. Chem. Int. Ed. Engl. **17**, 945 (1978).
27. BREWER, J. D., and J. A. ELIX: The Synthesis of Di-*O*-methylstrepsilin. Tetrahedron Letters **1969**, 4139.
28. — — Annelated Furans. VII. Synthetic Routes to Di-*O*-methylstrepsilin. Austral. J. Chem. **25**, 545 (1972).
29. BRODO, I. M.: Changing Concepts Regarding Chemical Diversity in Lichens. Lichenologist **10**, 1 (1978).
30. BROWN, C. J., D. E. CLARK, W. D. OLLIS, and P. L. VEAL: The Synthesis of the Depsidone, Diploicin. Proc. Chem. Soc. (London) **1960**, 393.
31. BRUUN, T.: Phenarctin, a Fully Substituted Depside from *Nephroma arcticum*. Acta Chem. Scand. **25**, 2831 (1971).
32. — Bourgeanic Acid in the Lichen *Stereocaulon tomentosum*. Acta Chem. Scand. **27**, 3120 (1973).
33. — Aliphatic Compounds in Some Lichens. Phytochem. **15**, 1261 (1976).
34. BRUUN, T., and A. LAMVIK: Haemoventosin. Acta Chem. Scand. **25**, 483 (1971).
35. BRUUN, T., and A. M. MOTZFELD: 5α,8α-Peroxyergosteryl Divaricatinate from *Haematomma ventosum*. Acta Chem. Scand. **29 B**, 274 (1975).
36. BRYAN, A. J., and J. A. ELIX: 2-*O*-Acetyltenuiorin, a New Tridepside from the Lichen *Pseudocyphellaria australiensis*. Austral. J. Chem. **29**, 1147 (1976).
37. BRYAN, A. J., J. A. ELIX, and S. NORFOLK: Synthesis of Orcinol Tridepsides and Aphthosin, an Orcinol Tetradepside. Austral. J. Chem. **29**, 1079 (1976).

38. Büchi, G., and P. G. Williard: The Total Synthesis of Mollicellin A. Heterocycles **11**, 437 (1978).

39. Bycroft, B. W., J. A. Knight, and J. C. Roberts: Studies in Mycological Chemistry. Part XV. Synthesis of 2,5-Dihydroxy-3-methyl-6-s-butyl-1,4-benzoquinone and its Bearing on the Structure of Nidulin. J. Chem. Soc. (London) **1963**, 5148.

40. Cagnaire, D., R. H. Marchessault, and M. Vincedon: N.m.r. of Lichenin. Tetrahedron Letters **1975**, 3953.

41. Cambie, R. C.: The Depsides from *Stereocaulon ramulosum* (Sw.) Räusch. N. Z. Journ. Sci. **11**, 48 (1968).

42. Chance, K. H., and D. L. Tibbetts: A Specific Method for the Identification of Usnic Acid. Bryologist **76**, 208 (1973).

43. Chawla, H. M., S. S. Chibber, and S. Niwas: Novel Photochemical Conversion of Pulvinic Acid to Leprapinic Acid. Tetrahedron Letters **1980**, 2089.

44. Chester, D. O., and J. A. Elix: The Identification of four New *meta*-Depsides in the Lichen *Ramalina asahinae*. Austral. J. Chem. **31**, 2745 (1978).

45. — — 2-O-Methylobtusatic Acid, a New Depside from the Lichen *Xanthoparmelia tusconensis*. Austral. J. Chem. **32**, 1399 (1979).

46. — — Three New Aliphatic Acids from Lichens of Genus *Parmelia* (Subgenus *Xanthoparmelia*). Austral. J. Chem. **32**, 2565 (1979).

47. — — A New Dibenzofuran and Diphenyl Ether from the Lichen *Gymnoderma melacarpum*. Austral. J. Chem. **33**, 1153 (1980).

48. — — Condidymic Acid, a New Dibenzofuran from the Lichen *Cladonia squamosula*. Austral. J. Chem. **34**, 1501 (1981).

49. — — New Metabolites from Australian Lichens. Austral. J. Chem. **34**, 1507 (1981).

50. Chester, D. O., J. A. Elix, and A. J. Jones: Lecideoidin and 3'-Dechlorolecideoidin, New Depsidones from an Australian *Lecidea* (Lichen). Austral. J. Chem. **32**, 1857 (1979).

51. Chin, W. J., R. E. Corbett, C. K. Heng, and A. L. Wilkins: Lichens and Fungi. Part XI. Isolation and Structural Elucidation of a New Group of Triterpenes from *Sticta coronata, S. colensoi,* and *S. flavicans*. J. Chem. Soc. (London) Perkin Trans. I **1973**, 1437.

52. Corbett, R. F., and S. D. Cumming: Lichens and Fungi. Part VII. Extractives of the Lichen *Sticta mougeotiana* var. *dissecta* Del. J. Chem. Soc. (London) C **1971**, 955.

53. Corbett, R. E., S. D. Cumming, and E. V. Whitehead: Lichens and Fungi. Part X. 14α-Taraxerane. J. Chem. Soc. (London) Perkin Trans. I **1972**, 2827.

54. Corbett, R. E., and H. L. Ding: Lichens and Fungi. Part VIII. 18α-Oleanane. J. Chem. Soc. (London) C **1971**, 1884.

55. Corbett, R. E., and C. K. Heng: Lichens and Fungi. Part IX. 17αH-Hopane, 17αH-Moretane and Derivatives. J. Chem. Soc. (London) C **1971**, 1885.

56. Corbett, R. E., C. K. Heng, and A. L. Wilkins: Lichens and Fungi. Part XIV. A Revised Structure for Retigerane Triterpenoids. Austral. J. Chem. **29**, 2567 (1976).

57. Corbett, R. E., and A. L. Wilkins: Lichens and Fungi. Part XII. Dehydration and Isomerisation of Stictane Triterpenoids. J. Chem. Soc. (London) Perkin Trans. I **1976**, 857.

58. — — Lichens and Fungi. Part XIII. Comparison of the Nuclear Magnetic Resonance and Mass Spectra of 17,21-Secohopane and 17,21-Secoflavicane Derivatives. J. Chem. Soc. (London) Perkin Trans. I **1976**, 1316.

59. — — Lichens and Fungi. XV. Revised Structures for Hopane Triterpenoids Isolated from the Lichen *Pseudocyphellaria mougeotiana*. Austral. J. Chem. **30**, 2329 (1977).

60. Cresp, T. M., P. Djura, M. V. Sargent, J. A. Elix, U. Engkaninan and D. P. H. Murphy: The Synthesis of Notatic Acid and 4-O-Methylhypoprotocetraric Acid. Austral. J. Chem. **28**, 2417 (1975).

61. CRESP, T. M., J. A. ELIX, S. KUROKAWA and M. V. SARGENT: The Structure of Two New Depsidones from the Lichen *Parmelia notata*. Austral. J. Chem. **25**, 2167 (1972).
62. CULBERSON, C. F.: Joint Occurrence of a Lichen Depsidone and its Probable Depside Precursor. Science **143**, 255 (1964).
63. — A Note on Chemical Strains of *Parmelia furfuracea*. Bryologist **68**, 435 (1965).
64. — Improved Conditions and New Data for the Identification of Lichen Products by a Standardized Thin-layer Chromatographic Method. J. Chromatogr. **72**, 113 (1972).
65. — Supplement to Chemical and Botanical Guide to Lichen Products. Bryologist **73**, 177 (1973).
66. — Conditions for the Use of Merck Silica Gel 60 F_{254} Plates in the Standardized Thin-layer Technique for Lichen Products. J. Chromatogr. **97**, 107 (1974).
67. — High-speed Liquid Chromatography of Lichen Extracts. Bryologist **75**, 54 (1975).
68. CULBERSON, W. L.: Analysis of Chemical and Morphological Variation in the *Ramalina siliquosa* Species Complex. Brittonia **19**, 333 (1967).
69. — The Behaviour of the *Ramalina siliquosa* Group in Portugal. Österreich Bot. Z. **116**, 85 (1969).
70. — The Chemistry and Systematics of Some Species of the *Cladonia cariosa* Group in North America. Bryologist **72**, 377 (1969).
71. — The *Parmelia perforata* Group: Niche Characteristics of Chemical Races, Separation by Parallel Evolution, and a New Taxonomy. Bryologist **76**, 20 (1973).
72. CULBERSON, C. F., and W. L. CULBERSON: Chemosyndromic Variation in Lichens. Systematic Botany **1**, 325 (1976).
73. — — β-Orcinol Derivatives in Lichens: Biogenetic Evidence from *Oropogon loxensis*. Exptl. Mycology **2**, 245 (1978).
74. CULBERSON, W. L., and C. F. CULBERSON: Habitat Selection by Chemically Differentiated Races of Lichens. Science **158**, 1195 (1967).
75. — — *Ramalina asahinae*, a New Boninic Acid Producing Species from Mexico. J. Jap. Bot. **51**, 374 (1976).
76. — — A New *Ramalina* with Two New Depsides. Occas. Pap. Farlow Herb. **16**, 37 (1981).
77. CULBERSON, C. F., W. L. CULBERSON, and T. L. ESSLINGER: Chemosyndromic Variation in the *Parmelia pulla* Group. Bryologist **80**, 125 (1977).
78. CULBERSON, C. F., W. L. CULBERSON, and A. JOHNSON: Second Supplement to Chemical and Botanical Guide to Lichen Products. St. Louis, Missouri: American Bryological and Lichenological Society. 1977.
79. CULBERSON, W. L., C. F. CULBERSON, and A. JOHNSON: A Standardized TLC Analysis of β-Orcinol Depsidones. Bryologist **84**, 16 (1981).
80. CULBERSON, C. F., and M. J. DIBBEN: 2-*O*-Methylperlatolic and 2′-*O*-methylperlatolic Acids: Two New Lichen Depsides from *Pertusaria*. Bryologist **75**, 362 (1972).
81. CULBERSON, C. F., and T. L. ESSLINGER: 4-*O*-Methylolivetoric and Loxodellic Acids. New Depsides from New Species of Brown *Parmeliae*. Bryologist **79**, 42 (1976).
82. CULBERSON, C. F., and M. E. HALE, JR.: 4-*O*-Demethylnotatic Acid, a New Depsidone in Some Lichens Producing Hypoprotocetraric Acid. Bryologist **76**, 77 (1973).
83. CULBERSON, C. F., and H. HERTEL: 2′-*O*-Methylanziaic Acid, a New Depside in *Lecidea diducens* and *Lecidea speirodes*. Bryologist **75**, 372 (1972).
84. — — Chemical and Morphological Analyses of the *Ledidea lithophila-plana* Group (Lecideaceae). Bryologist **82**, 189 (1979).
85. CULBERSON, C. F., and A. JOHNSON: A Standardized Two-dimensional Thin-layer Chromatographic Method for Lichen products. J. Chromatogr. **128**, 253 (1976).
86. CULBERSON, C. F., and H. KRISTINSSON: A Standardized Method for the Identification of Lichen Products. J. Chromatogr. **46**, 85 (1970).
87. CULBERSON, C. F., T. H. NASH III, and A. JOHNSON: 3-α-Hydroxybarbatic Acid, a New

Depside in Chemosyndromes of some *Xanthoparmeliae* with β-Orcinol Depsides. Bryologist **82**, 154 (1979).

88. CULLEN, L. J., and M. V. SARGENT: Depsidone Synthesis. XXII. An Alternative Synthesis of Gangaleoidin. Austral. J. Chem. **34**, 2701 (1981).

89. CZECZUGA, B.: Investigations on Carotenoids in Lichens I. The Presence of Carotenoids in Representatives of Certain Families. Nova Hedwigia **31**, 337 (1979).

90. — Investigations on Carotenoids in Lichens II. Members of the Usneaceae Family. Nova Hedwigia **31**, 349 (1979).

91. — Investigations on Carotenoids in Lichens IV. Representatives of the Parmeliaceae Family. Nova Hedwigia **32**, 105 (1980).

92. CZYGAN, F. C.: Carotinoid-Garnitus und -Stoffwechsel der Flechte *Haematomma ventosum* und ihres Phycobionten. Z. Pflanzanphysiol. **79**, 438 (1976).

93. DAHL, E., and H. KROG: Macrolichens of Denmark, Finland, Norway and Sweden. Oslo: Universitetsforlaget. 1973.

94. DAVIDSON, V. E., J. KEANE, and T. J. NOLAN: The Chemical Constituents of Lichens Found in Ireland. *Lecanora gangaleoides*. Part 3. The Constitution of Gangaleoidin. Sci. Proc. Roy. Dublin Soc. **23**, 143 (1943).

95. DEAN, F. M.: Naturally Occurring Oxygen Ring Compounds. London: Butterworths. 1963.

96. DEAN, F. M., D. S. DEORHA, A. D. T. ERNI, D. W. HUGHES, and J. C. ROBERTS: The Structure of Nidulin, a Metabolite of *Aspergillus nidulans*. J. Chem. Soc. (London) **1960**, 4829.

97. DEAN, F. M., A. D. T. ERNI, and A. ROBERTSON: The Chemistry of Fungi. Part XXVI. Dechloronornidulin. J. Chem. Soc. (London) **1956**, 3545.

98. DEAN, F. M., J. C. ROBERTS, and A. ROBERTSON: The Chemistry of Fungi. Part XXII. Nidulin and Nornidulin („Ustin"): Chlorine-containing Metabolic Products of *Aspergillus nidulans*. J. Chem. Soc. (London) **1954**, 1432.

99. DEVLIN, J. P., C. P. FALSHAW, W. D. OLLIS, and R. E. WHEELER: Phytochemical Examination of the Lichen *Lecanora rupicola*. J. Chem. Soc. (London) C **1971**, 1318.

100. DJURA, P.: Depsidone Synthesis. Ph. D. Thesis, University of Western Australia, 1977.

101. DJURA, P., and M. V. SARGENT: Depsidone Synthesis. III. Grayanic Acid. Austral. J. Chem. **29**, 899 (1976).

102. — — Depsidone Synthesis. VI. Colensoic Acid. Austral. J. Chem. **29**, 1069 (1976).

103. — — Depsidone Synthesis. IX. Nornotatic Acid and Notatic Acid. Austral. J. Chem. **30**, 1293 (1977).

104. — — Depsidone Synthesis. Part 11. Synthesis of Some Fungal Depsidones related to Nidulin. J. Chem. Soc. (London) Perkin Trans. I **1978**, 395.

105. DJURA, P., M. V. SARGENT, and P. D. CLARK: Depsidone Synthesis. X. Methoxy- and Hydroxy-colensoic Acids. Austral. J. Chem. **30**, 1545 (1977).

106. DJURA, P., M. V. SARGENT, J. A. ELIX, U. ENGKANINAN, S. HUNECK, and C. F. CULBERSON: Depsidone Synthesis. VIII. Isolation and Structural Determination of Hydroxy- and Methoxy-colensoic Acids. Synthesis of Methyl Methoxy-*O*-methylcolensoate. Austral. J. Chem. **30**, 599 (1977).

107. DJURA, P., M. V. SARGENT, and P. VOGEL: Depsidone Synthesis. Part II. Diploicin and Gangaleoidin. J. Chem. Soc. (London) Perkin Trans. I **1976**, 147.

108. EJIRI, H., U. SANKAWA, and S. SHIBATA: Graciliformin and its Acetates in *Cladonia graciliformis*. Phytochem. **14**, 277 (1975).

109. EJIRI, H., and S. SHIBATA: Zeorin from the Mycobiont of *Anaptychia hypoleuca*. Phytochem. **13**, 2871 (1974).

110. ELIX, J. A.: Annelated Furans. VIII. A Synthesis of Pannaric Acid. Austral. J. Chem. **25**, 1129 (1972).

111. — Synthesis of *para*-olivetol Depsides. Austral. J. Chem. **27**, 1767 (1974).

112. Elix, J. A.: 2'-*O*-Methylphysodic Acid and Hydroxyphysodic Acid: Two New Depsidones from the Lichen *Hypogymnia billardieri.* Austral. J. Chem. **28,** 849 (1975).

113. — Peculiarities of the Australasian Lichen Flora; Accessory Metabolites, Chemical and Hybrid Strains. J. Hattori Bot. Lab. **52,** 407 (1982).

114. Elix, J. A., and U. Engkaninan: The Structure of Galbinic Acid. A Depsidone from the Lichen *Usnea undulata.* Austral. J. Chem. **28,** 1793 (1975).

115. — — The Structure of Lividic Acid. A Depsidone from the Lichen *Parmelia formosana.* Austral. J. Chem. **29,** 203 (1976).

116. — — 3-Hydroxyphysodic Acid. Chemical Corroboration of the Structure of this Lichen Depsidone. Austral. J. Chem. **29,** 2693 (1976).

117. — — 4,5-Di-*O*-methylhiascic Acid, a New Tridepside from the Lichens *Parmelia pseudofatiscens* and *Parmelia horrescens.* Austral. J. Chem. **29,** 2701 (1976).

118. Elix, J. A., U. Engkaninan, A. J. Jones, C. L. Raston, M. V. Sargent, and A. H. White: Chemistry and Crystal Structure of Leprolomin, a Novel Diphenyl Ether from the Lichen *Psoroma leprolomum.* Austral. J. Chem. **31,** 2057 (1978).

119. Elix, J. A., and B. A. Ferguson: Synthesis of the Lichen Depside 2-*O*-Methylconfluentic Acid. Austral. J. Chem. **30,** 373 (1977).

120. — — Synthesis of the Lichen Depsides, Olivetoric Acid, Confluentic Acid and 4-*O*-Methylolivetoric Acid. Austral. J. Chem. **31,** 1041 (1978).

121. Elix, J. A., B. A. Ferguson, and M. V. Sargent: The Structure of Alectoronic Acid and Related Lichen Metabolites. Austral. J. Chem. **27,** 2403 (1974).

122. Elix, J. A., and V. K. Jayanthi: 5-*O*-Methylhiascic Acid, a New Tridepside from Australian Lichens. Austral. J. Chem. **30,** 2695 (1977).

123. — — 3-Methoxy-2,4-di-*O*-methylgyrophoric Acid, a Novel Tridepside from the Lichen *Parmelia subfatiscens.* Austral. J. Chem. **34,** 1153 (1981).

124. Elix, J. A., V. K. Jayanthi, and C. C. Leznoff: 2,4-Di-*O*-methylgyrophoric Acid and 2,4,5-Tri-*O*-methylhiascic Acid. New Tridepsides from *Parmelia damaziana.* Austral. J. Chem. **34,** 1757 (1981).

125. Elix, J. A., and L. Lajide: The Structure of Connorstictic Acid. A Depsidone from the Lichen *Lecidea aspidula.* Austral. J. Chem. **34,** 583 (1981).

126. — — 2'-*O*-Methyltenuiorin, 2''-*O*-methyltenuiorin and 2',2''-Di-*O*-methyltenuiorin. New Tridepsides from the Lichen *Pseudocyphellaria faveolata.* Austral. J. Chem. **34,** 2005 (1981).

127. Elix, J. A., and D. P. Murphy: Annelated Furans. XVIII. The Photocyclization of 2-Methoxyphenyl Phenyl Ethers. Austral. J. Chem. **28,** 1559 (1975).

128. Elix, J. A., H. W. Musidlak, T. Sala, and M. V. Sargent: Structure and Synthesis of some Lichen Xanthones. Austral. J. Chem. **31,** 145 (1978).

129. Elix, J. A., and S. Norfolk: Synthesis of *meta*-Divarinol and Olivetol Depsides. Austral. J. Chem. **28,** 399 (1975).

130. — — Synthesis of Para β-Orcinol Depsides. Austral. J. Chem. **28,** 1113 (1975).

131. — — Synthesis of β-Orcinol Meta-depsides. Austral. J. Chem. **28,** 2035 (1975).

132. Elix, J. A., and P. D. Tearne: Nordivaricatic Acid, a New Depside from the Lichen *Heterodea beaugleholei.* Austral. J. Chem. **30,** 2333 (1977).

133. Elix, J. A., and D. Tronson: Annelated Furans. XIV. Synthetic Analogues of Usnic Acid. Austral. J. Chem. **26,** 1093 (1973).

134. Elix, J. A., A. A. Whitton, and A. J. Jones: Triterpenes from the Lichen Genus *Physcia.* Austral. J. Chem. **35,** 641 (1982).

135. Erdtman, H., and C. A. Wachtmeister: Phenoldehydrogenation as a Biosynthetic Reaction. Festschrift A. Stoll, p. 144. Basel: Birkhäuser 1957.

136. Fahselt, D.: Gas-Liquid Chromatography of the Lichen Substance Usnic Acid. Bryologist **78,** 452 (1975).

137. FINLAY-JONES, P. F., T. SALA, and M. V. SARGENT: Depsidone Synthesis. Part 18. Dihydronidulin. J. Chem. Soc. (London) Perkin Trans. I **1981**, 874.

138. FITZPATRICK, L., T. SALA, and M. V. SARGENT: Further Total Synthesis of Chlorine Containing Lichen Xanthones. J. Chem. Soc. (London) Perkin Trans. I **1980**, 85.

139. FOO, L. Y., and D. J. GALLOWAY: Pseudodepsidones and Other Constituents from *Xanthoparmelia scabrosa.* Phytochem. **18**, 1977 (1979).

140. FOO, L. Y., and S. A. GWYN: The Identification of Norlobariol, a New Lichen Constituent from *Xanthoparmelia scabrosa* (Tayl.) Hale. Experientia **34**, 970 (1978).

141. FOX, C. H., E. Klein, and S. HUNECK: Colensoinsäure, ein neues Depsidon aus *Stereocaulon colensoi.* Phythochem. **9**, 2567 (1970).

142. FRANCK, B.: The Biosynthesis of the Ergochromes. In: P. S. STEYN, The Biosynthesis of Mycotoxins. A Study in Secondary Metabolism. London-New York: Academic Press. 1980.

143. GAREGG, P. J., B. LINDBERG, K. NILSSON, and C.-G. SWAHN: 1-*O*-β-D-Galactopyranosyl-D-ribitol from *Xanthoria parietina.* Acta Chem. Scand. **27**, 1595 (1973).

144. GAVIN, J., G. NICOLLIER, and R. TABACCHI: Composants Volatils de la „Mousse de Chêne" [*Evernia prunastri* (L.) Ach.]. Helv. Chim. Acta **61**, 352 (1978).

145. GLEIN, R. D., S. TRENBEATH, F. SUZUKI, and C. J. SIH: Regiospecific Syntheses of Islandicin and Digitopurpone Monomethyl Ethers. Chem. Commun. **1978**, 242.

146. GOH, E. M., and A. L. WILKINS: Structures of the Lichen Depsidones Granulatin and Chlorogranulatin. J. Chem. Soc. (London) Perkin Trans. I **1979**, 1656.

147. GOH, E. M., A. L. WILKINS, and P. T. HOLLAND: Structural Elucidation of a New Group of Secostictane Triterpenoids. J. Chem. Soc. (London) Perkin Trans. I **1978**, 1560.

148. GONZALEZ, A. G., J. D. MARTIN, and C. PEREZ: Three New Triterpenes from the Lichen *Xanthoria resendei.* Phytochem. **13**, 1547 (1974).

149. GREAM, G. E., and N. V. RIGGS: Chemistry of Australian Lichens. II. A New Depsidone from *Parmelia conspersa* (Ehrh.) Ach. Austral. J. Chem. **13**, 285 (1960).

150. HALE, M. E., Jr.: Chemical Strains of the Lichen *Parmelia furfuracea.* Amer. J. Bot. **43**, 456 (1956).

151. — A Synopsis of the Lichen Genus *Pseudevernia.* Bryologist **71**, 1 (1968).

152. HAMILTON, R. J., and M. V. SARGENT: Synthesis of Nephroarctin and Phenarctin. J. Chem. Soc. (London) Perkin Trans. I **1976**, 943.

153. HARRIS, T. M., and J. V. HAY: Biogenetically Modelled Syntheses of Heptaacetate Metabolites Alternariol and Lichexanthone. J. Amer. Chem. Soc. **99**, 1631 (1977).

154. HAUAN, E., and O. KJØLBERG: Studies on the Polysaccharides of Lichens. I. The Structure of a Water-soluble Polysaccharide in *Stereocaulon paschale* (L.) Fr. Acta Chem. Scand. **25**, 2622 (1971).

155. HAWKSWORTH, D. L.: Lichen Chemotaxonomy. In: Lichenology: Progress and Problems, p. 139. London: Academic Press. 1976.

156. HAWKSWORTH, D. L., and D. S. CHAPMAN: *Pseudevernia furfuracea* (L.) Zopf and its Chemical Races in the British Isles. Lichenologist **5**, 51 (1971).

157. HAY, J. V., and T. M. HARRIS: Biogenetic-type Syntheses of Heptaketide Natural Products: Alternariol and Lichexanthone. Chem. Commun. **1972**, 953.

158. HEBRARD, P., B. BODO, L. MOLHO, and D. MOLHO: L'Acide Hydroxy-5 Physodique: Nouvelle Depsidone de Lichens de Genre Hypogymnia. C. R. hebd. séances Acad. Sci, Ser. C **281**, 115 (1975).

159. HECKENDORF, A. H., and Y. SHIMUZU: Structure of a Polysaccharide of *Umbilicaria mammulata.* Phytochem. **13**, 2181 (1974).

160. HENDRICKSON, J. B., M. V. J. RAMSAY, and T. R. KELLY: A New Synthesis of Depsidones. Diploicin and Gangaleoidin. J. Amer. Chem. Soc. **94**, 6834 (1972).

161. HILL, D. J., and V. AHMADJIAN: Relationship between Carbohydrate Movement and the Symbiosis in Lichens with Green Algae. Planta **103**, 267 (1972).

162. HIRAYAMA, T., F. FUJIKAWA, I. YOSIOKA, and I. KITAGAWA: A New Depsidone Oxyphysodic Acid Isolated from a Lichen *Parmelia enteromorpha* (Ach.). Chem. Pharm. Bull. (Japan) **22**, 1678 (1974).

163. — — — — Constituents of the Lichen in the Genus *Hypogymnia* II. Vittatolic Acid, a New Optically Active Depsidone, and 2'-*O*-Methylphysodic Acid from *Hypogymnia vittata* (Ach.). Gas. Chem. Pharm. Bull. (Japan) **24**, 1602 (1976).

164. — — — — On the Constituents of the Lichen in the Genus *Menegazzia*. Menegazziaic Acid, a New Depsidone from *Menegazzia asahinae* (Yas. ex Zahlbr.) Sant. and *Menegazzia terebrata* (Hoffm.) Mass. Chem. Pharm. Bull. (Japan) **24**, 2340 (1976).

165. HIROSE, Y., M. KUROIWA, H. YAMASHITA, T. TANAKA, and T. MEGUMI: Chemical Studies on the Natural Anthraquinones. I. Synthesis of Munjistin, Emodin and 3-Hydroxy-2-methylanthraquinone. Chem. Pharm. Bull. (Japan) **21**, 2790 (1973).

166. HOOPER, J. W., W. MARLOW, W. B. WHALLEY, and (in part) A. D. BORTHWICK and R. BOWDEN: The Chemistry of Fungi Part LXV. The Structures of Ergochrysin A, Isoergochrysin A, and Ergoxanthin, and of Secalonic Acids A, B, C, and D. J. Chem. Soc. (London) C **1971**, 3580.

167. HOWE, M. L., and J. T. BARRETT: Studies on a Hemaglutinin from the Lichen *Parmelia michauxiana*. Biochem. Biophys. Acta **215**, 97 (1970).

168. HRANISAVLJEVIĆ-JAKOVLJEVIĆ, M., J. MILJKOVIĆ-STOJANOVIĆ, R. DIMITRIJEVIĆ, and V. M. MICOVIĆ: Water- and Alkali-Soluble Glucans from Oak Lichen. Carbohydrate Res. **39**, 115 (1975).

169. — — — — An Alkali-Soluble Polysaccharide, from the Oak Lichen *Cetraria islandica* (L.) Ach. Carbohydrate Res. **80**, 291 (1980).

170. HUNECK, S.: Lichen Substances. In: L. REINHOLD and Y. LIWSCHITZ (eds.), Progress in Phytochemistry, Vol. 1, p. 223. London-New York-Sydney: Interscience Publ. 1968.

171. — Chemie und Biosynthese der Flechtenstoffe. Fortschr. Chem. organ. Naturstoffe **29**, 209 (1971).

172. — Chemie der Flechteninhaltsstoffe. XCI. Chromonglucoside aus Flechten. J. prakt. Chem. **314**, 488 (1972).

173. — Chemie der Flechteninhaltsstoffe. XCIII. Struktur der (−)-Placodiolsäure. Tetrahedron **28**, 4011 (1972).

174. — 6-Hydroxymethyleugenetin, ein neues Chromon aus *Roccella fuciformis*. Phytochem. **11**, 1489 (1972).

175. — Lobodirin: Ein neues Chromonglucosid aus *Lobodirina cerebriformis*. Phytochem. **12**, 2497 (1973).

176. — Inhaltsstoffe von *Pyxine coccifera*. Phytochem. **15**, 799 (1976).

177. — Chemistry of some Yellow *Acarospora* Species. Lichenologist **12**, 239 (1980).

178. HUNECK, S., J. A. AKINNIYI, A. F. CAMERON, J. D. CONNOLLY, and A. G. MULHOLLAND: The Absolute Configurations of (+)-Usnic and (+)-Isousnic Acid. X-Ray Analyses of the (−)-α-Phenylethylamine Derivative of (+)-Usnic Acid and of (−)-Pseudoplacodiolic Acid, A New Dibenzofuran, from the Lichen *Rhizoplaca chrysoleuca*. Tetrahedron Letters **22**, 351 (1981).

179. HUNECK, S., and G. FOLLMANN: Mitteilungen über Flechteninhaltsstoffe. LXXXIV. Zur Phytochemie und Chemotaxonomie der Lecanoraceengattung *Haematomma*. J. Hattori Bot. Lab. **35**, 319 (1972).

180. — — Struktur der Glomellsäure. Phytochem. **12**, 2993 (1973).

181. HUNECK, S., and G. HÖFLE: Struktur und ^{13}C-NMR-Spektroskopie von chlorhaltigen Flechten Xanthonen. Tetrahedron **34**, 2491 (1978).

182. — — Structure of Acaranoic Acid and Acarenoic Acid. Phytochem. **19**, 2713 (1980).

183. HUNECK, S., G. HÖFLE, and C. F. CULBERSON: 3,5-Dichloro-2'-*O*-methylanziasäure, ein neues Depsid aus *Lecanora sulphurella*. Phytochem. **16**, 995 (1977).

184. HUNECK, S., and I. M. LAMB: 1'-Chloropannarin, A New Depsidone from *Argopsis friesiana:* Notes on the Structure of Pannarin and on the Chemistry of the Lichen Genus *Argopsis.* Phytochem. **14**, 1625 (1975).

185. HUNECK, S., and J. SANTESSON: 64. Mitteilung über Flechteninhaltsstoffe. Über die Inhaltsstoffe von *Lecanora rupicola* (L.) Zahlbr. und *Lecanora carpinea* (L.) Ach. ex. Vain. und die Strukturaufklärung sowie Synthese von 8-Chlor-5,7-dihydroxy-2,6-dimethylchromon. Z. Naturforsch. **24 B**, 750 (1969).

186. — — 65. Mitteilung über Flechteninhaltsstoffe. Die Inhaltsstoffe von *Lecidea carpathica* (Koerb.) Szat. und die Struktur des Thuringions, eines neuen Xanthons. Z. Naturforsch. **24 B**, 757 (1969).

187. HUNECK, S., and M. V. SARGENT: Depsidone Synthesis. V. The Chemistry of Psoromic Acid: A Reinvestigation. Austral. J. Chem. **29**, 1059 (1976).

188. HUNECK, S., and K. SCHREIBER: Wachstumsregulatorische Eigenschaften von Flechten- und Moos-Inhaltsstoffen. Phytochem. **11**, 2429 (1972).

189. — — Flechteninhaltsstoffe XCVIII. Struktur des Aspicilins. Tetrahedron **29**, 3687 (1973).

190. — — 2-*O*-Methylconfluentinsäure: ein neues Depsid aus *Lecidea fuscoatra.* Phytochem. **13**, 221 (1974).

191. — — 2'-*O*-Methylperlatolinsäure aus einer *Lecidea* spec. Phytochem. **14**, 1629 (1975).

192. HUNECK, S., K. SCHREIBER, G. HÖFLE, and G. SNATZKE: Neodihydromurol- und Murolsäure, zwei neue γ-Lactoncarbonsäuren aus *Lecanora muralis.* J. Hattori Bot. Lab. **45**, 1 (1979).

193. HUNECK, S., K. SCHREIBER, G. SNATZKE, and H.-W. FEHLHABER: Miriquidsäure, ein neues Depsid aus *Lecidea lilienstroemii* und *Lecidea leucophaea.* Z. Naturforsch. **26 B**, 1357 (1971).

194. HUNECK, S., K. SCHREIBER, and G. SUNDHOLM: Ovosäure, ein neues Tridepsid aus der Flechte *Parmelia substygia.* Phytochem. **19**, 885 (1980).

195. HUNECK, S., and G. SNATZKE: Die Absolutkonfiguration der (−)-2-Methylen-3-carboxy-18-hydroxynonadecansäure. J. Hattori Bot. Lab. **48**, 211 (1980).

196. HUNECK, S., W. STEGLICH, and G. HÖFLE: Canarion, ein neues Naphthochinon aus *Usnea canariensis.* Phytochem. **16**, 121 (1977).

197. HUNECK, S., G. SUNDHOLM, and G. FOLLMANN: 3-Chlorodivaricatsäure, ein neues Depsid aus *Thelomma*-Arten. Phytochem. **19**, 645 (1980).

198. JACKMAN, D. A., M. V. SARGENT, and J. A. ELIX: Structure of the Lichen Depsidone Pannarin. J. Chem. Soc. (London) Perkin Trans. I **1975**, 1979.

199. JONES, A. J., J. A. ELIX, and U. ENGKANINAN: The Structure of the Lichen Depsidone Hydroxyphysodic Acid. A Carbon-13 Magnetic Resonance Study. Austral. J. Chem. **29**, 1947 (1976).

200. JONGEN, R., T. SALA, and M. V. SARGENT: Depsidone Synthesis. Part 14. The Total Synthesis of Variolaric Acid. J. Chem. Soc. (London) Perkin Trans. I **1979**, 2588.

201. JUNG, M. E., and J. A. LOWE: Synthetic Approaches to Adriamycin Involving Diels-Alder Reactions of Photochemically Generated Bisketenes. Total Synthesis of Islandicin and Digitopurpone. J. Org. Chem. **42**, 2371 (1977).

202. KAMAL, A., Y. HAIDER, Y. A. KHAN, I. H. QURESHI, and A. A. QURESHI: Studies in the Biochemistry of Microorganisms. Part X. Isolation, Structure and Stereochemistry of Yasimin and other Metabolic Products of *Aspergillus unguis* Emile-Weil and Gaudin. Pakistan J. Sci. Ind. Res. **13**, 244 (1970).

203. KAMAL, A., Y. HAIDER, A. A. QURESHI, and Y. A. KHAN: Studies in the Biochemistry of Microorganisms. Part XII. Isolation and Structure of Haiderin, Rubinin, Shirin and Nasrin Metabolic Products of *Aspergillus unguis* Emile-Weil and Gaudin. Pakistan J. Sci. Ind. Res. **13**, 364 (1970).

204. KANEDA, M., Y. IITAKA, and S. SHIBATA: X-ray Studies of C_{25} Terpenoids. IV. The Crystal Structure of Retigeranic Acid *p*-Bromoanilide. Acta Crystallogr. **30**, 358 (1974).

15*

205. KANEDA, M., R. TAKAHASHI, Y. IITAKA, and S. SHIBATA: Retigeranic Acid, A Novel Sesterterpene Isolated from the Lichens of *Lobaria retigera* Group. Tetrahedron Letters **1972**, 4609.
206. KENDE, A. S., J. L. BELLETIRE, J. L. HERRMANN, and R. F. ROMANET: Regiospecific Construction of Polyoxygenated 9,10-Anthraquinones, Total Synthesis of Islandicin and Digitopurpone. Synth. Comm. **3**, 387 (1973).
207. KENDE, A. S., J. L. BELLETIRE, and E. L. HUME: Regiospecific Synthesis of 1,4,5-Trioxygenated Anthraquinones. A Total Synthesis of Islandicin. Tetrahedron Letters **1973**, 2935.
208. KEOGH, M. F.: Malonprotocetraric Acid from *Parmotrema conformatum*. Phytochem. **16**, 1102 (1977).
209. — New β-Orcinol Depsidones from *Xanthoparmelia quintaria* and a *Thelometra* Species. Phytochem. **17**, 1192 (1978).
210. KEOGH, M. F., and I. DURAN: A New Fatty Acid from *Usnea meridensis*. Phytochem. **16**, 1605 (1977).
211. KEOGH, M. F., and M. E. ZURITA: α-(15-Hydroxyhexadecyl)itaconic Acid from *Usnea aliphatica*. Phytochem. **16**, 134 (1977).
212. KNIGHT, D. W., and G. PATTENDEN: Syntheses of Permethylated Derivatives of Pinastric Acid and Gomphidic Acid, Pulvinic Acid Pigments of Lichen and Fungi. J. Chem. Soc. (London) Perkin Trans. I **1979**, 84.
213. KOMIYA, T., and S. SHIBATA: Polyols Produced by the Cultured Phyco- and Mycobionts of some *Ramalina* Species. Phytochem. **10**, 695 (1971).
214. KOTCHETKOV, N. K., A. J. KHORLIN, and A. F. BOCHOV: A New Method of Glycosylation. Tetrahedron **23**, 693 (1967).
215. KUROKAWA, S., and J. A. ELIX: Two New Australian *Parmeliae*. J. Japan Bot. **46**, 113 (1971).
216. KUSHNIR, E., A. TIETZ, and M. GALUN: "Oil Hyphae" of Endolithic Lichens and their Fatty Acid Composition. Protoplasma **97**, 47 (1978).
217. KUTNEY, J. P., W. H. BAARSCHERS, O. CHIN, Y. EBIZUKA, L. HURLEY, J. D. LEMAN, P. J. SALISBURY, I. H. SANCHEZ, and T. YEE: Studies in the Usnic Acid Series. VIII. The Biodegradation of (+)-Usnic Acid by *Mortierella isabellina*. Can. J. Chem. **55**, 2930 (1977).
218. KUTNEY, J. P., J. D. LEMAN, P. J. SALISBURY, I. H. SANCHEZ, T. YEE, and R. J. BANDONI: Studies in the Usnic Acid Series. VII. The Biodegradation of (+)-Usnic Acid by a *Pseudomnas* Species. Isolation, Structure Determination, and Synthesis of (+)-6-Desacetylusnic Acid. Can. J. Chem. **55**, 2336 (1977).
219. KUTNEY, J. P., and I. H. SANCHEZ: Studies in the Usnic Acid Series. I. The Condensation of (+)-Usnic Acid with Aliphatic and Aromatic Amines. Can. J. Chem. **54**, 2795 (1976).
220. — — Studies in the Usnic Acid Series. V. The Base Catalysed Usnic Acid-Isousnic Acid Rearrangement. Part III. (−)-Usnic Acid Isomethoxide Monoacetate. Can. J. Chem. **55**, 1079 (1977).
221. KUTNEY, J. P., and I. H. SANCHEZ: Studies in the Usnic Acid Series. VI. The Preparation of some Ether Derivatives of (+)-Usnic Acid. Can. J. Chem. **55**, 1085 (1977).
222. KUTNEY, J. P., I. H. SANCHEZ, and T. YEE: Mass Spectral Fragmentation Studies in Usnic Acid and Related Compounds. Org. Mass Spectrom. **8**, 129 (1974).
223. — — — Studies in the Usnic Acid Series. II. The Condensation of (+)-Usnic Acid with Hydroxylamine. Can. J. Chem. **54**, 3713 (1976).
224. — — — Studies in the Usnic Acid Series. III. The Base Catalysed Usnic Acid-Isousnic Acid Rearrangement. The Synthesis of (+)-Isousnic Acid. Can. J. Chem. **54**, 3721 (1976).
225. — — — Studies in the Usnic Acid Series. IV. The Base Catalysed Usnic Acid-Isousnic Acid Rearrangement. Part II. An Improved Synthesis of (+)-Isousnic Acid. Can. J. Chem. **55**, 1073 (1977).

226. LAM, J. K. K., M. V. SARGENT, J. A. ELIX, and D. O'N. SMITH: Synthesis of Valsarin and 5,7-Dichloroemodin. J. Chem. Soc. (London) Perkin Trans. I **1972**, 1466.

227. LELE, S. R., and B. D. HOSANGADI: Studies in Photoinduced Reactions on Depsidones: Synthesis of Dibenzofuran from 11H-Dibenzo[b,e] [1,4]dioxepin-11-one. Indian J. Chem. **16B**, 415 (1978).

228. LENTON, J. R., L. J. GOAD, and T. W. GOODWIN: Sterols of *Xanthoria parietina*: Evidence for Two Sterol "Pools" and the Identification of a Novel C_{28} Triene, Ergosta-5,8,22-trien-3β-ol. Phytochem. **12**, 1135 (1973).

229. — — — Sterols of the Mycobiont and Phycobiont Isolated from the Lichen *Xanthoria parietina*. Phytochem. **12**, 2249 (1973).

230. LINDBERG, B., B.-G. SILVANDER, and C. A. WACHTMEISTER: Studies of the Chemistry of Lichens. 19. Mannitol Glycosides in *Peltigera* Species. Acta Chem. Scand. **18**, 213 (1964).

231. MAASS, W. S. G.: Lichen Substances. IV. Incorporation of Pulvinic-^{14}C Acids into Calycin by the Lichen *Pseudocyphellaria crocata*. Can. J. Biochem. **48**, 1241 (1970).

232. — Lichen Substances V. Methylated Derivatives of Orsellinic Acid, Lecanoric Acid and Gyrophoric Acid from *Pseudocyphellaria crocata*. Can. J. Bot. **53**, 1031 (1975).

233. — Lichen Substances VII. Identification of Orsellinate Derivatives from *Lobaria linita*. Bryologist **78**, 178 (1975).

234. — The Phenolic Constituents of *Peltigera aphthosa*. Phytochem. **14**, 2487 (1975).

235. MAASS, W. S. G., A. G. MCINNES, D. G. SMITH, and A. TAYLOR: Lichen Substances. X. Physciosporin, a New Chlorinated Depsidone. Can. J. Chem. **55**, 2839 (1977).

236. MACDONALD, A. L., S. J. RETTIG, and J. TROTTER: Crystal and Molecular Structure of 2-Desacylusnic Acid. Can. J. Chem. **52**, 723 (1974).

237. MCEWEN, P. M., and M. V. SARGENT: Depsidone Synthesis. Part 20. Lecideoidin and Dechlorolecideoidin. J. Chem. Soc. (London) Perkin Trans. I **1981**, 883.

238. MAHANDRU, M. M., and O. L. GILBERT: Chemical Studies in *Fulgensia*: Structures of Two New Chlorodepsidones. Bryologist **82**, 302 (1979).

239. MIĆOVIĆ, V. M., M. HRANISAVLJEVIĆ-JAKOVLJEVIĆ, and J. MILJKOVIĆ-STOJANOVIĆ: Structural Study of Polysaccharides from the Oak Lichen *Evernia prunastri* (L.) Ach. Part I. An Alkali-soluble Galactomannan. Carbohydrate Res. **10**, 525 (1969).

240. MISHCHENKO, N. P., O. E. KRIVOSHCHENKOVA, O. B. MAKSIMOV, and L. S. STEPANENKO: Anthraquinones of the Lichen *Asahinea chrysantha*. Chem. Natural Compounds **16**, 117 (1980).

241. MOLHO, D., B. BODO, W. L. CULBERSON, and C. F. CULBERSON: A Chemically Distinctive New *Ramalina* from Fiji. Bryologist **84**, 396 (1981).

242. MOLHO, L., B. BODO, and D. MOLHO: L'Acide *O*-Methyl-4'-norsekikaique, Nouveau meta-Depside Isole d'un Lichen du Genre *Ramalina*. Phytochem. **18**, 2049 (1979).

243. MORONEY, S. E., K. J. RONALDSON, A. L. WILKINS, T. G. A. GREEN, and P. W. JAMES: Depsidone Constituents from the Quintaria Group of *Nephroma* Species. Phytochem. **20**, 787 (1981).

244. MOSBACH, K.: Biosynthesis of Lichen Substances, Ch. 16. In: The Lichens (V. AMADJIAN and M. E. HALE, JR., eds.). New York and London: Academic Press, 1973.

245. MURPHY, D., J. KEANE, and T. J. NOLAN: Chemical Constituents of Lichens Found in Ireland. Constitution of Variolaric Acid. Sci. Proc. Roy. Dublin Soc. **23**, 71 (1943).

246. NAKANISHI, T., H. YAMAUCHI, T. FUJIWARA, and K. TOMITA: The Crystal Structure of 6-*O*-*p*-Bromobenzoyl Zeorin. Tetrahedron Letters **1971**, 1157.

247. NAKANO, H., T. KOMYA, and S. SHIBATA: Anthraquinones of the Lichens of *Xanthoria* and *Caloplaca* and their Cultivated Mycobionts. Phytochem. **11**, 3505 (1972).

248. NEELAKANTAN, S., R. PADMASANI, and T. R. SESHADRI: A Note on the Synthesis of Diploicin Methyl Ether. Current Sci. (India) **33**, 365 (1964).

249. — — — Halogenation of the Depsides, Lecanoric Acid and Atranorin. Indian J. Chem. **2**, 478 (1964).

250. Neelakantan, S., and N. Thillaichidambaran: A New Synthesis of Thiophanic Acid. Current Sci. (India) **42**, 21 (1973).
251. Nicollier, G., M. Rebetez, R. Tabacchi, H. Gerlach, and A. Thalmann: Synthése de l'Evernine. Helv. Chim. Acta **61**, 2899 (1978).
252. Nicollier, G., M. Rebetez, and R. Tabacchi: Identification et Synthése de Nouveaux Depsides Isoles de la Mousse de Chêne [*Evernia prunastri* (L.) Ach.). Helv. Chim. Acta **62**, 711 (1979).
253. Nicollier, G., and R. Tabacchi: Isolement et Identification de l'Evernine dans la Mousse de Chêne [*Evernia prunastri* (L.) Ach.]. Helv. Chim. Acta **59**, 2979 (1976).
254. Nishikawa, Y., K. Michishita, and G. Kurono: Studies on the Water Soluble Constituents of Lichens. I. Gas Chromatographic Analysis of Low Molecular Weight Carbohydrates. Chem. Pharm. Bull. (Japan) **21**, 1014 (1973).
255. Nishikawa, Y., K. Ohki, K. Takahashi, G. Kurono, F. Fukuoka, and M. Emori: Studies on the Water Soluble Constituents of Lichens. II. Antitumor Polysaccharides of *Lasallia, Usnea,* and *Cladonia* Species. Chem. Pharm. Bull. (Japan) **22**, 2692 (1974).
256. Nishikawa, Y., T. Takeda, S. Shibata, and F. Fukuoka: Polysaccharides in Lichens and Fungi. III. Further Investigation on Structures and the Antitumour Activity of the Polysaccharides from *Gyrophora esculenta* Miyoshi and *Lasallia papulosa* (Ach.) Llano. Chem. Pharm. Bull. (Japan) **17**, 1910 (1969).
257. Nishikawa, Y., K. Yoshimoto, R. Horiuchi (née Murakami), K. Michishita, M. Okabe, and F. Fukuoka: Change in Antitumor Effect Caused by Modifications of Pustulan Type and Lichenan Type Glucans. Chem. Pharm. Bull. (Japan) **27**, 1065 (1979).
258. Nolan, T. J., J. Algar, E. P. McCann, W. A. Manahan, and N. Nolan: The Chemical Constituents of Lichens Found in Ireland. *Buellia canescens.* Part 3. The Constitution of Diploicin. Sci. Proc. Roy. Dublin Soc. **24**, 319 (1948).
259. Norrestam, R., M. von Glehn, and C. A. Wachtmeister: Three-Dimensional Structure of Usnic Acid. Acta Chem. Scand. **28B**, 1149 (1974).
260. Nuno, M.: On the Isolation of Chemical Ingredients of *Usnea baileyi.* J. Japan Bot. **46**, 294 (1971).
261. — Miscellaneous Notes on *Cladonia* (1). Misc. Bryol. Lichenol. **6**, 126 (1973).
262. Nuno, M., Y. Kuwada, and K. Kamiya: The Structure of Nephroarctin. Chem. Commun. **1969**, 78.
263. O'Donovan, D. G., G. Roberts, and M. F. Keogh: Structure of the β-Orcinol Depsidones, Connorstictic and Consalazinic Acids. Phytochem. **19**, 2497 (1980).
264. Ollis, W. D.: Personal Communication.
265. Pattenden, G.: Natural 4-Ylidenebutenolides and 4-Ylidenetetronic Acids. Fortschr. Chem. organ. Naturstoffe **35**, 133 (1978).
266. Poelt, J., and S. Huneck: *Lecanora vinetorum* nova spec., ihre Vergesellschaftung, ihre Ökologie und ihre Chemie. Österr. Bot. Z. **115**, 411 (1968).
267. Pring, B. E., and N. E. Stjernström: Complex Dibenzofurans. XI. The Mechanism of the Acid-Catalysed Dehydration of 2,2′-Dihydroxybiphenyls. Acta Chem. Scand. **22**, 538 (1968).
268. Ramaut, J. L., and J. Thonar: Inhibition de la Germination de Différentes Graines d'Angiospermes par *Evernia prunastri* (L.) Ach. I. Quimica Anales Real Soc. Españ. Fis. Quim. **68**, 575 (1972).
269. — — Inhibition de la Germination de Différentes Graines d'Angiospermes par *Evernia prunastri* (L.) Ach. 2ème partie Quimica Anales Real Soc. Españ. Fis. Quim. **68**, 597 (1972).
270. Rana, N. M., M. V. Sargent, and J. A. Elix: Structure of the Lichen Depsidone Variolaric Acid. J. Chem. Soc. (London) Perkin Trans. I **1975**, 1992.
271. Rao, P. S., and T. R. Seshadri: Chemical Investigation of Indian Lichens: Part XXIX. Structural Studies on Retigeradiol. Indian J. Chem. **6**, 398 (1968).

272. RICHARDSON, D. H. S.: Photosynthesis and Carbohydrate Movement. In: The Lichens (V. AHMADJIAN and M. E. HALE, eds.), p. 249. New York: Academic Press. 1973.
273. RONALDSON, K. J., and A. L. WILKINS: The Structure of Amphistictinic Acid a Triterpenoid Constituent of the Lichen *Pseudocyphellaria amphisticta*. Austral. J. Chem. **31**, 215 (1978).
274. SAFE, S., L. M. SAFE, and W. S. G. MAASS: Sterols of Three Lichen Species: *Lobaria pulmonaria, Lobaria scrobiculata* and *Usnea longissima*. Phytochem. **14**, 1821 (1975).
275. SALA, T., and M. V. SARGENT: Depsidone Synthesis. Part 14. The Total Synthesis of Psoromic Acid: Isopropyl Ethers as Useful Phenolic Protective Groups. J. Chem. Soc. (London) Perkin Trans. I **1979**, 2593.
276. — — Depsidone Synthesis. Part 16. Benzophenone-Grisa-3′,5′-diene-2′,3-dione-Depsidone Interconversion: a New Theory of Depsidone Biosynthesis. J. Chem. Soc. (London) Perkin Trans. I **1981**, 855.
277. — — Depsidone Synthesis. Part 19. Some β-Orcinol Depsidones. J. Chem. Soc. (London) Perkin Trans I **1981**, 877.
278. SALA, T., M. V. SARGENT, and J. A. ELIX: Depsidone Synthesis. Part 15. New Metabolites of the Lichen *Buellia canescens* (Dicks.) De Not: Novel Phthalide Catabolites of Depsidones. J. Chem. Soc. (London) Perkin Trans. I **1981**, 849.
279. SANKAWA, U., Y. EBIZUKA, and S. SHIBATA: Biosynthetic Incorporation of Emodin and Emodinanthrone into the Anthraquinoids of *Penicillium brunneum* and *P. islandicum*. Tetrahedron Letters **1973**, 2125.
280. SANTESSON, J.: Chemical Studies on Lichens. 16. The Xanthones of *Lecanora straminea*. I. Arthothelin and Thiophanic Acid. Ark. Kemi **30**, 449 (1969).
281. — Chemical Studies on Lichens. 17. The Xanthones of *Lecanora straminea*. II. 2,7-Dichloronorlichexanthone. Ark. Kemi **30**, 455 (1969).
282. — Chemical Studies on Lichens. 18. The Xanthones of *Lecanora straminea*. III. Norlichexanthone, 2-Chloronorlichexanthone, and 2,4-Dichloronorlichexanthone. Ark. Kemi **30**, 461 (1969).
283. — Chemical Studies on Lichens. 20. The Xanthones of some Crustaceous Lichens. Ark. Kemi **31**, 57 (1969).
284. — Chemical Studies on Lichens. 21. Two Novel Chlorinated Lichen Xanthones. Ark. Kemi **31**, 121 (1969).
285. SANTESSON, J., and G. SUNDHOLM: Chemical Studies on Lichens. 14. Syntheses and Chlorinations of Norlichexanthone. Ark. Kemi **30**, 427 (1969).
286. SARGENT, M. V., and P. VOGEL: Depsidone Synthesis. IV. Caloploicin. Austral. J. Chem. **29**, 907 (1976).
287. SARGENT, M. V., P. VOGEL, and J. A. ELIX: Structure of the Lichen Depsidone Gangaleoidin. J. Chem. Soc. (London) Perkin Trans. I **1975**, 1986.
288. SARGENT, M. V., P. VOGEL, J. A. ELIX, and B. A. FERGUSON: Depsidone Synthesis. VII. Vicanicin and Norvicanicin. Austral. J. Chem. **29**, 2263 (1976).
289. SEO, S., U. SANKAWA, Y. OGIHARA, Y. IITAKA, and S. SHIBATA: Structure of (−)-Flavoskyrin, a Colouring Matter of *Penicillium islandicum* Sopp NRRL 1175. Tetrahedron Letters **1972**, 735.
290. — — — — — Studies on Fungal Metabolites — XXXII. A Renewed Investigation on (−)-Flavoskyrin and its Analogues. Tetrahedron **29**, 3721 (1973).
291. SEO, S., U. SANKAWA, and S. SHIBATA: Dimerisation of 3,4-Dihydro-1,9,10(2H)-anthracenetrione Derivatives. Tetrahedron Letters **1972**, 731.
292. SHIBATA, S.: Polysaccharides of Lichens. J. Nat. Res. Council Sri Lanka **1**, 183 (1973).
293. SHIBATA, S., Y. NISHIKAWA, T. TAKEDA, and M. TANAKA: Polysaccharides in Lichens and Fungi. I. Antitumour Active Polysaccharides of *Gyrophora esculenta* Miyoshi and *Lasallia papulosa* (Ach.) Llano. Chem. Pharm. Bull. (Japan) **16**, 2362 (1968).

294. Shibata, S., Y. Nishikawa, M. Tanaka, F. Fukuoka, and M. Nakanishi: Antitumour Activities of Lichen Polysaccharides. Z. Krebsforsch. **71**, 102 (1968).

295. Shimada (née Miyoshi), S., T. Saitoh, U. Sankawa, and S. Shibata: New Depsidones from *Lobaria oregana*. Phytochem. **19**, 328 (1980).

296. Shimada (née Miyoshi), S., T. Saitoh, Y. Namiki, U. Sankawa, and S. Shibata: New Siphulin Derivatives from the Lichen *Siphula ceralites*. Phytochem. **19**, 467 (1980).

297. Sierankiewicz, J., and S. Gatenbeck: A New Depsidone from *Aspergillus nidulans*. Acta Chem. Scand. **26**, 455 (1972).

298. — — The Biosynthesis of Nidulin and Trisdechloronornidulin. Acta Chem. Scand. **27**, 2710 (1973).

299. Solberg, Y. J.: Studies on the Chemistry of Lichens. VIII. An Examination of the Free Sugars and Ninhydrin-positive Compounds of Several Norwegian Lichen Species. Lichenologist **4**, 271 (1970).

300. — Studies on the Chemistry of Lichens. X. Chemical Investigation of the Lichen Species *Xanthoria parietina* (L.) Th. Fr. Bryologist **74**, 144 (1971).

301. — Studies on the Chemistry of Lichens, XII. Chemical Investigation of the Lichen Species *Xanthoria parietina* (L.) Th. Fr. −2. Bryologist **77**, 203 (1974).

302. — Flechteninhaltsstoffe, XIII. Physodol, ein neues Depsidon aus *Hypogymnia physodes*. Z. Naturforsch. **29 B**, 364 (1974).

303. — Studies on the Chemistry of Lichens, XI. Chemical Investigation of Five Norwegian *Alectoria* Species. Acta Chem. Scand. **29 B**, 145 (1975).

304. — Studies on the Chemistry of Lichens, XIV. Chemical Investigation of the Lichen Species *Anaptychia fusca, Peltigera canina,* and *Omphalodiscus spodochrous*. Z. Naturforsch. **30 C**, 445 (1975).

305. — Studies on the Chemistry of Lichens, XVI. Chemical Investigation of the Lichen Species *Alectoria ochroleuca, Stereocaulon vesuvianum* var. *pulvinatum* and *Icmadophila ericetorum*. Z. Naturforsch. **32 C**, 182 (1977).

306. Steglich, W., and K. F. Jedtke: Neue Anthrachinonfarbstoffe aus *Solorina crocea*. Z. Naturforsch. **31 C**, 197 (1976).

307. Stodola, F. H., R. F. Vesonder, D. I. Fennell, and D. Weisleder: A New Depsidone from *Aspergillus unguis*. Phytochem. **11**, 2107 (1972).

308. Sundholm, E. G.: Total Synthesis of Lichen Xanthones. Revision of Structures. Tetrahedron **34**, 577 (1978).

309. — ^{13}C NMR Spectra of Lichen Xanthones. Temperature Dependent Collapse of Long-range Couplings to Hydrogen-bonded Hydroxyl Protons. Acta Chem. Scand. **32 B**, 177 (1978).

310. — Synthesis and ^{13}C NMR Spectra of some 5-Chloro Substituted Lichen Xanthones. Acta Chem. Scand. **33 B**, 475 (1979).

311. Sundholm, E. G., and S. Huneck: ^{13}C NMR-Spectra of Lichen Depsides, Depsidones and Depsones. 1. Compounds of the Orcinol Series. Chemica Scripta **16**, 197 (1980).

312. — — ^{13}C NMR-Spectra of Lichen Depsides, Depsidones and Depsones. 2. Compounds of the β-Orcinol Series. Chemica Scripta **18**, 233 (1981).

313. Takahashi, R., H.-C. Chiang, N. Aimi, O. Tanaka, and S. Shibata: The Structures of Retigeric Acids A and B from Lichens of the *Lobaria retigera* Group. Phytochem. **11**, 2039 (1972).

314. Takahashi, R., and Y. Iitaka: The Crystal and Molecular Structure of *p*-Bromophenacyl Retigerate A. Acta Crystallogr. **28 B**, 764 (1972).

315. Takahashi, K., and M. Takani: Usnic Acid. IX. The Pyrolysis of Tetrahydrodesoxyusnic Acid. Chem. Pharm. Bull. (Japan) **19**, 2079 (1971).

316. — — Usnic Acid. X. The Pyrolysis of Tetrahydrodesoxyusnic and Dihydrousnic Acids. Chem. Pharm. Bull. (Japan) **20**, 1230 (1972).

317. — — Usnic Acid. XIII. The Pyrolysis of the Oxidation Product of Dihydrousnic Acid. Chem. Pharm. Bull. (Japan) **26,** 526 (1978).

318. — — Usnic Acid. XIV. The Photo-oxidation of Usnic Acid. Chem. Pharm. Bull. (Japan) **26,** 3585 (1978).

319. — — Usnic Acid. XV. Alkaline Degradation of Usnic Acid. Chem. Pharm. Bull. (Japan) **28,** 177 (1980).

320. TAKAHASHI, K., M. TAKANI, and A. FUKUMATO: Usnic Acid. XI. Photolysis of Dihydrousnic and Methyldihydrousnic Acids. Chem. Pharm. Bull. (Japan) **22,** 115 (1974).

321. TAKAHASHI, K., M. TAKANI, and Y. WADA: Usnic Acid. XVI. Alkaline Degradation of Dihydrousnic Acid. Chem. Pharm. Bull. (Japan) **28,** 1590 (1980).

322. TAKAHASHI, K., T. TAKEDA, and S. SHIBATA: Polysaccharides of Lichen Symbionts. Chem. Pharm. Bull. (Japan) **27,** 238 (1979).

323. TAKAHASHI, K., T. TAKEDA, S. SHIBATA, M. INOMATA, and F. FUKUOKA: Polysaccharides of Lichens and Fungi. VI. Antitumor Active Polysaccharides of Lichens of *Stictaceae.* Chem. Pharm. Bull. (Japan) **22,** 404 (1974).

324. TAKAHASHI, K., and Y. TANAKA: Usnic Acid. XII. The Oxidation of Dihydrousnic Acid. Chem. Pharm. Bull. (Japan) **23,** 623 (1975).

325. TAKAHASHI, R., O. TANAKA, and S. SHIBATA: Ergosterol Peroxide from *Peltigera* Species. Phytochem. **11,** 1850 (1972).

326. TAKANI, M., and K. TAKAHASHI: Usnic Acid. VIII. The Dienone-Phenol Rearrangement of 9-*O*-Acetyltetrahydrodesoxyusnic and Dihydrousnic Acids. Chem. Pharm. Bull. (Japan) **19,** 2072 (1971).

327. TAKEDA, T., M. FUNATSU, S. SHIBATA, and F. FUKUOKA: Polysaccharides of Lichens and Fungi. V. Antitumor Active Polysaccharides of Lichens of *Evernia, Acroscyphus* and *Alectoria* spp. Chem. Pharm. Bull. (Japan) **20,** 2445 (1972).

328. TAKEDA, T., Y. NISHIKAWA, and S. SHIBATA: A New α-Glucan from the Lichen *Parmelia caperata* (L.) Ach. Chem. Pharm. Bull. (Japan) **18,** 1074 (1970).

329. TAKEDA, N., S. SEO, Y. OGIHARA, U. SANKAWA, I. IITAKA, I. KITAGAWA, and S. SHIBATA: Studies on Fungal Metabolites — XXXI. Anthraquinoid Colouring Matters of *Penicillium islandicum* Sopp and some other Fungi. (−)-Luteoskyrin, (−)-Rubroskyrin, (ǀ)-Rugulosin and their Related Compounds. Tetrahedron **29,** 3703 (1973).

330. TOKUZEN, R.: Comparison of Local Cellular Reaction to Tumor Grafts in Mice Treated with Some Plant Polysaccharides. Cancer Res. **31,** 1590 (1971).

331. WEINSTOCK, J., J. E. BLANK, H.-J. OH, and B. M. SUTTON: A Regiospecific Synthesis of Substituted Vulpinic Acids. J. Organ. Chem. (USA) **44,** 673 (1979).

332. WOJCIECHOWSKI, Z. A., L. J. GOAD, and T. W. GOODWIN: Sterols of *Pseudevernia furfuracea.* Phytochem. **12,** 1433 (1973).

333. YAMAMOTO, Y., and A. WATABABE: Fatty Acid Composition of Lichens and their Phyco- and Mycobionts. J. Gen. Appl. Microbiol. **20,** 83 (1974).

334. YANG, D.-M., N. TAKEDA, Y. IIKATA, U. SANKAWA, and S. SHIBATA: Structure of Eumitrins A_1, A_2 and B. The Yellow Pigments of the Lichen, *Usnea baileyi* (Stirt.) Zahlbr. Tetrahedron **29,** 519 (1973).

335. YOKOTA, I., and S. SHIBATA: A Polysaccharide of the Lichen, *Stereocaulon japonicum.* Chem. Pharm. Bull. (Japan) **26,** 2668 (1978).

336. YOKOTA, I., S. SHIBATA, and H. SAITÔ: A ^{13}C n.m.r. Analysis of Linkages in Lichen Polysaccharides: An Approach to Chemical Taxonomy of Lichens. Carbohydrate Res. **69,** 252 (1979).

337. YOSIOKA, I.: On the Structure of Pannarin, a Component of the Lichen Genus *Pannaria.* J. Pharmac. Soc. Japan **61,** 332 (1941).

338. YOSIOKA, I., K. HINO, M. FUJIO, and I. KITAGAWA: The Structure of Caloploicin, a New Lichen Trichlorodepsidone. Chem. Pharm. Bull. (Japan) **21,** 1547 (1973).

339. Yosioka, I., A. Matsuda, and I. Kitagawa: Pyxinic Acid, a Novel Lichen Triterpene with 3β-Hydroxyl Function. Tetrahedron Letters **1966,** 613.

340. Yosioka, I., T. Nakanishi, S. Izumi, and I. Kitagawa: Structure of a Lichen Pigment Entothein and its Identity with Secalonic Acid A, a Major Ergot Pigment. Chem. Pharm. Bull. (Japan) **16,** 2090 (1968).

341. Yosioka, I., T. Nakanishi, M. Yamaki, and I. Kitagawa: The Structures of Leucotylic Acid and Methyl Isoleucotylate, an Acid-Induced Isomer of Methyl Leucotylate. Chem. Pharm. Bull. (Japan) **20,** 487 (1972).

342. Yosioka, I., T. Nakanishi, H. Yamauchi, and I. Kitagawa: Lichen Triterpenoids. III. The Final Conclusion on the Stereochemistry of Zeorin and its Correlation with Leucotylin. The Structure of Isoleucotylin. Chem. Pharm. Bull. (Japan) **20,** 147 (1972).

343. Yosioka, I., M. Yamaki, T. Nakanishi, and I. Kitagawa: The Structure of Methyl Isoleucotylate, an Acid Isomerised Product of Methyl Leucotylate. Tetrahedron Letters **1966,** 2227.

344. Yosioka, I., H. Yamauchi, and I. Kitagawa: Diacetylpyxinol, a Triterpene Alcohol from a Lichen: *Pyxine endochrysina* Nyl. Tetrahedron Letters **1969,** 4241.

345. — — — Neutral Triterpenoids of *Pyxine endochrysina*. Chem. Pharm. Bull. (Japan) **20,** 502 (1972).

346. Yosioka, I., H. Yamauchi, K. Murata, and I. Kitagawa: Colouring Substance of a Lichen *Cetraria ornata*. Chem. Pharm. Bull. (Japan) **20,** 1082 (1972).

(Received April 4, 1983)

Paralytic Shellfish Poisons

By Y. Shimizu, Department of Pharmacognosy and Environmental
Health Sciences, College of Pharmacy, University of Rhode Island,
Kingston, Rhode Island, U.S.A.

Contents

I. Introduction

Paralytic shellfish poisons (PSP) are the toxins responsible for acute and
often fatal poisonings caused by the consumption of certain shellfish. The
phenomenon has been known since prehistoric times. A thorough docum-
entation of the incidents can be found in HALSTEAD's treatise on poisonous

marine organisms (*1*). Saxitoxin (**1**) was the first toxin recognized in toxic shellfish, but recently a series of new toxins has been discovered in various specimens. The toxins have particular importance as pharmacological tools because of their unique action against sodium channels in the excitable membrane. Earlier accounts of toxin research were reviewed by SCHANTZ (*2*) and SHIMIZU (*3*).

II. Occurrence and Isolation

A. Occurrence

SOMMER and MEYER were the first to suggest that dinoflagellates are the primary sources of the toxins discovered in shellfish (*4, 5*). To date, *Gonyaulax (=Protogonyaulax) catenella, G. tamarensis (=G. tamarensis* var. *excavata* or *G. excavata)* (*6* and references therein), and *Pyrodinium bahamense* var. *compressa* (*7, 8*) have been identified as toxin-producing organisms. In addition to these dinoflagellates, a toxic strain of freshwater blue-green alga, *Aphanizomenon flos-aquae* is also known to contain related compounds (*9—11*). The accumulation of the toxins in shellfish is the result of filter-feeding of these organisms especially during the outbreak of blooms (red tides). Thus filter-feeding bivalves are generally considered to become toxic. However, the PSP toxins are also found in certain species of crabs (*12—16*) and carnivorous snails (*17*) probably as a result of secondary toxin transfer in the food chain. The toxicity of shellfish is normally transitory and gradually diminishes as the causative organisms disappear from the water. One exception is Alaska butter clam, *Saxidomus giganteus,* which tends to retain the toxicity in the siphon for a length of time (*18*).

The occurrence of paralytic shellfish poisons was once considered to be a problem confined mostly to northern temperate waters. It is now found widespread in both northern and southern hemispheres including subtropical to tropical waters (*19*).

B. Isolation

Saxitoxin (**1**) was the first toxin brought to a pure form (*20*). The purification procedure was based upon the toxin's ability to bind tightly to weakly acidic resins. A single chromatography on a carboxylate resin such as Amberlite IRC-50 was capable of raising the specific toxicity of the extract several hundred times. Repeated chromatography on carboxylate resins and subsequent chromatography on acidic alumina afforded pure saxitoxin as an amorphous dihydrochloride.

References, pp. 261—264

This isolation procedure was, however, not applicable to the other PSP toxins. The small molecular water-soluble toxins were difficult to separate from the overwhelming mass of other ingredients in the shellfish extracts. A breakthrough came when it was found that the toxins could be adsorbed on small matrix gels such as Sephadex-15 or Bio-Gel P-2 at pH around 6.0 and eluted with a dilute acetic acid solution (*21, 22*). The crude toxin mixture was then separated into individual toxins by chromatography on weakly acidic resin such as Bio-Rex 70 by acetic acid gradient elution (*21*). Toxins with positive net charges can be purified by this procedure (*23*) (Scheme I). Toxins having zero or negative net charges are not held tightly by the aforementioned gels. These toxins have to be purified by repeated chromatography on Bio-Gel P-2 or preparative thin-layer chromatography (*24, 25*).

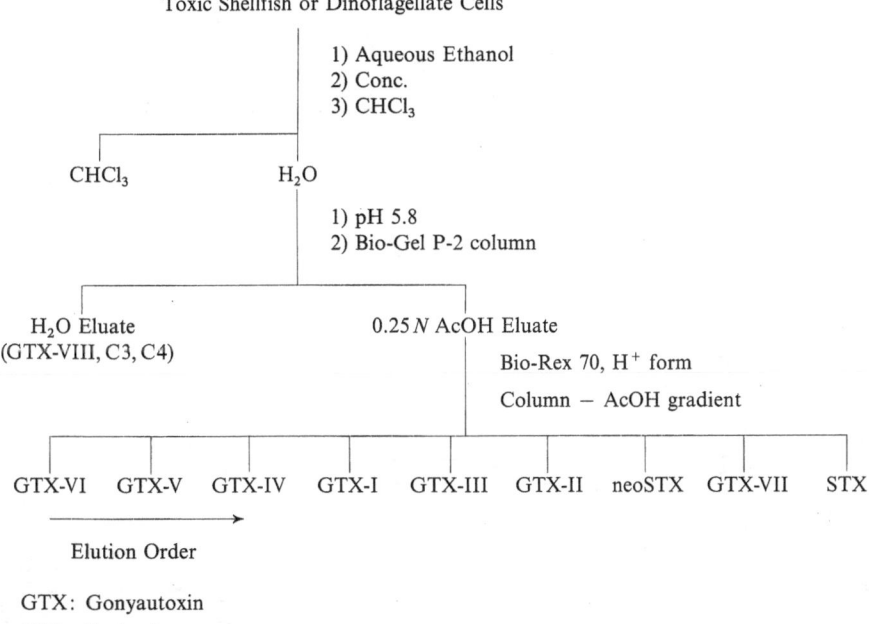

Toxic Shellfish or Dinoflagellate Cells

1) Aqueous Ethanol
2) Conc.
3) CHCl₃

CHCl₃   H₂O

1) pH 5.8
2) Bio-Gel P-2 column

H₂O Eluate   0.25 N AcOH Eluate
(GTX-VIII, C3, C4)

Bio-Rex 70, H⁺ form

Column — AcOH gradient

GTX-VI   GTX-V   GTX-IV   GTX-I   GTX-III   GTX-II   neoSTX   GTX-VII   STX

Elution Order

GTX: Gonyautoxin
STX: Saxitoxin

Scheme I. General isolation procedure for paralytic shellfish toxins (*26*)

To date twelve toxins including saxitoxin have been isolated from various sources. The isolated toxins and their chromatographic and electrophoretic behavior are summarized in Table 1. Those toxins whose origins were identified as *Gonyaulax* spp. were named serially gonyautoxin-I — -VIII (abbreviated GTX-I — -VIII or GTX_1 — GTX_8). Neosaxitoxin

Table 1. *Paralytic Shellfish Toxins and Their Chromatographic and Electrophoretic Behaviors*

Toxin	TLC Rf a	TLC Rf b	Electrophoresis Rm c	Elution Order	Major Sources	References for First Reports
Saxitoxin (1)	0.62	0.51	1.00	9	*Gonyaulax* spp. shellfish	(20)
Neosaxitoxin (2)	0.70	0.54		7	*Gonyaulax* spp. *Aphanizomenon flos-aquae* shellfish, crabs, etc.	(26)
Gonyautoxin-I (3)	0.90	0.70	0.15	4	*Gonyaulax* spp. shellfish	(21)
Gonyautoxin-II (4)	0.81	0.65	0.56	6	*Gonyaulax* spp. shellfish	(21)
Gonyautoxin-III (5)	0.69	0.61	0.28	5	*Gonyaulax* spp. shellfish	(21)
Gonyautoxin-IV (6)	0.81	0.65	0	3	*Gonyaulax* spp. shellfish	(26)
Gonyautoxin-V (=B1) (7)	0.61	0.52	0.28	2	*Gonyaulax* spp. *Pyrodinium bahamense* var. *compressa* shellfish	(25, 26)
Gonyautoxin-VI (=B2) (8)	0.57		0.08	1	*Gonyaulax* spp. *Pyrodinium bahamense* var. *compressa* shellfish	(25, 27)
Gonyautoxin-VII (9)	0.44		0.97	8	Scallop (*Placopecten magellanicus*)	(28)
Gonyautoxin-VIII (=C2) (10)	0.88		−0.40	—	*Gonyaulax* spp. shellfish	(24, 25, 29)
Epigonyautoxin-VIII (=C1) (11)				—	*Gonyaulax* spp. shellfish	(24, 25)
C3 (12)				—	*Gonyaulax* spp.	(30)
C4 (13)				—	*Gonyaulax* spp.	(30)

a Silica gel 60 (pyridine : ethyl acetate : water : acetic acid = 75 : 25 : 30 : 15).
b Silica gel GF (t-butyl alcohol : acetic acid : water = 2 : 1 : 1).
c Tris buffer pH 8.7, 0.2 mA/cm, 200 V (31).

References, pp. 261—264

(2) was named differently because it was first discovered in Alaska butter clam (26) together with saxitoxin although it was later found in *Gonyaulax* organisms. Gonyautoxin-VII has been so far found only in a toxic scallop sample exposed to *Gonyaulax* blooms (28). Later some of the toxins were reisolated and reported under different codes or as derivatives of saxitoxin.

In general, most strains of *Gonyaulax tamarensis* and *G. catenella* contain mixtures of gonyautoxins, neosaxitoxin and saxitoxin and their side-chain sulfated derivatives (32). In a few strains, however, saxitoxin or neosaxitoxin appears to be the sole toxic constituent (33). The toxin profiles in toxic shellfish basically reflect those of the causative organisms. Thus toxin components in shellfish are normally quite heterogeneous and saxitoxin constitutes only a minor portion except for a few cases (20).

III. Structure and Chemistry

Structurally the toxins can be divided into three groups: 1) toxins without sulfate conjugation: saxitoxin (1), neosaxitoxin (2), and gonyautoxin-VII (9); 2) toxins with single sulfate conjugation in the 11-hydroxyl group: gonyautoxin-I (3), -II (4), -III (5), and -IV (6); 3) toxins with sulfate conjugation in the side chain carbamoyl group: gonyautoxin-V (7), -VI (8), -VIII (10), C3 toxin (12), and C4 toxin (13).

A. Saxitoxin

Since saxitoxin (1) was purified in 1957 (20), it has been the subject of extensive investigation. Although the final structure of saxitoxin (1) was established by X-ray crystallography (34, 35), the chemistry of saxitoxin for which we are indebted mostly to RAPOPORT's group presents some intriguing aspects.

The molecular formula of amorphous saxitoxin dihydrochloride was first proposed as $C_{10}H_{17}N_7O_4 \cdot 2\,HCl$ by elemental analysis (20, 36). Fourteen years later a question arose regarding the appropriateness of the formula (37). Thus when a thoroughly dried sample was analyzed, it showed a formula $C_{10}H_{15}N_7O_3 \cdot 2\,HCl$, which is one molecule of water less than proposed originally. It was difficult to determine if the difference was due to a molecule of solvation or derived from part of the structure. Nonetheless, the "thoroughly dried sample" was shown to possess the original toxicity. It was also demonstrated that only one ^{18}O atom was incorporated by $H_2^{18}O$ exchange and that it was removed again by thorough drying. As the structure of saxitoxin is now known to have a hydrated keto group, the whole episode can be explained by the keto-keto hydrate conversion shown in Scheme II with preferential attack and departure of the $^{18}OH^-$ group only from one side (possibly from the α-side) (14).

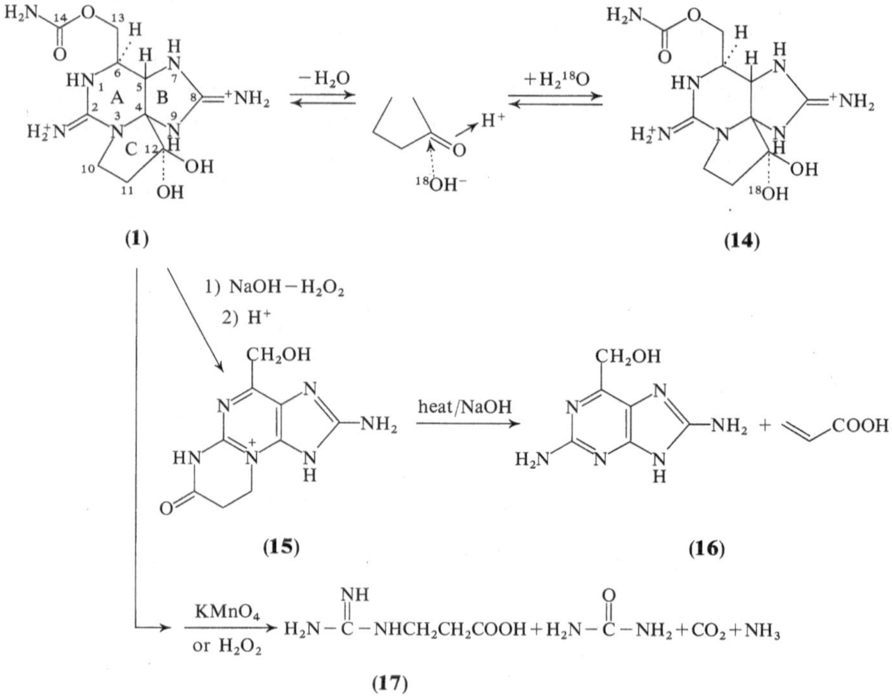

Scheme II. Degradation and ^{18}O-exchange reactions of saxitoxin

Saxitoxin (1) was found to be extremely unstable under alkaline conditions in the presence of oxygen, a reaction leading to the formation of a fluorescent degradation product (36). RAPOPORT's group oxidized saxitoxin (1) with $NaOH-H_2O_2$, and obtained a fluorescent product (15) [λ_{max} 324 nm (21,000), 236 (12,800)] in high yield (38). The propionyl moiety attached to a quarternary nitrogen in (15) can be easily removed by heating in an alkaline medium as a result of retro-Michael addition (or Hoffmann degradation). The structure of the resulting aminopurine derivative (16) was established by synthesis (38). This degradation confirmed the presence of two guanidinium groups in the saxitoxin molecule, a hypothesis previously based on the observation that permanganate oxidation of saxitoxin afforded fragments such as guanidopropionic acid (17), urea, carbon dioxide and ammonia (36). In 1971, RAPOPORT's group proposed structure (18) for saxitoxin on the basis of the degradation studies and ^{1}H-NMR data (37). Although this formula was eventually proved to be incorrect, it was quite close to the correct structure (1). The only major difference between (18) and (1) was in the juncture of the third ring formed by a propionyl moiety. The reason that C-5 was chosen as the point of ring fusion was the appearance in the ^{1}H-NMR spectrum of the H-5 signal as a near singlet. In retrospect, the dihedral angle between H-5 and H-6 is as such that the coupling between the two hydrogens is very small $(0-1$ Hz depending on the measurement conditions) (34). This unexpected lack of coupling was a misfortune particularly because in earlier degradation studies the same group had isolated 8-methyl-2-oxo-2,4,5,6-tetrahydro-pyrrole[1,2-C]pyrimidine (19) which shows the correct ring juncture (39).

(18)   (19)

In the ^{1}H-NMR study, it was found that in D_2O the C-11 methylene protons in (1) easily exchange with deuterium atoms even at near neutral pH (37). In the deuterated compound (20), the signals for the C-10 methylene protons appear as a clear AB pattern (Table 2). Oxidation of the dideuterated compound (20) afforded α-dideuterated guanidopropionic acid (17).

The structure of saxitoxin (1) uncovered by X-ray crystallography of saxitoxin p-bromobenzenesulfonate (34) shows that the 12-carbonyl function exists in the hydrated form. The hydrated form is stabilized by the two electron-withdrawing guanidinium groups linked to the adjacent C-4; that is to say, the keto group opts for the hydrated form to avoid the double charge situation on the two neighboring carbons, C-4 and C-12. The unusually easy enolization of the keto group demonstrated by the facile deuterium exchange may also be an indication of the tendency of the keto function to avoid the keto form (Scheme III). In fact, saxitoxin (1) can exist in the keto form at high pH once one of the guanidinium groups (imidazoline) is deprotonated (vide infra).

Table 2. *NMR Data of Paralytic Shellfish Toxins*

No. C–H	2	8	14	4	5	6	10	11	12	13
Saxitoxin (1)										
¹³C NMR	157.9*(s)	156.1*(s)	159.0*(s)	82.6(s)	57.3(d)	53.2(d)	43.0(t)	33.1(t)	98.9(s)	63.3(t)
¹H NMR[a]					4.77 (d, J=1)	3.87 (J=9,5)	3.85 (d, J=10) 3.57 (d, J=10)	2.37 (m)		4.27 (dd, J=11,9) 4.05 (dd, J=11,5)
Neosaxitoxin (2)										
¹³C NMR	158.8*(s)	158.1*(s)		82.2(s)	56.9(d)	64.4(d)	43.7(t)	31.9(?)	98.6(s)	61.1(t)
¹H NMR					4.83 (s)	4.15 (J=6,6)	3.80 (J=10) 3.58 (J=10)	2.44 (m)		4.43 (J=11,6) 4.28 (J=11,6)
Gonyautoxin-I (3)										
¹³C NMR	158.9*	158.4*		82.6	57.8	64.8	51.6	78.1	97.8	61.4
¹H NMR					4.89ᵇ (s)	4.80 (J=7,6)	4.10 (J=12) 3.93 (J=12,5)	4.74ᵇ (J=5)		4.38 (J=11,7) 4.18 (J=11,6)
Gonyautoxin-II (4)										
¹³C NMR	158.0*(s)	156.2*(s)	159.1*(s)	81.5(s)	57.9(d)	53.2(d)	50.9(d)	77.6(d)	97.5(s)	63.3(t)
¹H NMR					4.77ᵇ (J=1)	3.77 (J=10,6)	4.09 (J=12) 3.91 (J=12,5)	4.90ᵇ (J=5)		4.20 (J=11,10) 3.94 (J=11,6)

Gonyautoxin-III (5)									
^{13}C NMR	160.6*	159.2*	156.6*	82.6	58.1	53.5	48.0	76.3	97.8	57.8
^{1}H NMR					4.79^c (J=1)	3.79 (J=10,6)	4.14 (J=11,9) 3.55 (J=11,7)	4.93^c (J=9,7)		4.28 (J=12,10) 4.02 (J=12,6)
Gonyautoxin-IV (6)									
^{1}H NMR					4.57 (br s)	4.04 (t,J=6)	4.11 (J=10,8) 3.53 (J=10,7)	4.88 (t,J=10)		4.37 (J=12,6) 4.21 (J=12,6)
Gonyautoxin-V (7)									
^{1}H NMR					4.61 (s)	3.86 (m)	3.62 (J=10) 3.81 (J=10)	D exchanged		4.13 (J=10,2) 4.13 (J=10,5)
Gonyautoxin-VIII (10)									
^{13}C NMR	158.3*	156.1*	154.3*	81.9	57.5	53.0	47.7	76.0	97.5	64.0
^{1}H NMR					4.78 (s)	3.81 (J=10,5)	3.54 (J=11,8) 4.12 (J=11,8)	4.11 (J=12,5)		4.32 (J=12,10) 4.11 (J=12,5)

^{a} Data taken on a 100 MHz instrument (37). All other proton data on a 270 MHz instrument (28). Discrepancies in chemical shifts seen in some papers seem to be due to difference in selection and assignment of the standard. Here δ values from DSS in D_2O are shown in accordance with the first report on saxitoxin (37).

^{b} Measured at 85° C.

^{c} Measured at 55° C.

* Assignments can be interchanged.

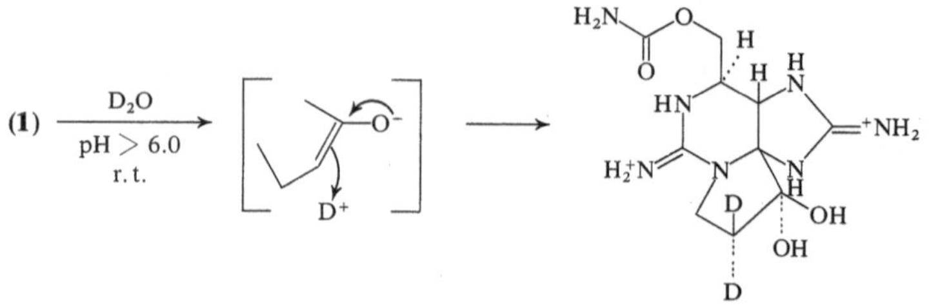

(20)

Scheme III. Deuterium exchange of saxitoxin

Two pKa's, 11.5 and 8.1 (or 11.28 and 8.24) were reported for saxitoxin
(1) *(36, 37)*. One of them, that at 11.5, is undoubtedly attributable to one of
the guanidinium groups. The origin of the other pKa at 8.24, had long been
a subject of discussion. The value was initially considered to be too low for a
guanidine group. Titrations in less polar media gave higher pKa values
which is normally an indication of proton dissociation from oxygen *(37)*.
The hemicyclitol structure in **(18)** was proposed in consideration of this
(37). After the presence of a keto hydrate in the molecule was revealed, the
dissociation of one of the hydrate hydroxyl groups was also considered *(34)*.
Finally, the pKa was assigned to the C-8 guanidine group based on ^{13}C- *(40)*
and ^{1}H-NMR *(41)* studies of saxitoxin **(1)** in solution at different pH's. The
largest chemical shift changes were observed for H-5, C-4 and C-5 of **(1)** and
its derivatives in the pH range corresponding to the pKa 8.24. Moreover,
the NMR studies revealed the presence of significant proportions of the
keto form at higher pH's. Thus the molecule of saxitoxin exists as an
equilibrium among three species **(1)**, **(21)** and **(22)** (Scheme IV). Also, it was
found that the chemical shift of H-5 in the keto form is independent of pH
indicating that the imidazoline moiety in the keto form is not protonated, or
that only the deprotonated molecules are allowed to take the keto form **(22)**
in solution *(41)*.

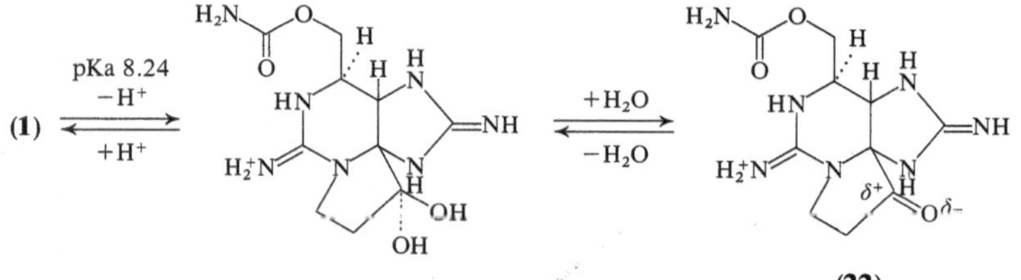

Scheme IV. Molecular forms of saxitoxin in solutions

The unusually depressed pKa of the imidazoline moiety suggests that the C-8 guanidinium ion is not fully resonance stabilized. Because the C-4 signal exhibits a large pH dependent chemical shift, ROGERS and RAPOPORT speculated that there is bond fixation between C-8 and N-9 (40). However, the X-ray data rather show that the C-8, N-9 bond is longer than the other two C-N bonds (34), implying insufficient participation of N-9 in the resonance (41). As the cause of the depressed pKa, the interaction of two neighboring guanidinium groups was once considered but excluded (41), and there is an indication that such anomaly may be intrinsic to condensed five-membered 2-aminoimidazolines as exemplified by another anomalous guanidinium compound phakellin (23) with a pKa of 7.7 (42, 43).

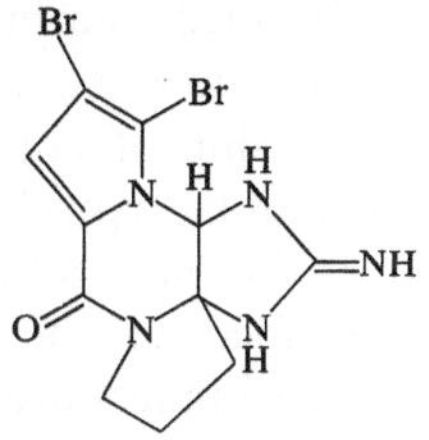

(23)

Catalytic hydrogenation of saxitoxin (1) in hydrochloric acid (36) affords α-dihydrosaxitoxin (24) accompanied by a small amount of β-dihydrosaxitoxin (25), whereas reduction with $NaB(CN)H_3$ or catalytic hydrogenation in ethanol afford a mixture of almost equal amounts of (24) and (25) (40, 41) (Scheme V). The stereochemistry of both isomers was unequivocally established on the basis of the coupling constants in the ^{1}H-NMR spectrum (41).

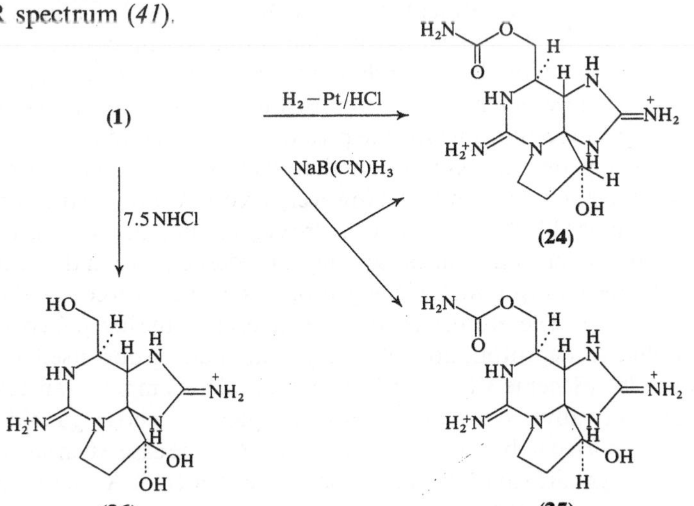

Scheme V. Hydrogenation and hydrolysis of saxitoxin (1)

Saxitoxin is a very stable compound under strongly acidic conditions. Only upon heating in 7.5 N HCl at 100° for several hours is the carbamate moiety hydrolyzed to afford decarbamoylsaxitoxin (**26**) which retains about 70% of the original toxicity (*43*). On the other hand, saxitoxin decomposes rapidly at basic pH's, although in the absence of oxygen it is stable enough to withstand NMR measurements (*40*).

B. Neosaxitoxin

Neosaxitoxin (**2**) can be obtained as a highly hygroscopic powder or gum (*26*). It was eluted just prior to co-existing saxitoxin (**1**) from a Bio-Rex 70 column, and a close relationship of both compounds was expected. However, unlike saxitoxin (**1**), neosaxitoxin (**2**) on NaOH − H$_2$O$_2$ oxidation did not afford a good yield of the purinyl-propionic acid derivative (**15**) (*44*). The difference was ascribed to the presence of an additional substituent on the nucleus which hinders aromatization.

Both ^{1}H- and ^{13}C-NMR spectra of neosaxitoxin (**2**) closely paralleled those of saxitoxin (**1**) with some differences in chemical shifts of C-6 and nearby carbons and protons (*45*). Microtitration of (**2**) gave two pKa's, 6.75 and 8.65, and possibly there is another pKa larger than 11.5 which is not directly titratable. A pH dependency study of proton chemical shifts indicated that the first pKa of 6.75 is associated with a functional group located in the proximity of C-6 and the second pKa of 8.65 with the C-8 guanidinium group. Treatment of neosaxitoxin (**2**) with Zn-acetic acid afforded saxitoxin (**1**), indicating that the additional functional group is removable by reduction. Based on these observations, the structure of 1 (N)-hydroxysaxitoxin was tentatively proposed for neosaxitoxin (**2**) (*45*). The functional group responsible for the pKa of 6.75 was assumed to be the 1-hydroxyl group on the assumption that such a hydroxyguanidine dissociates to the N-oxide form (*27*). However, a recent titration study of δ-N-hydroxyarginine (**28**) showed that its hydroxyguanidinium group has a pKa of 7.2 and that deprotonation causes only a modest change in the chemical shift of the neighboring methylene protons as also observed for the H-6 proton of (**2**) (*46*). Therefore, it is now more likely that the low pKa of 6.75 is actually due to deprotonation of C-2 guanidinium group itself to give structure (**29**) (Scheme VI). Another comparable example is an alkoxy-guanidine derivative, canavanine (**30**), the pKa of whose guanidinium group was reported to be 7.1 (*47*). If that were really the case, such structures as ^{2}N-hydroxysaxitoxin (**31**) should also be reconsidered. As in the case of saxitoxin, final confirmation of the structure of neosaxitoxin must probably await X-ray analysis of a crystalline derivative.

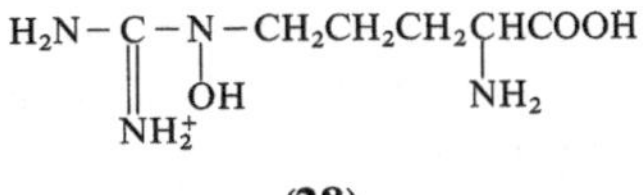

$$H_2N-\underset{\underset{NH_2^+}{\overset{\displaystyle \|}{\underset{}{}}}}{\overset{\displaystyle OH}{C}}-N-CH_2CH_2CH_2\underset{\underset{NH_2}{|}}{C}HCOOH$$

(28)

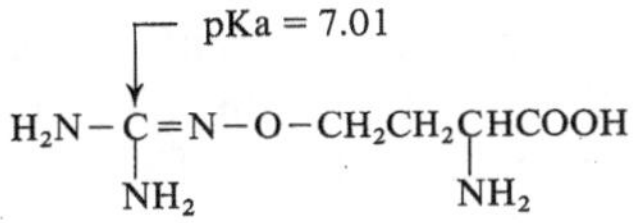

pKa = 7.01

$$H_2N-\underset{\underset{NH_2}{|}}{C}=N-O-CH_2CH_2\underset{\underset{NH_2}{|}}{C}HCOOH$$

(30)

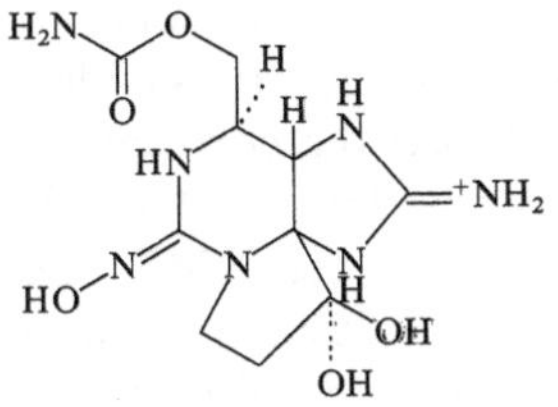

(31)

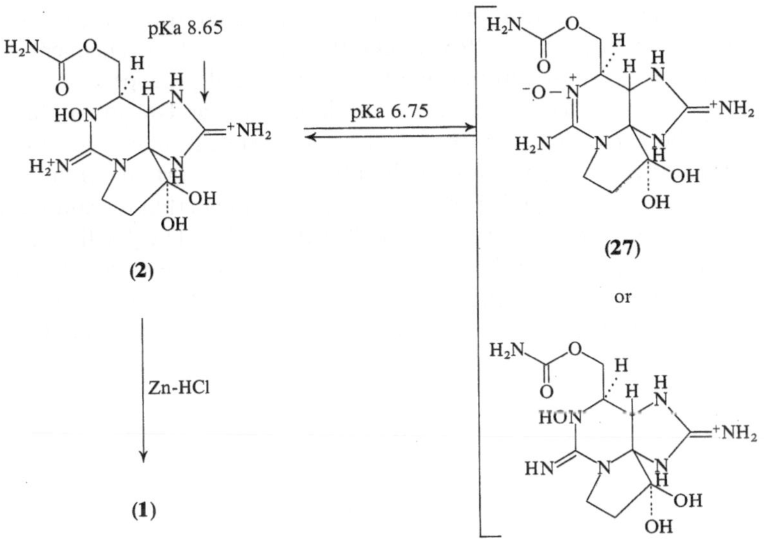

(2)

pKa 8.65

pKa 6.75

(27)

or

(29)

Zn-HCl

(1)

Scheme VI. Dissociation and reduction of neosaxitoxin (**2**)

Note added in proof: The alternative structure (**31**) for neosaxitoxin has recently been ruled out as a result of ¹⁵N-NMR study of neosaxitoxin (HORI, A., and Y. SHIMIZU: Biosynthetic ¹⁵N-Enrichment and ¹⁵N N.M.R. Spectra of Neosaxitoxin and Gonyautoxin-II: Application to Structure Determination. Chem. Commun. **1983**, 790).

C. Gonyautoxin-II and Gonyautoxin-III

Gonyautoxin-II (**4**) is the major toxic constituent of many toxic shellfish samples and organisms (*49*). The first studies aimed at structure elucidation were carried out with a minute amount of toxin isolated from softshell clams, *Mya arenaria,* which had been exposed to a red tide bloom along the New England coast in 1972 (*49*). NaOH−H₂O₂ oxidation of gonyautoxin-II (**4**) afforded two highly fluorescent products, (**32**) and (**33**). The UV spectra of both products closely resembled that of the oxidation product of saxitoxin (**1**). In acidic solutions, both (**32**) and (**33**) formed the lactams (**34**) and (**35**) (Scheme VII). Oxidation of saxitoxin (**1**) under identical condition afforded also the corresponding oxidation products (**36**) (*50*) and (**37**), which can be converted to (**15**) and (**40**) on acid treatment. From the mass spectrum and the NMR spectra (**32**) was assumed to be a hydroxyl derivative of (**36**) and the product (**33**) was assumed to be its hydrolyzed urea derivative. Significantly, (**32**) and (**33**) are optically active as shown by absorption in the CD spectra. Since (**36**) and (**37**) do not possess a chiral center, the optical activity must be derived from the additional substituent. Both ^{1}H- and ^{13}C-NMR spectra of (**4**) suggested the presence of a substituent at C-11 of saxitoxin. On the basis of these chemical and spectroscopic data the structure of (**4**) was initially proposed as 11α-hydroxysaxitoxin (*49*). Due to the lack of the material and the toxin's failure to give the molecular ion by various mass spectrometric techniques including the plasma ionization method the molecular formula of the toxin itself had never been established, and it was later shown that the compound is actually a conjugate containing a sulfate group (*51*). The position of this group was speculated to be on the 11-hydroxyl; this was subsequently confirmed by the observation that removal of the O-sulfate group by reductive cleavage with zinc-hydrochloric acid affords saxitoxin (**1**) (*52*) (Scheme VIII).

(**36**) X = NH₂
(**37**) X = OH

Gonyautoxin-III (**5**) is the 11β-epimer of (**4**); epimerization which takes place even near neutral pH is greatly enhanced by the presence of trace amounts of bases such as acetate (*49*). At equilibrium the ratio of (**4**) to (**5**) is 7:3. The stereochemical assignment was based on the magnitude of coupling constants involving H-10 and H-11. The five-membered ring is slightly puckered with C-12 out of the plane (*34*). The dihedral angle between 10α-H and 11β-H is close to 90° (*35, 37, 41, 53*), and only a small coupling is observed between the two protons in the 11α-O-sulfate isomer (**4**) (*28, 51*).

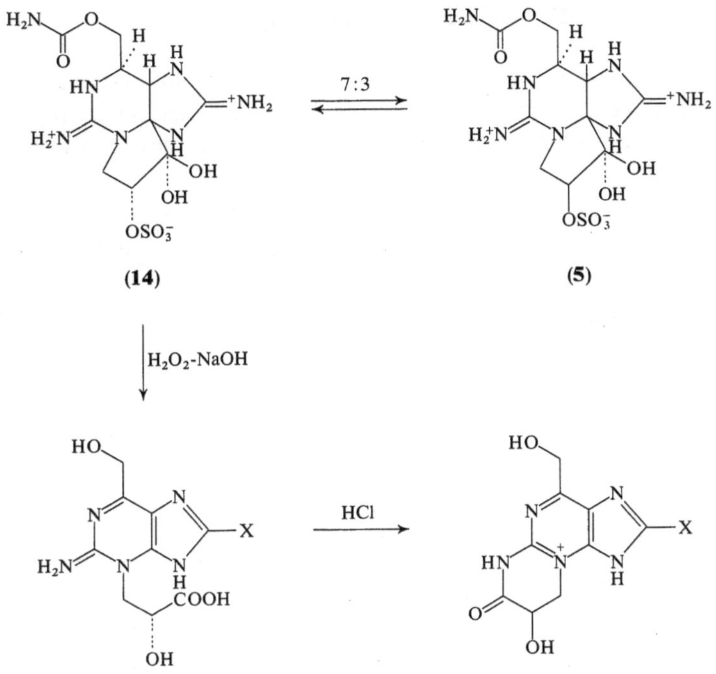

(14) (5)

(32) $X = NH_2$
(33) $X = OH$

(34) $X = NH_2$
(35) $X = OH$

Scheme VII. Reactions of gonyautoxin-II (**4**) and gonyautoxin-III (**5**)

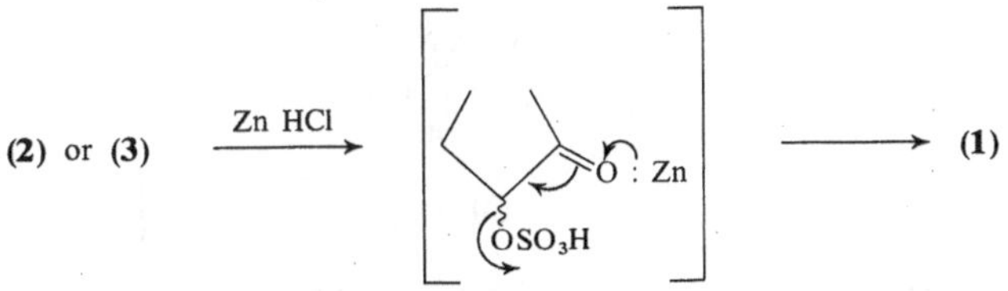

Scheme VIII. Possible mechanism for reductive conversion of gonyautoxin-II (**4**) and gonyautoxin-III (**5**) to saxitoxin

The facile equilibration of (**4**) and (**5**) in basic media raises a question about the optical activity observed with the $NaOH-H_2O_2$ oxidation products, (**32**) and (**33**). One possible explanation would be that it stemmed from the unequal ratio of the two epimers at equilibrium, another that one of the stereoisomers is preferentially oxidized.

D. Gonyautoxin-I and Gonyautoxin-IV

Gonyautoxin-I (3) and gonyautoxin-IV (6) were also first found in *Mya arenaria* and *Gonyaulax tamarensis* (20). Later these toxins were reported in a number of other species and in some cases they constituted the

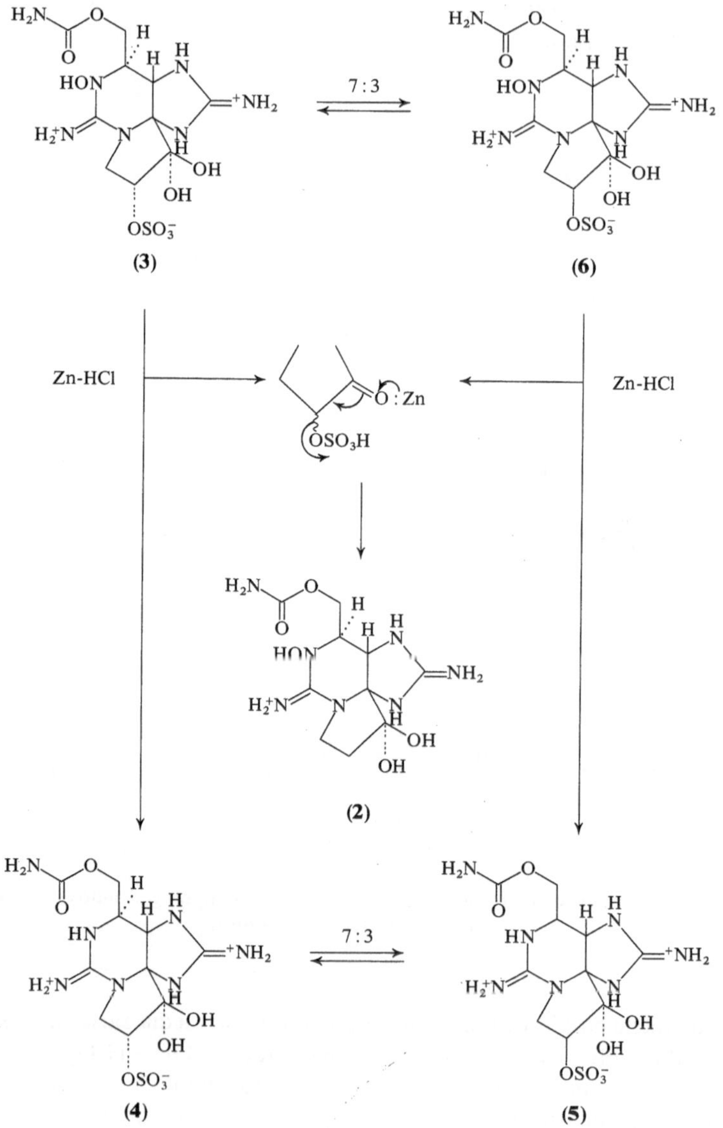

Scheme IX. Stereochemical relationship of gonyautoxin-I (3), -II (4), -III (5), and -IV (6)

predominant portions of the total toxin mixtures (*54*). The two toxins are epimeric and their relationship is similar to that of gonyautoxin-II (**4**) and (**5**). The ^{1}H- and ^{13}C-NMR spectra of (**3**) are essentially composites of the spectra of gonyautoxin-II (**4**) and neosaxitoxin (**2**); this suggested that gonyautoxin-I (**3**) was 11α-hydroxyneosaxitoxin sulfate and gonyautoxin-IV (**6**) its 11β-epimer (*27, 55—59*). Confirmation for this assumption was provided by conversion of (**3**) to a mixture of neosaxitoxin (**2**) and gonyautoxin-II (**4**). Similarly, gonyautoxin-IV (**6**) was converted by Zn-HCl reduction to a mixture of neosaxitoxin (**2**) and gonyautoxin-III (**5**) (*52*) (Scheme IX). Thus the structures of (**3**) and (**5**) depend on the unequivocal establishment of the structure of neosaxitoxin.

E. Gonyautoxin-V and Gonyautoxin-VI (B1 and B2 Toxins)

Both gonyautoxin-V (**7**) and gonyautoxin-VI (**8**) were first isolated as minor components in *Gonyaulax* cells and shellfish (*26, 27*). Both toxins have characteristically low specific toxicities. Oxidation of gonyautoxin-V (**7**) with $NaOH - H_2O_2$ gave high yields of fluorescent products (**36**) and (**37**) identical with those from saxitoxin (**1**) (*45*). Meanwhile, working with an Alaskan strain of *Gonyaulax* sp., HALL *et al.* isolated a series of latent toxins which can be converted to more virulent known toxins. Two of them, designated B1 and B2, were found to be identical with gonyautoxin-V (**7**) and gonyautoxin-VI (**8**) respectively (*25*). The ^{1}H-NMR spectrum of (**7**) was almost identical with that of saxitoxin (**1**). Mild acid hydrolysis of (**7**) and (**8**) afforded saxitoxin (**1**) and neosaxitoxin (**2**) respectively (Scheme X).

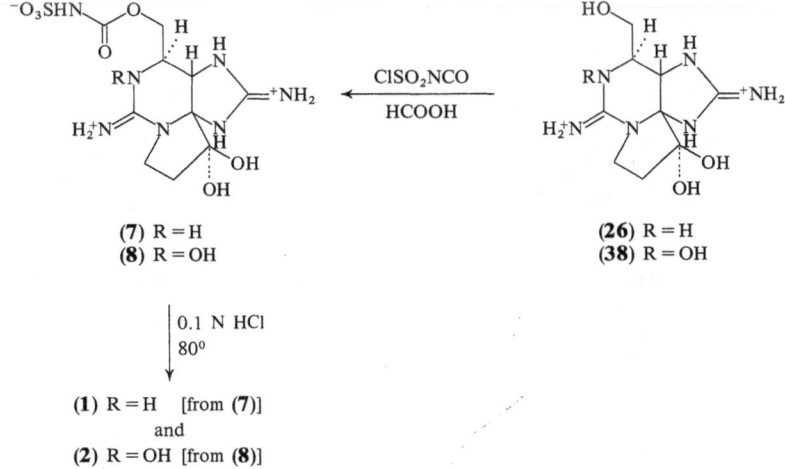

Scheme X. Hydrolysis and partial synthesis of gonyautoxin-V (**7**) and gonyautoxin-VI (**8**)

When the structure of gonyautoxin-VIII (**10**) was established *(vide infra)*, it was immediately suspected that (**7**) and (**8**) were analogous carbamoyl-N-sulfated derivatives of saxitoxin and neosaxitoxin (*8, 60—63*). The structures were confirmed by partial synthesis of (**7**) and (**8**) from decarbamoylsaxitoxin (**26**) and decarbamoylneosaxitoxin (**38**) respectively on treatment with chlorosulfonyl isocyanate in formic acid (*60*) according to a previously reported procedure (*64*).

F. Gonyautoxin-VII

Gonyautoxin-VII (**9**) was isolated from toxic scallops collected in the Bay of Fundy (*28*). The compound does not possess sulfate conjugation and was eluted very close to saxitoxin in Bio-Rex 70 ion-exchange chromatography. It was first considered to be identical with decarbamoylsaxitoxin (**26**), but the identity was excluded by direct comparison (*28*).

G. Gonyautoxin-VIII

Gonyautoxin-VIII (**10**) does not bind tightly to Bio-Rex 70 gel and due to its weak toxicity remained unnoticed in the earlier isolation studies. The compound was recognized as a radioactive peak associated with weak toxicity in cultured *Gonyaulax* organisms fed with radioactive arginine (*24, 61*) and was later proved to be identical with C2, one of the latent toxins found in an Alaskan strain of *Gonyaulax* sp. (*25*).

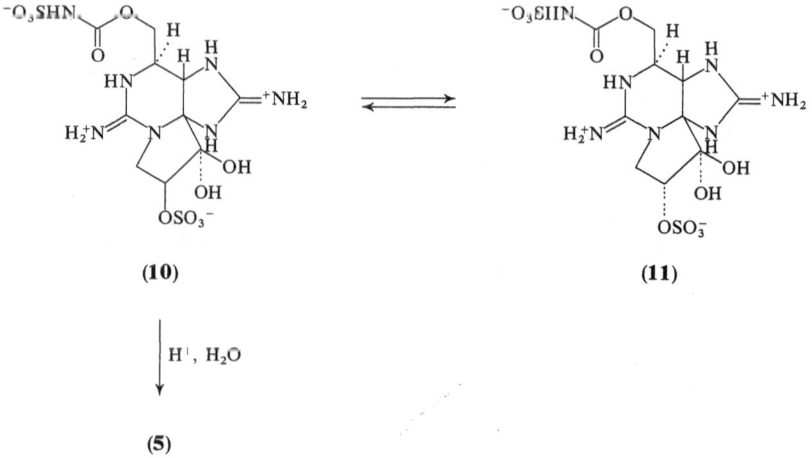

Scheme XI. Reactions of gonyautoxin-VIII (**10**)

The ^{1}H- and ^{13}C-NMR spectra of (**10**) are almost identical with those of gonyautoxin-III (**5**) except for small differences in the chemical shifts of C-13 and the attached protons. Mild acid hydrolysis liberated one mole of inorganic sulfate and gonyautoxin-III (**5**); hence it was postulated to be carbamoyl-N-sulfogonyautoxin-III (*24*). Subsequently, the structure was unequivocally established by X-ray crystallography (*65*). The substance easily isomerizes at slightly elevated pH's and forms an equilibrium mixture of (**10**) and the 11α-isomer (= C1) (**11**) corresponding to gonyautoxin-II (**4**) (Scheme XI).

Gonyautoxin-VIII is the first toxin which was proved to possess sulfate conjugation on the carbamoyl group. It is also the first toxin in the series which carries a negative net charge at or near physiological pH as shown by its mobility on electrophoresis (*29*).

H. C3 and C4 Toxins

These two toxins were found in the latent toxins from the Alaskan *Gonyaulax* strain which can be converted to more potent toxins by mild acid treatment (*30*). Since all the known toxins seem to exist also in the form of the corresponding carbamoyl-N-sulfated derivatives, they are speculated to be conjugated gonyautoxin-I (**12**) and -IV (**13**).

(12)   (13)

IV. Synthesis

So far the only total synthesis of saxitoxin has been accomplished by KISHI and his coworkers at Harvard University (*64*).

The group first synthesized the model compound (**39**), to test their strategy which was to build the A/C ring system and then to form the B-ring in the final step (*66*) (Scheme XII).

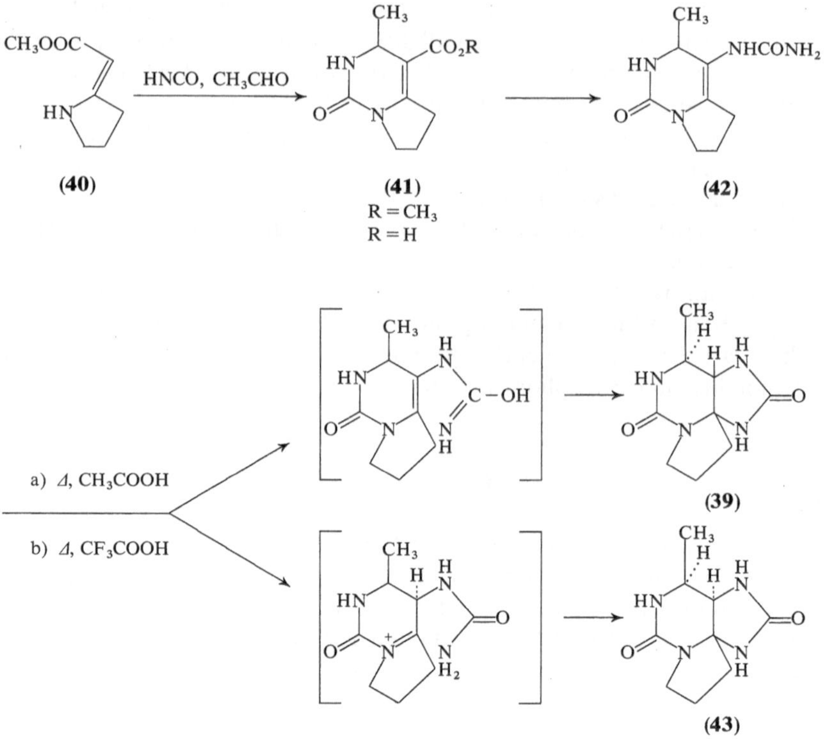

Scheme XII. Synthesis of saxitoxin model compounds

Condensation of the vinylogous carbamate (**40**) with acetaldehyde and isocyanic acid afforded the pyrimidine derivative (**41**) which was converted to the urea derivative (**42**) by Curtius rearrangement in 65% overall yield. Treatment of (**42**) with acetic acid at 50° afforded exclusively the desired product (**39**), whereas treatment with trifluoroacetic acid gave a mixture of (**39**) and stereoisomer (**43**). The stereochemistry was established by comparing the coupling constants involving H-5 and H-6 with those of saxitoxin (**1**). Different cyclization mechanisms were suggested for the acetic acid and trifluoroacetic acid catalyzed cyclization to account for the different stereochemistry of the products.

To synthesize saxitoxin (**1**) itself, the ketal (**47**) was prepared in five steps starting with methyl 2-oxo-4-phthalimidobutyrate (**44**) and then subjected to the cyclization using benzyloxyacetaldehyde and silicon tetraiso-thiocyanate. The resulting bicyclic thiourea (**48**) was then converted to thiourea urea (49) in a sequence analogous to that used for the preparation of (42). Because the acidic conditions used for the cyclization of (42) were not applicable to the ketal (49), it was converted to the thioketal (50). Cyclization of (50) to the desired tricyclic compound (51) was effected by warming in a 9 : 1 mixture of acetic acid and trifluoroacetic acid at 50°. As in the case of the model experiment, cyclization in neat trifluoroacetic acid afforded predominantly the undesirable isomer (52). The tricyclic thiourea urea (51) was converted to guanidine (53) by treatment with Meerwein reagent followed by displacement with ammonia. After debenzylation the thioketal (54) was converted to known decarbamoylsaxitoxin (26) by a standard procedure. The synthesis of d,l-saxitoxin was completed by treatment of (26) with chlorosulfonyl isocyanate in formic acid followed by hydrolysis (Scheme XIII).

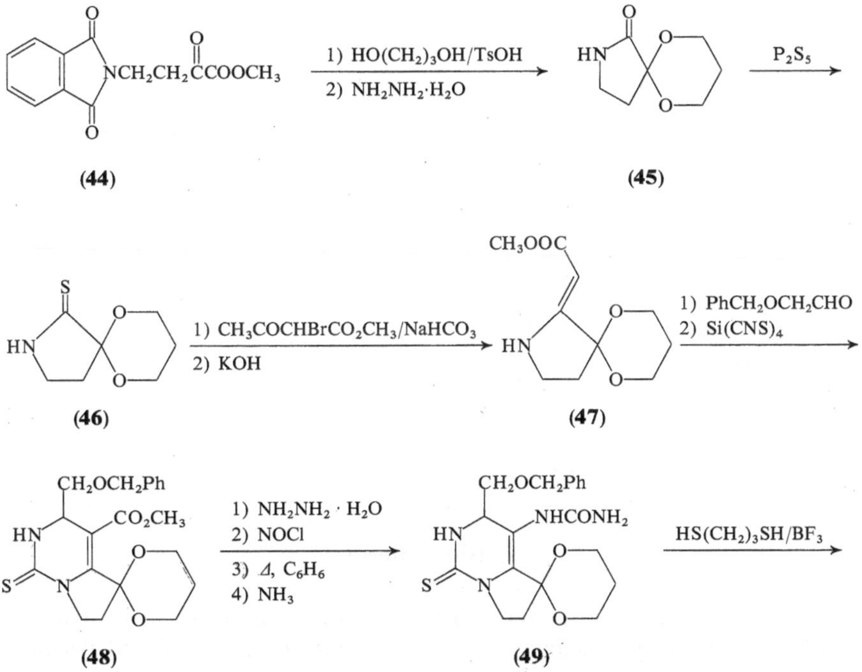

Scheme XIII. Total synthesis of saxitoxin (1)

Scheme XIII (continued)

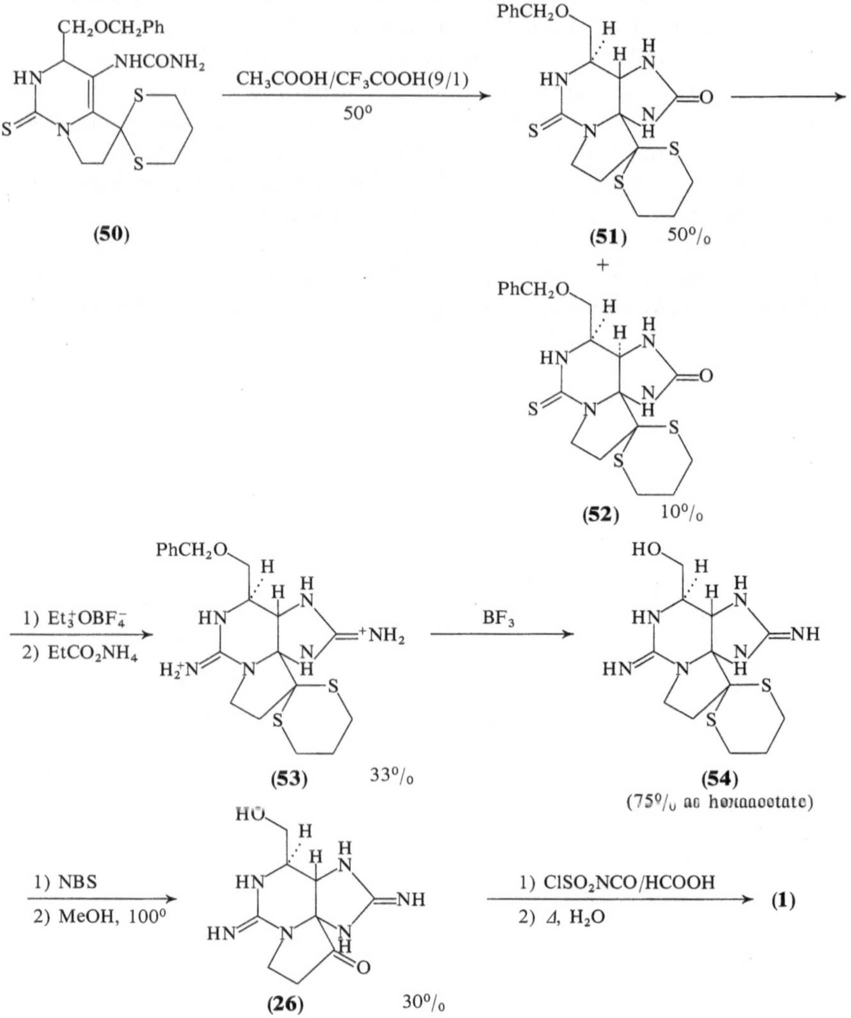

(50)

$$\xrightarrow[50^\circ]{CH_3COOH/CF_3COOH(9/1)}$$

(51)   50%

+

(52)   10%

1) $Et_3^+OBF_4^-$
2) $EtCO_2NH_4$

(53)   33%

$$\xrightarrow{BF_3}$$

(54)
(75% as hexaacetate)

1) NBS
2) MeOH, 100°

(26)   30%

1) $ClSO_2NCO/HCOOH$
2) Δ, H_2O **(1)**

V. Biosynthesis

Little is known about the biosynthesis of paralytic shellfish toxins, although one may speculate about the origin of the condensed perhydropurine skeleton.

In an earlier experiment, several ^{14}C-labeled precursors were fed to a culture of *Gonyaulax catenella* (*67*). Some incorporation into the crude toxin fraction was observed, but the results were inconclusive. In a recent report, [guanido-^{14}C]-labeled arginine (55) was fed to a culture of *Gonyaulax tamarensis* (*61*). Radioactive gonyautoxin-III (6) was isolated and after degradation about one-third of the total radioactivity was found in the carbamate carbon (Scheme XIV). The result appears to support the contention that guanido and ureido groups are generally derived from the ornithine-urea cycle.

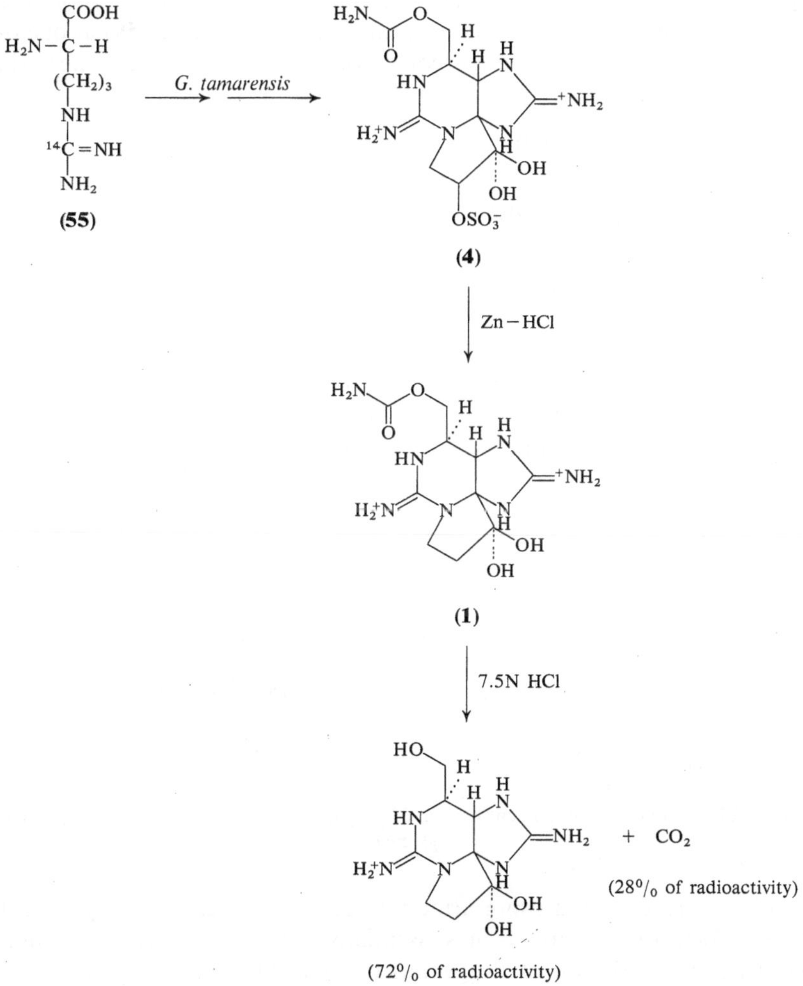

Scheme XIV. Degradation of gonyautoxin-III labeled by feeding of [guanido-^{14}C]-arginine

Feeding of [2-^{13}C]-labeled glycine (**56**) to the culture of *Gonyaulax tamarensis* resulted in rather random incorporation of the labeled carbon into the toxin molecules as judged from the ^{13}C-NMR spectrum of gonyautoxin-II (**4**) (*61*). Interestingly, extra enrichment was observed for C-11 and C-12. One plausible explanation for the unexpected enrichment in two neighboring carbons from the single-labeled precursor could be the incorporation of glycine *via* the TCA cycle. In such a scheme, glycine is brought into the TCA cycle by the way of the glyoxalic acid pathway. In the cycle, the assymetry of the labeling pattern will be lost and the label will appear on both C-2 and C-3 of succinate (**58**). The results seem to support a biosynthetic pathway to the toxins from α-ketoglutaric acid (**59**) or one of its derivatives whose C-4 and C-5 correspond to C-11 and C-12 of the toxin molecule (Scheme XV).

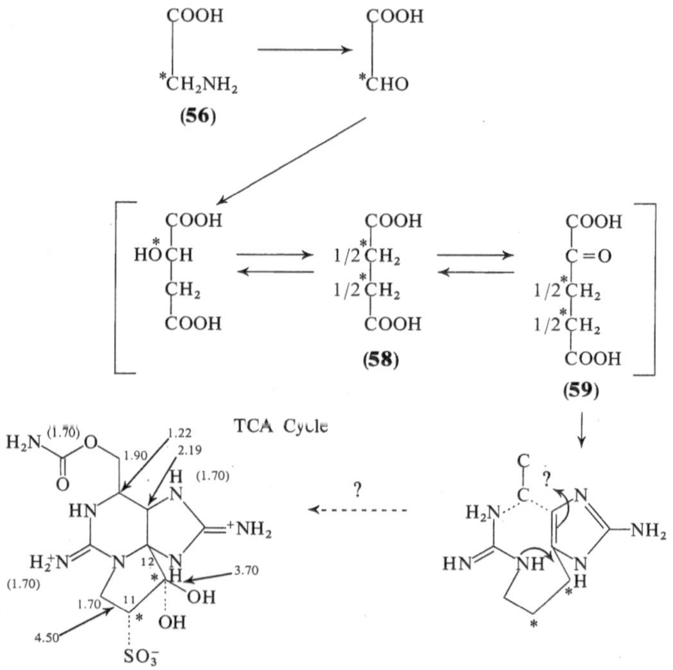

Scheme XV. Relative ^{13}C-enrichment pattern of gonyautoxin-II (**4**) derived from the [2-^{13}C]-glycine

Biosynthetic studies using armoured dinoflagellates are fraught with many difficulties. The organism is essentially phototrophic and shows high selectivity in utilization of exogenous organic substances in addition to complications which arise from compartmentalization of certain biochemical reactions.

References, pp. 261—264

VI. Pharmacology

The toxicity of paralytic shellfish toxins is traditionally expressed in mouse units (mu) where one mouse unit is defined as the amount required to kill a 20 g mouse within 15 minutes by intraperetoneal injection (68). Saxitoxin has a specific toxicity of 5500 mu/mg or 2045 mu/μmol. The specific toxicities of other toxins vary depending upon the researchers, probably due to the difficulty associated with weighing and due to differences in the assay conditions. Table 3 shows some selected numbers which have been reported.

Table 3. *Specific and Related Mouse Toxicity of PSPs.* (*69*)

	Mouse Unit/mole	Related Toxicity
Saxitoxin	2045	1
Neosaxitoxin	1038	0.50
Gonyautoxin-I	1638	0.80
Gonyautoxin-II	793	0.39
Gonyautoxin-III	2234	1.09
Gonyautoxin-IV	673	0.33
Gonyautoxin-V	354	0.17
Gonyautoxin-VI*	180	0.09
Gonyautoxin-VIII	280	0.14
11-Epigoniautoxin-VIII**	17	8.3×10^{-3}

* KOEHN et al. (*60*).
** WICHMANN et al. (*58*).

The action mechanism of saxitoxin has been most thoroughly studied. The toxin blocks the influx of sodium ions through the excitable nerve membranes and prevents the formation of action potentials (*70, 71*). All other paralytic shellfish toxins seem to act in the same manner (*72—74*). According to the model proposed by HILLE, saxitoxin acts as a plug in the sodium channel by blocking the flow of sodium ions (Fig. 1) (*75*).

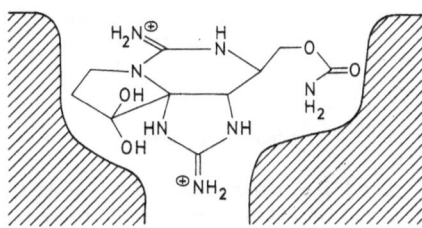

Fig. 1. Hypothetic plugging model for saxitoxin-sodium channel interaction (*75*)

Structural variations by adding bulky dissociable functional groups at the 11α, 11β, and 1(N) positions seem to have no critical influence on the toxicity. This is also true in the isolated nerve systems (73—75). The finding is difficult to explain by the plugging model proposed by Hille (75). Recently, it has been suggested that the effect is due to a rather shallow penetration by the toxin molecule or that the latter acts entirely on the surface (61, 74). For the binding force hemiketal formation between the hydrated 12-keto group and a functionality on the receptor was considered as well as ionic interactions and hydrogen bonding (75).

However, the latest data show that dihydrosaxitoxin, especially the 12α-isomer (24), also possesses activity against the isolated nerve system (61, 74, 76). Comparison of the molecules of saxitoxin and tetrodotoxin (60), another marine toxin with a similar activity, suggested that hydrogen bonding with the 12α- and 12β-hydroxyl and ion pairing between the guanidinium group and an anionic site on the membrane are probably major factors in binding of the toxin (61, 74). Interestingly, the presence of sulfate esters in the side chain results in a drastic loss of activity. The finding appeared to contradict the earlier belief that the side chain plays a small role in the activity, an assumption based on the fact that decarbamoylsaxitoxin maintains 70% of the original activity (43). In the new binding model, however, the side chain may extend very close to the membrane surface and the negatively charged sulfonate group may interact repulsively with an anionic site scattered on the membrane surface (61, 63) (Fig. 2).

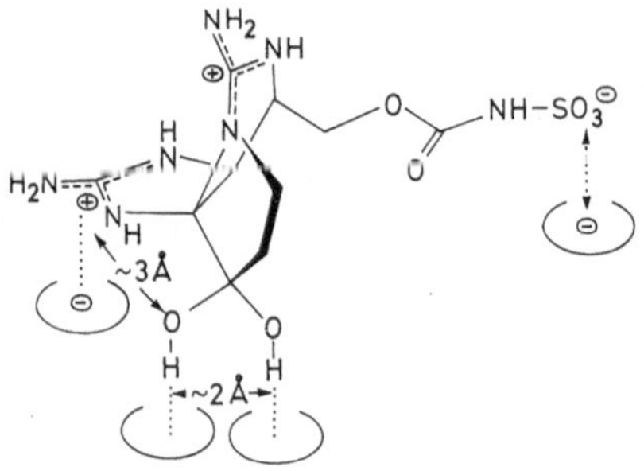

Fig. 2. Perspective view of possible interaction between paralytic shellfish toxin and excitable membrane surface (61, 63)

Acknowledgements

I would like to thank Drs. P. B. Reichardt and S. Hall of the University of Alaska for furnishing preliminary data and Dr. T. Fukai and Miss S. Stoner for help in preparing the manuscript. Unpublished data cited in this review were taken from the work supported by PHS grants GM28754 and GM24425 which are also acknowledged.

References

1. HALSTEAD, B. W.: Poisonous and Venomous Marine Animals of the World. Princeton, N.J.: Darwin. 1978.
2. SCHANTZ, E. J.: The Dinoflagellate Poisons in Microbial Toxins, Vol. II — Algal and Fungal Toxins. New York: Academic Press. 1972.
3. SHIMIZU, Y.: Dinoflagellate Toxins in Marine Natural Products, Chemical and Biological Perspectives Vol. I, p. 1—42. New York: Academic Press. 1979.
4. MEYER, K. F., H. SOMMER, and P. SCHOENHOLZ: Mussel Poisoning. J. Prev. Med. 2, 365 (1928).
5. SOMMER, H., and K. F. MEYER: Paralytic Shellfish Poisoning. Arch. Pathol. 24, 560 (1937).
6. SHIMIZU, Y.: Compounds from Microalgae — Their Influence on the Field of Marine Natural Products in Recent Advances in Phytochemistry 13, 199 (1979) and references therein.
7. KAMIYA, H., and Y. HASHIMOTO: Occurrence of Saxitoxin and Related Toxins in Palauan Bivalves. Toxicon 16, 303 (1978).
8. HARADA, T., Y. OSHIMA, and T. YASUMOTO: Structure of Two Paralytic Shellfish Toxins. Gonyautoxins V and VI, Isolated from a Tropical Dinoflagellate, Pyrodinium bahamense var. compressa. Agric. Biol. Chem. 46, 1861 (1982).
9. JACKIM, E., and J. GENTILE: Toxins of a Blue-Green Alga: Similarity to Saxitoxin. Science 162, 915 (1968).
10. ALAM, M., Y. SHIMIZU, M. IKAWA, and J. J. SASNER: Reinvestigation of the Toxin from Aphanizomenon flos-aquae by High Performance Chromatographic Methods. J. Environ. Sci. Health A 13, 493 (1978).
11. IKAWA, M., K. WEGENER, T. L. FAXALL, and J. J. SASNER, JR.: Comparison of the Toxins of the Blue-Green Alga Aphanizomenon flos-aquae with the Gonyaulax Toxins. Toxicon 20, 747 (1982).
12. HASHIMOTO, Y., S. KONOSU, T. YASUMOTO, A. INOUE, and T. NOGUCHI: Occurrence of Toxic Crabs in Ryukyu and Amami Islands. Toxicon 5, 85 (1967).
13. NOGUCHI, T., S. KONOSU, and Y. HASHIMOTO: Identity of Crab Toxin with Saxitoxin. Toxicon 5, 325 (1969).
14. YASUMOTO, T., Y. OSHIMA, and T. KONTA: Analysis of Paralytic Shellfish Toxins of Xanthid Crabs in Okinawa. Suisan Gakkaishi 47, 957 (1981).
15. KOYAMA, K., T. NOGUCHI, Y. UEDA, and K. HASHIMOTO: Occurrence of Neosaxitoxin and Other Paralytic Shellfish Poisons in Toxic Crabs Belonging to the Family Xanthidal. Suisan Gakkaishi 47, 965 (1981).
16. SHIOMI, K., H. INAOKA, H. YAMANAKA, and T. KIKUCHI: Occurrence of a Large Amount of Gonyautoxins in a Xanthid Crab Atergatis Floridus from Chiba. Suisan Gakkaishi 48, 1407 (1982).
17. KOTAKI, H., Y. OSHIMA, and T. HASHIMOTO: Analysis of Paralytic Shellfish Toxins of Marine Snails. Suisan Gakkaishi 47, 943 (1981).
18. SCHANTZ, E. J., and H. W. MAGNUSSON: Observations on the Origin of the Paralytic Poison in Alaska Butter Clam. J. Protozool. 11, 239 (1964).
19. TAYLOR, L., and H. H. SELIGER (ed.): Toxic Dinoflagellate Blooms. New York: Elsevier North Holland. 1979.
20. SCHANTZ, E. J., J. D. MOLD, D. W. STANGER, J. SHAVEL, F. J. RIEL, J. P. BOWDEN, J. M. LYNCH, R. S. WYLER, B. R. RIEGEL, and H. SOMMER: Paralytic Shellfish Poison. VI. A Procedure for the Isolation and Purification of the Poison from Toxic Clams and Mussel Tissues. J. Amer. Chem. Soc. 79, 5230 (1957).
21. SHIMIZU, Y., M. ALAM, Y. OSHIMA, and W. E. FALLON: Presence of Four Toxins in Red Tide Infested Clams and Cultured Gonyaulax tamarensis Cells. Biochem. Biophys. Res. Comm. 66, 731 (1975).

22. BUCKLEY, L. J., M. IKAWA, and J. J. SASNER, JR.: Isolation of *Gonyaulax tamarensis* Toxins from Softshell Clams *(Mya arenaria)* and a Thin-Layer Chromatographic-Fluorometric Method for Their Detection. J. Agric. Food Chem. **24,** 107 (1976).

23. SHIMIZU, Y.: Red Tide Toxins: Assay and Isolation of the Toxic Components in Handling of Natural Products — Extraction and Isolation of Bioactive Compounds, pp. 151. Kodansha Scientific (1977).

24. KOBAYASHI, M., and Y. SHIMIZU: Gonyautoxin VIII, A Cryptic Precursor of Paralytic Shellfish Poisons. Chem. Commun. 827 (1981).

25. HALL, S., P. B. REICHARDT, and R. A. NEVE: Toxins Extracted from an Alaskan Isolate of *Protogonyaulax* sp. Biochem. Biophys. Res. Comm. **97,** 649 (1980).

26. OSHIMA, Y., L. J. BUCKLEY, M. ALAM, and Y. SHIMIZU: Heterogeneity of Paralytic Shellfish Poisons. Three New Toxins from Cultured *Gonyaulax tamarensis* Cells, *Mya arenaria,* and *Saxidomus giganteus.* Comp. Biochem. Physiol. **57c,** 31 (1977).

27. OSHIMA, Y., W. E. FALLON, Y. SHIMIZU, T. NOGUCHI, and Y. HASHIMOTO: Toxins of the *Gonyaulax* sp. and Infested Bivalves in Owase Bay. Suisan Gakkaishi **42,** 851 (1976).

28. HSU, C. P., A. MARCHAND, Y. SHIMIZU, and G. G. SIMS: Paralytic Shellfish Toxins in Sea Scallops, *Placopecten magellanicus* in the Bay of Fundy. J. Fish. Res. Board Can. **36,** 32 (1979).

29. SHIMIZU, Y.: Red Tide Toxins — Paralytic Shellfish Poisons Produced by *Gonyaulax* Organisms. Kagaku to Seibutsu **18,** 792 (1980).

30. HALL, S.: Private communication.

31. FALLON, W. E., and Y. SHIMIZU: Electrophoretic Analysis of Paralytic Shellfish Toxins. J. Environ. Sci. Health **A 12,** 455 (1977).

32. MARANDA, L., and Y. SHIMIZU: Distribution Mapping and Typing of Toxic Dinoflagellates in Sea Grant Report, University of Rhode Island 160 (1981).

33. SCHANTZ, E. J., J. M. LYNCH, G. VAYVADA, K. MATSUMOTO, and H. RAPOPORT: The Purification and Characterization of the Poison Produced by *Gonyaulax catenella* in Axenic Culture. Biochemistry **5,** 1191 (1966).

34. SCHANTZ, E. J., V. E. GHAZAROSSIAN, H. K. SCHNOES, F. M. STRONG, J. P. SPRINGER, J. O. PEZZANITE, and J. CLARDY: The Structure of Saxitoxin. J. Amer. Chem. Soc. **97,** 1238 (1975).

35. BORDNER, J., W. E. THIESSEN, H. A. BATES, and H. RAPOPORT: The Structure of a Crystalline Derivative of saxitoxin. The Structure of Saxitoxin. J. Amer. Chem. Soc. **97,** 6008 (1975).

36. SCHANTZ, E. J., J. D. MOLD, W. L. HOWARD, J. P. BOWDEN, D. W. STANGER, J. M. LYNCH, O. P. WINTERSTEINER, J. D. DUTCHER, D. R. WALTERS, and B. RIEGEL: Some Chemical and Physical Properties of Purified Clams and Mussel Poisons. Can. J. Chem. **39,** 2117 (1961).

37. WONG, J. L., R. OESTERLIN, and H. RAPOPORT: The Structure of Saxitoxin. J. Amer. Chem. Soc. **93,** 7344 (1971).

38. WONG, J. L., M. S. BROWN, K. MATSUMOTO, R. OESTERLIN, and H. RAPOPORT: Degradation of Saxitoxin to a Pyrimido[2,1-b]purine. J. Amer. Chem. Soc. **93,** 4633 (1971).

39. SCHUETT, W., and H. RAPOPORT: Saxitoxin, the Paralytic Shellfish Poison. Degradation to a Pyrrolpyrimidine. J. Amer. Chem. Soc. **84,** 2266 (1962).

40. ROGERS, R. S., and H. RAPOPORT: The pK_a's of Saxitoxin. J. Amer. Chem. Soc. **102,** 7335 (1980).

41. SHIMIZU, Y., C. P. HSU, and A. GENENAH: Structure of Saxitoxin in Solution and Stereochemistry of Dihydrosaxitoxin. J. Amer. Chem. Soc. **103,** 605 (1981).

42. SHARMA, G., and B. MAGDOFT-FAIRCHILD: Natural Products of Marine Sponges. 7. The Constitution of Weakly Basic Guanidine Compounds. Dibromophakellin and Monobromophakellin. J. Org. Chem. **42,** 4118 (1977).

43. GHAZAROSSIAN, V. E., E. J. SCHANTZ, H. K. SCHNOES, and F. M. STRONG: A Biologically Active Acid Hydrolysis Product of Saxitoxin. Biochem. Biophys. Res. Comm. **68**, 776 (1976).
44. BUCKLEY, L. J., Y. OSHIMA, and Y. SHIMIZU: Construction of a Paralytic Shellfish Toxin Analyzer and its Application. Anal. Biochem. **85**, 157 (1978).
45. SHIMIZU, Y., C. P. HSU, W. E. FALLON, Y. OSHIMA, I. MIURA, and K. NAKANISHI: Structure of Neosaxitoxin. J. Amer. Chem. Soc. **100**, 6791 (1978).
46. SHIMIZU, Y., A. HORI, and M. GHAZALA: Unpublished data.
47. BOYAR, A., and R. E. MARSH: 1. Canavanine, a Paradigm for the Structures of Substituted Guanidines. J. Amer. Chem. Soc. **104**, 1995 (1982).
48. ALAM, M., Y. OSHIMA, and Y. SHIMIZU: About Gonyautoxin-I, -II, -III, and -IV. Tetrahedron Letters **23**, 321 (1981).
49. SHIMIZU, Y., L. J. BUCKLEY, M. ALAM, Y. OSHIMA, W. E. FALLON, H. KASAI, I. MIURA, V. P. GULLO, and K. NAKANISHI: Structure of Gonyautoxin-II and -III from the East Coast Toxic Dinoflagellate, *Gonyaulax tamarensis*. J. Amer. Chem. Soc. **98**, 5414 (1976).
50. BATES, H. A., and H. RAPOPORT: A Chemical Assay for Saxitoxin, the Paralytic Shellfish Poison. J. Agric. Food Chem. **23**, 237 (1975).
51. BOYER, G. L., E. J. SCHANTZ, and H. K. SCHNOES: Characterization of 11-hydroxysaxitoxin Sulphate, a Major Toxin in Scallops Exposed to Blooms of the Poisonous Dinoflagellate *Gonyaulax tamarensis*. Chem. Commun. 889 (1978).
52. SHIMIZU, Y., and C. P. HSU: Confirmation of the Structures of Gonyaulax I—VI by Correlation with Saxitoxin. Chem. Commun. 314 (1981).
53. NICCOLAI, N., H. K. SCHNOES, and W. H. GIBBONS: Study of the Stereochemistry Relaxation Mechanisms, and Internal Motions of Natural Products Utilizing Proton Relaxation Parameters, Solution and Crystal Structures of Saxitoxin. J. Amer. Chem. Soc. **80**, 1502 (1980).
54. SHIMIZU, Y., W. E. FALLON, J. C. WEKELL, D. GERBER, JR., and E. J. GAUGLITZ, JR.: Analysis of Toxic Mussels (*Mytilus* sp.) from the Alaskan Inside Passage. J. Agric. Food Chem. **26**, 878 (1978).
55. SHIMIZU, Y.: Red Tide Toxins: Symposium Lecture No. 337. The American Chemical Society — Chemical Society of Japan Chemistry Congress, Honolulu, Hawaii (1979).
56. — Shellfish Toxins and Their Origins. Plenary Section Abstract of the 99th Annual Meeting of Pharmaceutical Society of Japan, pp. 87—89 (1979).
57. SHIMIZU, Y., and M. YOSHIOKA: Transformation of Paralytic Shellfish Toxins as Demonstrated in Scallop Homogenates. Science **212**, 546 (1981).
58. WICHMANN, C. F., G. L. BOYER, C. L. DIVAN, E. J. SCHANTZ, and H. K. SCHNOES: Neurotoxins of *Gonyaulax excavata* and Bay of Fundy Scallops. Tetrahedron Letters **22**, 1941 (1981).
59. NOGUCHI, T., Y. UEDA, K. HASHIMOTO, and H. SETO: Isolation and Characterization of Gonyautoxin-1 from the Toxic Digestive Gland of Scallop. Suisan Gakkaishi **47**, 1227 (1981).
60. KOEHN, F. E., S. HALL, C. F. WICHMANN, H. K. SCHNOES, and P. B. REICHARDT: Dinoflagellate Neurotoxins Related to Saxitoxin: Structure and Latent Activity of Toxins B1 and B2. Tetrahedron Letters **23**, 2247 (1982).
61. SHIMIZU, Y.: Recent Progress in Marine Toxin Research. Pure & Appl. Chem. **54**, 1973 (1982).
62. NISHIO, S., T. NOGUCHI, Y. ONOUE, J. MARUYAMA, and K. HASHIMOTO: Isolation and Properties of Gonyautoxin-5, an Extremely Low-toxic Component of Paralytic Shellfish Poison. Suisan Gakkaishi **48**, 959 (1982).
63. SHIMIZU, Y., M. KOBAYASHI, A. GENENAH, and Y. OSHIMA: Isolation of Side-Chain Sulfated Saxitoxin Analogs — Their Significance in Interpretation of the Mechanism of Action. Tetrahedron, in press.

64. TANINO, H., T. NAKATA, T. KANEKO, and Y. KISHI: A Stereospecific Total Synthesis of d,l-Saxitoxin. J. Amer. Chem. Soc. **99,** 2818 (1977).
65. WICHMANN, C. F., W. P. NIEMCZURA, H. K. SCHNOES, S. HALL, P. B. REICHARDT, and S. D. DARLING: Structures of Two Novel Toxins from *Protogonyaulax*. J. Amer. Chem. Soc. **103,** 6977 (1981).
66. TAGUCHI, H., Y. YAZAWA, J. F. ARNETT, and Y. KISHI: A Promising Cyclization Reaction to Conduct the Saxitoxin Ring System. Tetrahedron Letters 627 (1977).
67. PROCTOR, N. H., S. L. CHAN, and A. J. TREVOR: Production of Saxitoxin by Cultures of *Gonyaulax catenella*. Toxicon **13,** 1 (1975).
68. Association of Official Analytical Chemists: Paralytic Shellfish Poison, Biological Method. In: Official Methods of Analysis, 12th ed. (rev.), Washington, D.C., Assoc. Offic. Anal. Chem. **28,** 319 (1975).
69. GENENAH, A. A., and Y. SHIMIZU: Specific Toxicity of Paralytic Shellfish Poisons. J. Agric. Food Chem. **29,** 1289 (1981).
70. NARAHASHI, T.: Mechanism of Action of Tetrodotoxin and Saxitoxin on Excitable Membranes. Fed. Proc. (Am. Soc. Exp. Biol.) **31,** 1124 (1972).
71. KAO, C. Y.: Tetrodotoxin, Saxitoxin and Their Significance in the Study of Excitation Phenomenon. Phrm. rev. **18,** 997 (1966).
72. WALKER, S., and C. Y. KAO: Structure-Activity Relations of Saxitoxin Analogs. Fed. Proc. **39,** 380 (1980).
73. STRICHARTZ, G.: Relative Potencies of Several Derivatives of Saxitoxin: Electrophysiological and Toxin-binding Studies. Biophys. J. **33,** 209a (1981).
74. KAO, C. Y., and S. E. WALKER: Active Groups of Saxitoxin and Tetrodotoxin as Deduced from Actions of Saxitoxin Analogues on Frog Muscle and Squid Axon. J. Physiol. **323,** 619 (1982).
75. HILLE, B.: The Receptor for Tetrodotoxin and Saxitoxin: A Structural Hypothesis. Biophys. J. **15,** 615 (1975).
76. STRICHARTZ, G.: Private communication.

(Received February 4, 1983)

Author Index

Page numbers printed in *italics* refer to References

Subject Index

By

A. Siegel, Wien

Monatshefte für Chemie
Chemical Monthly

ISSN 0026-9247

Editorial Board: **E. Hengge**, Graz; **A. Neckel**, Wien; **K. Schlögl**, Wien (Managing Editor); **U. Schmidt**, Stuttgart; **H. Tuppy**, Wien

Papers published in 1983/84:

Subscription Information:

1984: Volume 115 (12 issues):
DM 658,—, öS 4.610,—, plus carriage charges

Springer-Verlag Wien New York

Mikrochimica Acta

An International Journal for Physical and Chemical Methods of Analysis

Editorial Board:

E. S. Etz, Washington, D. C.; **M. Grasserbauer,** Wien; **T. S. Ma,** New York; **A. Mizuike,** Nagoya; **W. Simon,** Zürich; **G. Tölg,** Dortmund; **M. K. Zacherl,** Wien (Managing Editor)

Presenting the latest results from all areas of analytical chemistry, "Mikrochimica Acta" is a journal of some tradition, published regularly since 1923, when it was founded by Nobel Prize winner F. Pregl. It has pioneered the present trend in analytical chemistry. In contrast to many of the highly specialized journals, "Mikrochimica Acta" covers a variety of topics, such as

- Elemental Analysis
- Trace Analysis
- Surface Analysis
- Chromatographic Analysis
- Molecular and Atomic Spectroscopic Techniques
- Computerized Analysis
- Electrochemical Analysis
- Sampling Methods
- Standard Reference Materials and Methods
- Enrichment Techniques
- Microchemistry
- Environmental Sciences
- Life Sciences
- Agricultural and Food Chemistry
- Forensic Analysis
- Material Sciences
- Geochemistry

Subscription Information:

1984. Volumes I—III (6 issues each):
Price per vol. DM 328,—, öS 2350,—, plus carriage charges

Prices for back volumes upon request

Springer-Verlag Wien New York